ADVANCES IN PROTEIN CHEMISTRY

Volume 63

Membrane Proteins

ADVANCES IN PROTEIN CHEMISTRY

EDITED BY

FREDERIC M. RICHARDS
Department of Molecular Biophysics
and Biochemistry
Yale University
New Haven, Connecticut

DAVID S. EISENBERG
Department of Chemistry and Biochemistry
University of California, Los Angeles
Los Angeles, California

JOHN KURIYAN
Department of Molecular Biophysics
Howard Hughes Medical Institute
Rockefeller University
New York, New York

VOLUME 63

Membrane Proteins

EDITED BY

DOUGLAS C. REES
Division of Chemistry and Chemical Engineering
Howard Hughes Medical Institute
California Institute of Technology
Pasadena, California

ACADEMIC PRESS
An imprint of Elsevier Science

Amsterdam Boston London New York Oxford Paris
San Diego San Francisco Singapore Sydney Tokyo

This book is printed on acid-free paper. ∞

Academic Press
An Elsevier Science Imprint.
525 B Street, Suite 1900, San Diego, California 92101-4495, USA
http://www.academicpress.com

Academic Press
84 Theobald's Road, London WC1X 8RR, UK
http://www.academicpress.com

International Standard Book Number: 0-12-034263-4

PRINTED IN THE UNITED STATES OF AMERICA
03 04 05 06 07 08 09 9 8 7 6 5 4 3 2 1

CONTENTS

Membrane Protein Assembly *in Vivo*

GUNNAR VON HEIJNE

Construction of Helix-Bundle Membrane Proteins

AARON K. CHAMBERLAIN, SALEM FAHAM, SARAH YOHANNAN, AND JAMES U. BOWIE

Transmembrane β-Barrel Proteins

GEORG E. SCHULZ

Length, Time, and Energy Scales of Photosystems

CHRISTOPHER C. MOSER, CHRISTOPHER C. PAGE, RICHARD J. COGDELL, JAMES BARBER, COLIN A. WRAIGHT, AND P. LESLIE DUTTON

Structural Clues to the Mechanism of Ion Pumping in Bacteriorhodopsin

HARTMUT LUECKE AND JANOS K. LANYI

The Structure of *Wolinella succinogenes* Quinol:Fumarate Reductase and Its Relevance to the Superfamily of Succinate:Quinone Oxidoreductases

C. Roy D. Lancaster

Structure and Function of Quinone Binding Membrane Proteins

Momi Iwata, Jeff Abramson, Bernadette Byrne, and So Iwata

Prokaryotic Mechanosensitive Channels

Pavel Strop, Randal Bass, and Douglas C. Rees

The Voltage Sensor and the Gate in Ion Channels

Francisco Bezanilla and Eduardo Perozo

Rhodopsin Structure, Dynamics, and Activation: A Perspective from Crystallography, Site-Directed Spin Labeling, Sulfhydryl Reactivity, and Disulfide Cross-Linking

Wayne L. Hubbell, Christian Altenbach, Cheryl M. Hubbell, and H. Gobind Khorana

The Glycerol Facilitator GlpF, Its Aquaporin Family of Channels, and Their Selectivity

Robert M. Stroud, Peter Nollert, and Larry Miercke

PREFACE

The first glimpse of the three-dimensional structure of a membrane protein was revealed in 1975, when Henderson and Unwin published their landmark analysis of bacteriorhodopsin at 7 Å resolution (Henderson and Unwin, 1975). This study established the presence of seven rodlike features, identified as α-helices, in the membrane spanning region of bacteriorhodopsin, consistent with expectations based on the behavior of polypeptides in nonaqueous solvents (see Singer, 1962). To place this development in context with the overall status of protein structure determinations at that time, X-ray crystal structures were available for ~80 water-soluble proteins (Matthews, 1976) when Henderson and Unwin's work appeared. During the intervening quarter century since this initial peek at bacteriorhodopsin, the structures of ~60 membrane proteins have been determined (see http://www.mpibp-frankfurt.mpg.de/michel/public/memprotstruct.html). Correcting for the ~25-year offset, the rate of membrane protein structure determinations (Figure 1) over the 15-year period following the first high resolution structure of the photosynthetic reaction center (Deisenhofer *et al.*, 1985) closely parallels the progress observed for water-soluble proteins after the myoglobin structure (Kendrew *et al.*, 1960). As we approach +20 years since the solution of the reaction center structure, it is not a very bold extrapolation to predict that the number of solved membrane protein structures is poised to explode, much as the number of water-soluble protein structures did in the 1980s, ~20 years after the myoglobin structure. For example, since the chapters for this review were commissioned, new structures have appeared for the chloride channel (Dutzler *et al.*, 2002), formate dehydrogenase (Jormakka *et al.*, 2002), the vitamin B_{12} transporter (Locher *et al.*, 2002), the multidrug efflux transporter AcrB

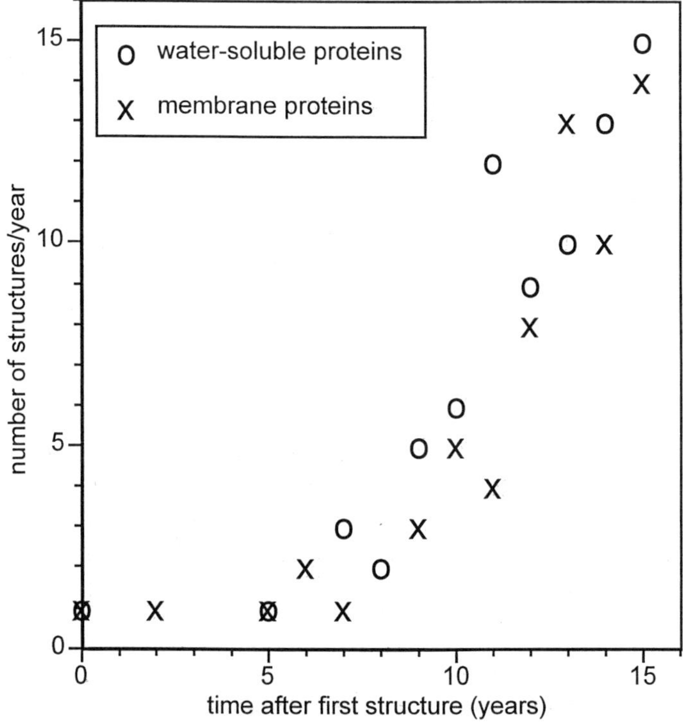

FIG. 1. The number of water soluble (o) and membrane proteins (x) solved per year, relative to the date of the first structure determination in each class, myoglobin (Kendrew *et al.*, 1960), and the photosynthetic reaction center (Deisenhofer *et al.*, 1985), respectively. The data for water-soluble proteins are from Matthews (1976), while those for membrane proteins are from the website http://www.mpibp-frankfurt.mpg.de/michel/public/memprostruct.html. Adjusting for the ~25 year difference, it is evident that progress in the structure determination of membrane proteins mirrors that experienced for water-soluble proteins. For perspective, there were 18,838 total available structures in the Protein Data Bank as of October 3, 2002, with 3298 structures deposited in 2001 (see http://www.rcsb.org/pdb/holdings_table.html).

(Murakami *et al.*, 2002), and the MscS mechanosensitive channel (Bass *et al.*, 2002), among others.

Although membrane protein structure will not likely become a mature field for another decade or two, one consequence of this recent surge in activity is that systematic coverage of all known structures is no longer possible. The eleven chapters in this volume review recent developments for selected membrane proteins from a variety of perspectives that emphasize the blending of structural and functional approaches, with the objective of establishing a comprehensive mechanistic framework for a particular system.

The first three chapters address issues of general relevance to all membrane proteins, namely, their biosynthesis and the basic structural principles underlying their construction. Gunnar von Heijne describes the cellular mechanisms responsible for "functionalization" of the lipid bilayer with proteins and the possible constraints imposed on membrane protein structure by this process. A detailed understanding of these mechanisms is not only of biological interest, but also of practical significance since the overexpression of eukaryotic membrane proteins looms as one of the greatest challenges to their structural characterization. James Bowie and co-workers address the energetic principles, including lipid–protein interactions, that drive the formation and stability of helix-bundle membrane proteins. As described in this chapter, the thermodynamic stability of membrane proteins and the energetic consequences of protein folding in a nonaqueous environment represent some of the challenging and outstanding problems in this field. While many membrane proteins consist of α-helices, those found in the outer membranes of bacteria are based on β-barrels of remarkable diversity. Georg Schulz provides an analysis of their construction principles, highlighting the implications of these observations for the engineering of channels with novel properties.

Due to their high natural abundance, it is no coincidence that significant progress has been made in the characterization of membrane protein complexes that mediate energy transduction processes such as photosynthesis and respiration. P. Leslie Dutton and coauthors develop a comprehensive framework for analyzing energy and electron transfers in photosystems centered around the spatial organization of cofactors into chains. These considerations again have significant implications for the design principles of both biological and engineered systems. The "other" photosynthetic system, bacteriorhodopsin, has not only played a central role in the structural analysis of membrane proteins, but also in deciphering the basic features of energy transduction processes. Hartmut Luecke and Janos Lanyi describe the exciting recent progress and the challenges in establishing the mechanism of proton pumping by bacteriorhodopsin in molecular detail. The next two chapters describe the structure and mechanism of quinone binding respiratory complexes. Quinones play a central role in membrane bioenergetics, serving as diffusible, lipid-soluble carriers of electrons and protons that link many of the photosynthetic and respiratory systems in electron transfer chains. C. R. D. Lancaster reviews the *Wolinella* fumarate quinol reductase that, with the structure of the *Escherichia coli* enzyme (Iverson *et al.*, 1999), has structurally defined the family of succinate:quinone oxidoreductases. This family is of central importance to the energy metabolism of many organisms, as various members can function in aerobic respiration (as complex II), the Krebs cycle, and as terminal acceptors during anaerobic respiration. So Iwata and co-workers describe the structure and function of two respiratory complexes, cytochrome bc_1 (or complex III) and cytochrome bo_3 (or ubiquinol oxidase) that can couple electron transfer to proton translocation at the heart of energy conservation during respiration. These structures, together with the structures of cytochrome *c* oxidase or complex IV (Iwata *et al.*, 1995; Tsukihara *et al.*, 1995), ATP synthase (Abrahams *et al.*, 1994; Stock *et al.*, 1999), and photosystem I (Jordan *et al.*, 2001), provide a nearly complete structural characterization of the major photosynthetic and aerobic respiratory complexes, missing only the structure of complex I (NADH dehydrogenase) and the extension of the photosystem II structure (Zouni *et al.*, 2001) to high resolution.

Channels and receptors mediate the flow of matter and information across the membrane bilayer that are fundamental to many biological processes. The last four chapters of this volume address systems that highlight the rich functional diversity of membrane proteins in these capacities. Microorganisms must be able to adapt to rapid changes in their environment, such as sudden drops in external osmolarity that can lead to swelling and lysis. To protect against this, stretch-activated (mechanosensitive) channels of large conductance, first identified by Kung and co-workers (Martinac *et al.*, 1987), are present in the cell membrane that appear to serve as safety valves to reduce the possibility of cell rupture under these conditions. Advances in the structure and mechanism of these mechanosensitive channels are reviewed in the chapter by my group. In eukaryotic organisms, voltage-gated channels mediate signaling processes that are of tremendous physiological and pharmacological significance. Francisco Bezanilla and Eduardo Perozo present an analysis of structural and dynamic properties of the voltage sensor derived from exquisite biophysical and biochemical studies. An important development will be the high resolution structure of a voltage-gated channel to define the structural organization of the voltage sensor, to complement the studies of MacKinnon in establishing the structural basis of ion selectivity (Doyle *et al.*, 1998). G-Protein-coupled receptors are key elements of signal transduction pathways in eukaryotes and represent important pharmacological targets. Rhodopsin, that initiates the visual response, is the paradigm of this receptor family. Wayne Hubbell and coauthors describe the properties of rhodopsin as established from crystallographic and biophysical studies, particularly the site-directed spin labeling approach developed by his group. Of particular interest is the analysis of the light-mediated changes in the structure and dynamics of rhodopsin. In the final chapter, Robert Stroud and co-workers detail the structure and selectivity mechanism of the glycerol facilitator GlpF that is a member of the aquaporin family of channels. By mediating the flow of water and a few other small solutes such as glycerol across the relatively impermeable cell membrane, aquaporins are critical to the maintenance of the appropriate osmotic pressure balance in prokaryotic and eukaryotic cells.

At the start of the twenty-first century, the pace of membrane protein structure determinations is clearly accelerating (Figure 1). With the exceptions of rhodopsin (Palczewski *et al.*, 2000) and the calcium ATPase (Toyoshima *et al.*, 2000), however, eukaryotic channels, transporters, and receptors are conspicuously absent from the list of known membrane protein structures. These two exceptions, as proteins of naturally high abundance, highlight the current reality that no structure has been determined for an overexpressed eukaryotic membrane protein. This situation reflects the present difficulties in the reliable overexpression of membrane proteins, particularly those of eukaryotic organisms. Just as the development ~20 years ago of overexpression systems for water-soluble proteins revolutionized the structure determinations of this class of proteins, advances in membrane protein expression will be essential to successful realization of the goal of routine structural analysis of membrane proteins.

In this era of proliferating reviews, investigators have many opportunities to satisfy such urges, particularly in a field such as membrane protein structure and function. Consequently, I would particularly like to thank the authors of this volume for the time commitment and effort required to prepare their contributions.

Douglas C. Rees

References

Abrahams, J. P., Leslie, A. G., Lutter, R., and Walker, J. E. (1994). *Nature* **370,** 621–628.

Bass, R. B., Strop, P., Barclay, M., and Rees, D. C. (2002). *Science* **298,** 1582–1587.

Deisenhofer, J., Epp, O., Miki, K., Huber, R., and Michel, H. (1985). *Nature* **318,** 618–624.

Doyle, D. A., Cabral, J. M., Pfuetzner, R. A., Kuo, A., Gulbis, J. M., Cohen, S. L., Chait, B. T., and MacKinnon, R. (1998). *Science* **280,** 69–77.

Dutzler, R., Campbell, E. B., Cadene, M., Chait, B. T., and MacKinnon, R. (2002). *Nature* **415,** 287–294.

Henderson, R. and Unwin, P. N. T. (1975). *Nature* **257,** 28–32.

Iverson, T. M., Luna-Chavez, C., Cecchini, G., and Rees, D. C. (1999). *Science* **284,** 1961–1966.

Iwata, S., Ostermeier, C., Ludwig, B., and Michel, H. (1995). *Nature* **376,** 660–643.

Jordan, P., Fromme, P., Witt, H. T., Klukas, O., Saenger, W., and Krauß, N. (2001). *Nature* **411,** 909–917.

Jormakka, M., Törnroth, S., Byrne, B., and Iwata, S. (2002). *Science* **295,** 1863–1868.

Kendrew, J. C., Dickerson, R. E., Strandberg, B. E., Hart, R. B., Davies, D. R., Phillips, D. C., and Shore, V. C. (1960). *Nature* **185,** 422–427.

Locher, K. P., Lee, A. T., and Ress, D. C. (2002). *Science* **296,** 1091–1098.

Martinac, B., Buechner, M., Delcour, A. H., Adler, J., and Kung, C. (1987). *Proc. Natl. Acad. Sci. USA* **84,** 2297–2301.

Matthews, B. W. (1976). *Ann. Rev. Phys. Chem.* **27,** 493–523.

Murakami, S., Nakashima, R., Yamashita, E., and Yamaguchi, A. (2002). *Nature* **419,** 587–593.

Palczewski, K., Kumasake, T., Hori, T., Behnke, C. A., Motoshima, H., Fox, B. A., Trong, I. L., Teller, D. C., Okada, T., Stenkamp, R. E., Yamaoto, M., and Miyano, M. (2000). *Science* **289,** 739–745.

Singer, S. J. (1962). *Adv. Prot. Chem.* **17,** 1–68.

Stock, D., Leslie, A. G. W., and Walker, J. E. (1999). *Science* **286,** 1700–1705.

Toyoshima, C., Nakasako, M., Nomura, H., and Ogawa, H. (2000). *Nature* **405,** 647–655.

Tsukihara, T., Aoyama, H., Yamashita, E., Tomizaki, T., Yamaguchi, H., Shinazawa-Itoh, K., Nakashima, R., Yaono, R., and Yoshikawa, S. (1995). *Science* **69,** 1069–1074.

Zouni, A., Witt, H.-T., Kern, J., Fromme, P., Krauß, N., Saenger, W., and Orth, P. (2001). *Nature* **409,** 739–743.

MEMBRANE PROTEIN ASSEMBLY *IN VIVO*

By GUNNAR VON HEIJNE

Department of Biochemistry and Biophysics, Stockholm University,
SE-10691 Stockholm, Sweden

I. Introduction

Although much current interest is focused on the structure–function relationships of membrane proteins, a good understanding of the cellular processes responsible for the insertion of proteins into lipid bilayer membranes is a necessary prerequisite for assessing how much the accessibly "structure space" is constrained not by the bilayer per se but rather by the idiosyncrasies of the machineries that have evolved to guide the insertion process. In most cases, membrane proteins use the same targeting and insertion mechanisms used to sort soluble proteins between different cellular compartments but add an extra level of complexity: How are membrane proteins recognized as proteins that should be only partly translocated, leaving the transmembrane segments of the polypeptide chain spanning the bilayer? Is the basis for this recognition the same in different translocation machineries, and, if not, how are such differences translated into different constraints on the allowable structures in any given cellular membrane?

In this article, I will review the current knowledge of how membrane proteins are handled by different targeting–translocation machineries from the perspective of possible structural constraints imposed by these machineries. As will become clear, this point of view has not been much elaborated up to now, and there are not many clear examples of structural constraints that go beyond those imposed by the lipid bilayer itself. Nevertheless, I will argue that such constraints do exist and that they cannot be ignored if we wish to fully understand the principles underlying

ADVANCES IN PROTEIN CHEMISTRY, Vol. 63

0065-3233/03 $35.00

membrane protein structure. For in-depth reviews of protein targeting mechanisms in general, a good up-to-date source is Dalbey and von Heijne (2002).

II. Overview of Membrane Protein Assembly Pathways in Prokaryotic and Eukaryotic Cells

Over the past 30 years, the study of intracellular protein sorting has grown to a large and diversified field. A host of different sorting pathways have been found in both prokaryotic and eukaryotic cells, and most of these can handle both soluble and membrane-bound proteins.

The most well understood pathway is the one that delivers secretory and membrane proteins to the endoplasmic reticulum (ER) membrane in eukaryotic cells and to the inner membrane in bacteria. In both kinds of cells, the pivotal role is played by the so-called Sec61 (in eukaryotes) or SecYEG (in prokaryotes) translocon, a multisubunit translocation channel that provides a conduit for soluble proteins to cross the membrane. The same translocon also serves to integrate membrane proteins into the lipid bilayer.

Bacteria harbor additional inner membrane translocation machineries, such as the "twin-arginine translocation" (Tat) system, the YidC system, and the so-called type II translocation systems that are dedicated to one or a small number of substrates and ensure their delivery into the extracellular medium.

The Tat and YidC systems are also found in certain organelles in eukaryotic cells. The former is present in the thylakoid membrane in plant cells, and the latter is found both in the thylakoid membrane (where it is called the Albino3 system) and in the inner mitochondrial membrane (where it is called the Oxa1p system). Both chloroplasts and mitochondria appear to have unique systems for importing proteins across their outer and inner membranes, and these systems also handle membrane proteins.

Finally, peroxisomes contain another unique machinery for protein import and membrane protein assembly.

From this list of targeting–translocation machineries it is clear that membrane proteins from different subcellular compartments are not handled in the same way and thus may be expected to be under different constraints as concerns the requirements for membrane insertion. However, very few comparative data are available and the only system for which we have a detailed understanding of the insertion process and how it constrains the final structure of the protein is the eukaryotic ER. The main focus of the chapter will thus be on ER targeting and insertion of membrane proteins, though the other systems will also be briefly reviewed.

III. MEMBRANE PROTEIN ASSEMBLY IN THE ER

For both secretory and membrane proteins, the initial event associated with ER targeting is the binding of the N-terminal signal peptide—or the most N-terminal transmembrane segment—to the signal recognition particle (SRP) (Johnson and van Waes, 1999). SRP binding is thought to slow down the rate of elongation, giving the ribosome–nascent chain–SRP complex more time to find an empty translocation site on the ER membrane. These sites are composed of the SRP receptor complex, the basic Sec61 translocon (an oligomer of the Sec61α, β, and γ subunits), the TRAM protein, the oligosaccharyl transferase (OST) complex, and the signal peptidase (SPase) complex. On binding to the SRP receptor, SRP dissociates from the ribosome, allowing it to dock to the Sec61 translocon and to insert the N-terminal part of the nascent chain (including the signal peptide) into the translocation channel. As the elongation rate picks up again, the nascent chain is extruded cotranslationally through the long conduit thus assembled from the nascent chain tunnel in the ribosomal large subunit and the Sec61 transmembrane channel (Beckmann *et al.*, 2001) The signal peptide is cleaved by SPase at an early stage of translocation, and Asn-linked oligosaccharides are attached to the growing nascent chain as it passes in the vicinity of the OST.

How is this process modified for membrane proteins? The major difference is that the ribosome–Sec61 translocon complex somehow can recognize sufficiently hydrophobic segments in the nascent chain, either when they enter the Sec61 channel, or possibly already when they move through the ribosomal large subunit tunnel. The recognition of a hydrophobic transmembrane segment by the ribosome–translocon complex leads alternately to the channel closing at its lumenal end and opening at the cytoplasmic ribosome–Sec61 junction or the opposite—closing of the cytoplasmic junction, opening at the lumenal end (Hamman *et al.*, 1998; Liao *et al.*, 1997). In the simplest model based on these observations (Fig. 1), the signal peptide (or the most N-terminal transmembrane segment) thus opens the lumenal translocon "gate," allowing free passage of the immediate downstream part of the nascent chain across the membrane, the next transmembrane segment closes the lumenal gate

FIG. 1. Model for membrane protein insertion into the ER membrane. In stage I, an N-terminal signal peptide (not shown) has already initiated translocation across the membrane.

and opens up an exit route toward the cytoplasmic side for the following part of the nascent chain, the succeeding transmembrane segment again opens the lumenal gate, etc., effectively "stitching" the transmembrane segments into the membrane one by one.

Obviously, the transmembrane segments must at some point be expelled laterally from the translocation channel into the surrounding lipid bilayer. The available information on this crucial step is limited, but it has been shown by rather sophisticated cross-linking techniques that a transmembrane segment appears to follow an ordered exit pathway where it is first found in the vicinity of the Sec61α protein, then TRAM, and finally lipids (Do *et al.*, 1996). Some transmembrane segments move to a lipid-exposed location almost immediately on entering the Sec61 channel, whereas others may remain within the channel for a longer time, possibly until a downstream transmembrane segment appears whereupon the two may exit *en bloc* (Heinrich *et al.*, 2000). It has even been claimed that multiple transmembrane segments or even the whole protein may assemble within the translocation channel before moving away from the translocon (Borel and Simon, 1996a,b). A better definition of the exit step in the assembly process will be crucial for our understanding of membrane protein folding *in vivo.*

If our knowledge of the molecular mechanisms responsible for the recognition and membrane integration of transmembrane segments is thus quite limited, the situation is significantly better when we ask about the substrate, i.e., the characteristics of the nascent chain that have an impact on the assembly process. Starting with the hydrophobic transmembrane segment itself, many studies have shown that there is a minimum hydrophobicity threshold required for membrane integration (Kuroiwa *et al.*, 1991; Sääf *et al.*, 1998); to a first approximation, the precise amino acid sequence is thus unimportant (but see below). As an example, a stretch of 7–8 consecutive Leu residues is sufficient for the formation of a transmembrane topology. The threshold hydrophobicity (measured over a 20-residue window) appears to be close to that of poly(Ala) (Kuroiwa *et al.*, 1991), as has also been suggested by calculations based on the insertion of synthetic peptides into lipid bilayers (Jayasinghe *et al.*, 2001). Nevertheless, it is not possible to cleanly discriminate all known transmembrane segments from nontransmembrane segments based on threshold hydrophobicity alone, suggesting that there is more to the story than simple greasiness.

Transmembrane segments are connected to one another by loops (anything from short turns to large domains) of the polypeptide chain, and these have been found to have a strong influence on the assembly and final topology of membrane proteins. In multispanning membrane proteins, loops tend to be short [typically ~10 residues (Tusnady and Simon, 1998)]. For short loops, there is a very strong asymmetry in the content of positively charged residues (Lys and Arg), with cytoplasmic ("inside") loops being 2- to 4-fold richer in such residues than extracytoplasmic ("outside") loops (Sipos and von Heijne, 1993; von Heijne, 1986; Wallin and von Heijne, 1995). This so-called positive inside rule holds for proteins from nearly all organisms and intracellular membranes (Gavel *et al.*, 1991; Gavel and von Heijne, 1992; van de Vossenberg *et al.*, 1998; Wallin and von Heijne, 1998), and the only clear exception found so far is the class of mitochondrial inner membrane proteins that are imported by using a "stop-transfer" mechanism (see below). Mitochondrial inner membrane proteins that are synthesized within the organelle or those that are imported using the "conservative sorting" mechanism do seem to follow the positive inside rule, however.

Numerous examples now exist where positively charged residues have been shown to influence the topology of both single- and multispanning membrane proteins. It has even been shown that a strongly hydrophobic segment can be prevented from inserting across the membrane if it is flanked by positively charged residues on both ends (so-called "frustrated" topologies (Gafvelin *et al.*, 1997; Gafvelin and von Heijne, 1994)), and, vice versa, that a polar segment of the polypeptide chain can be forced into a transmembrane disposition by flanking hydrophobic segments that both have the same orientational preference (Ota *et al.*, 1998). Recently, examples of "topology evolution" have been found, where two homologous proteins with the same number of transmembrane segments adopt opposite orientations in the membrane as a result of different distributions of positively charged residues in their loops (Sääf *et al.*, 1999, 2001).

Despite much work the molecular basis for the positive inside rule is still not clear, although both the membrane potential and lipid composition have been shown to affect the degree to which positively charged residues affect topology (Andersson and von Heijne, 1994; van Klompenburg *et al.*, 1997). Given the ubiquitous nature of the positive inside rule—it applies across all domains of life, in different organelles, and in microorganisms from very different ecological niches—it seems unlikely that it can be fully explained by the thermodynamics of protein–lipid interactions. Rather, it probably finds its basis in a kinetic difference between the rate of translocation of positively versus negatively charged polypeptide segments across biological membranes, perhaps modulated by the intrinsic properties of different kinds of translocons.

Positively charged residues were the first topological determinants to be discovered, but they are not the only ones (Goder and Spiess, 2001). Thus, at least for single-spanning membrane proteins with an N-terminal signal-anchor sequence, the length of the hydrophobic stretch also helps determine its orientation. Long hydrophobic segments favor an N_{out}–C_{in} orientation, and short segments have the opposite preference (Wahlberg and Spiess, 1997). Finally, the N_{out}–C_{in} orientation for signal-anchor sequences is prevented if the N-terminal tail preceding the hydrophobic segment folds too fast (Denzer *et al.*, 1995).

As noted above, more than one transmembrane segment may be present within the translocon at any one time, suggesting the possibility that transmembrane segments may affect one anothers topological preferences. Indeed, many studies have demonstrated that downstream transmembrane segments can influence the insertion of upstream segments (Monné *et al.*, 1999a; Nilsson *et al.*, 2000). In general, it seems reasonable that the possibilities for such interhelix effects should be maximal when the two transmembrane segments are located very close to each other in the primary sequence. Indeed, such closely spaced transmembrane segments may in fact act as one "insertion unit." The term "helical hairpin" (Engelman and Steitz, 1981) is often used to describe this situation.

We have studied the sequence determinants for helical hairpin formation during the insertion of a model membrane protein into the ER membrane. To simplify the problem, we engineered a 40-residue long poly(Leu) stretch into a membrane protein that inserts readily into ER-derived microsomes when expressed *in vitro* (Fig. 2A). Asn-X-Thr acceptor sites for N-linked glycosylation were used as topological markers, as they can only be modified when located in the

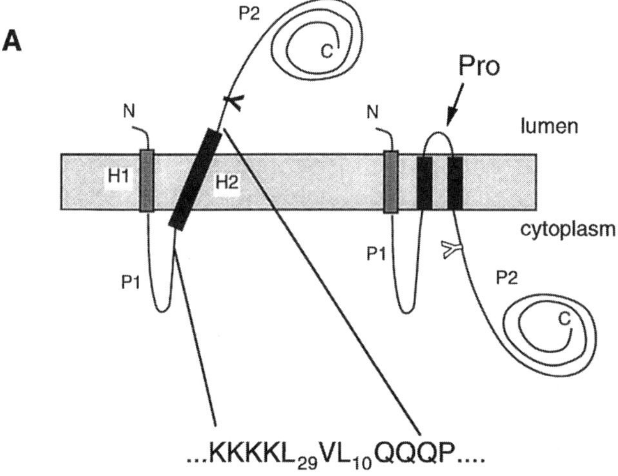

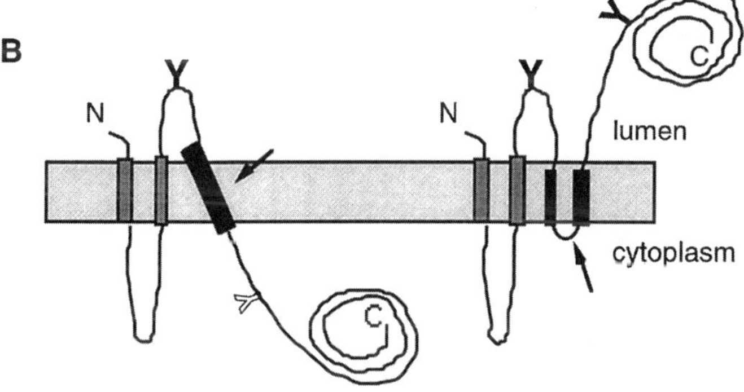

FIG. 2. Model proteins used to study helical hairpin formation. (A) The H2 transmembrane segment in the leader peptidase protein was replaced by a 40-residue long hydrophobic segment composed of 39 leucines and one valine. An Asn-Ser-Thr glycosylation acceptor site (Y) was used as a topological marker (this site can only be glycosylated if present in the ER lumen). A single Leu → Pro mutation in the middle of the poly(Leu) segment is enough to convert it into a helical hairpin. (B) Same as in A, but with the poly(Leu) segment inserted into the normally lumenal P2 domain. In this case, two glycosylation acceptor sites were used to determine the topology of constructs where different amino acids were inserted in the middle of the poly(Leu) segment (arrow). Filled Y's indicate glycosylated acceptor sites, unfilled Y's indicate nonglycosylated acceptor sites.

lumen of the microsomes. The poly(Leu) stretch was found to insert as a single transmembrane segment, even though it is twice as long as "normal" transmembrane helices. However, single Leu → Pro mutations efficiently converted the single transmembrane segment to a helical hairpin when more than 15 residues distant from either end of the poly(Leu) stretch (Nilsson and von Heijne, 1998). Similar effects were seen when charged and polar residues were substituted for Leu, but not when apolar and hydrophobic residues were introduced (Monné *et al.*, 1999b; Monné *et al.*, 1999c). We also found that the poly(Leu) stretch has to be longer than 30 residues to allow Pro-induced helical hairpin formation (Monné *et al.*, 1999c). Finally, blocks of charged residues placed immediately downstream of the poly(Leu) stretch also promote the formation of a helical hairpin topology (Hermansson *et al.*, 2001).

Interestingly, it appears that it is "easier" to induce the formation of a helical hairpin with the tight turn on the lumenal side of the ER membrane than one with the opposite orientation (cytoplasmic turn); i.e., whereas a single Pro is enough to convert the 40-residues long poly(Leu) stretch to a helical hairpin with a lumenal turn, three consecutive prolines are needed for a helical hairpin with a cytoplasmic turn to form (Sääf *et al.*, 2000) (Fig. 2B). If one only considers simple protein–lipid interactions there is no obvious thermodynamic reason why this should be so; instead, we favor the view that this reflects a constraint on helical hairpin structure imposed by the Sec machinery.

Finally, there is one particular class of membrane proteins for which special rules seem to apply: the so-called tail-anchored proteins. Tail-anchored proteins have only a single hydrophobic segment close to their C-terminus that inserts into the ER membrane with the C-terminus penetrating to the lumenal side. One consequence of this is that there is no N-terminal targeting information, and the cotranslational SRP pathway hence cannot be brought into play. Instead, tail-anchored proteins are targeted posttranslationally to the ER and do not appear to use the normal Sec61 translocon. The mechanistic details of the targeting and membrane insertion processes are largely unknown; however, in contrast to the SRP-Sec61 pathway, tail-anchored proteins depend on ATP rather than GTP for membrane insertion (Kutay *et al.*, 1995). Interestingly, the "minimum hydrophobicity" of the transmembrane segment is similar to what is required for membrane insertion via the SRP–Sec61 pathway (Whitley *et al.*, 1996).

In conclusion, the available studies on membrane protein assembly into the ER suggest that the ribosome–translocon channel is where hydrophobic segments are first recognized, eventually leading to their expulsion into the lipid bilayer as transmembrane helices. Orientational preferences are largely encoded within the regions immediately flanking the hydrophobic stretches, even if the length of the hydrophobic stretch can play a role as well. Pairs of closely spaced transmembrane segments—"helical hairpins"—may behave as "insertion units" that are recognized *en bloc* by the translocation machinery. Finally, although most of the general structural aspects of helix bundle membrane proteins appear reasonably consistent with the basic principles of protein–lipid thermodynamics, there are some details where constraints imposed by the translocation machinery seem to be responsible: the positive-inside rule is one such instance (and, *inter alia,* "frustrated" topologies where hydrophobic segments are prevented from crossing the membrane as well as topologies where polar segments are forced into a transmembrane disposition), and the different requirements for the formation of helical hairpins with lumenal versus cytoplasmic loops are another.

IV. Membrane Protein Assembly in *Escherichia coli*

As far as is known, the basic rules for membrane protein insertion into the inner membrane of *E. coli* are quite similar to those discussed in the preceding section. Thus, most inner membrane proteins are thought to be targeted to the membrane by the bacterial SRP homologue, and inserted via the SecYEG translocon (de Gier and Luirink, 2001). Whether membrane insertion is co- or posttranslational is not really clear, however. For inner membrane proteins with large periplasmic domains, the SecA ATPase seems to be required in order to translocate these large parts but possibly not for shorter loops (Andersson and von Heijne, 1993; Gafvelin and von Heijne, 1994; Sääf *et al.*, 1995).

A new inner membrane protein, YidC, has been shown to be critical for the assembly process (Samuelson *et al.*, 2000). YidC is found associated with the SecYEG translocon, but may also exist as a separate complex (Scotti *et al.*, 2000). Homologues to YidC are present in yeast mitochondria (the Oxa1p protein) and thylakoids (the Albino3 protein), where they are also involved in membrane protein assembly (Hell *et al.*, 1998; Luirink *et al.*, 2001; Sundberg *et al.*, 1997). Interestingly, mitochondria from the yeast *Saccharomyces cerevisiae* lack the SecYEG components, strongly suggesting that YidC can function in the absence of a Sec machinery (Glick and von Heijne, 1996). Indeed, even in *E. coli* there are proteins such as the M13 procoat protein that appear not to use SecYEG but only depend on YidC (Samuelson *et al.*, 2001).

Concerning how hydrophobic transmembrane segments are recognized by the SecYEG–YidC translocon, not much is known. A cross-linking study using a protein with a single transmembrane segment has revealed an apparently ordered process where the hydrophobic segment is first found in the vicinity of SecY, then YidC, and finally lipids (Urbanus *et al.*, 2001; van der Laan *et al.*, 2001). This is very similar to what has been found for the Sec61–TRAM complex in the ER (see above) and suggests that YidC and TRAM, although not related by sequence similarity, may nevertheless perform similar functions. This impression is reinforced by recent work on a multispanning inner membrane protein showing that at least three consecutive transmembrane segments can be cross-linked simultaneously to YidC (Beck *et al.*, 2001). At the same time, these segments form lipid cross-links, suggesting that YidC can interact with multiple transmembrane segments in a nascent protein and that it also allows the transmembrane segments to contact lipids. Thus, either the transmembrane segment can partition dynamically between the YidC complex and surrounding lipids, or lipids can penetrate deep into the YidC complex.

In addition to the SecYEG–YidC translocon, *E. coli* also has a machinery that can translocate folded polypeptides to the periplasmic space. Signal peptides that target proteins to this machinery contain a diagnostic Arg-Arg motif (Berks *et al.*, 2000); hence the name twin arginine translocation (Tat) system. The Tat system is also found in the thylakoid membrane of chloroplasts. Most Tat substrates identified so far are soluble proteins, but at least one inner membrane protein has been reported to use this pathway (Molik *et al.*, 2001). It will be very interesting to see whether the requirements in terms of hydrophobicity and flanking charged residues are the same for Tat-targeted and SecYEG-targeted membrane proteins.

Finally, a few words on the assembly of outer membrane β-barrel proteins. These proteins are made with an N-terminal signal peptide that targets them for translocation across the inner membrane via the SecYEG translocon. Aided by periplasmic chaperones such as Skp (de Cock *et al.*, 1999; Harms *et al.*, 2001; Schäfer *et al.*, 1999), the protein then folds and inserts into the outer membrane, possibly in a concerted folding–insertion reaction. Folding and insertion can be reconstituted *in vitro* (Tamm *et al.*, 2001), although the kinetics seen so far are orders of magnitude slower than *in vivo*. It is completely unclear how the orientation of the β-barrel proteins across the outer membrane is controlled; the regularities noted so far are (i) short loops face the periplasm, long loops the extracellular side, and (ii) the N- and C-termini tend to be located on the periplasmic side (Schulz, 2000), but whether these characteristics are just correlations or actually help determine the orientation is not known.

V. Membrane Protein Assembly in Mitochondria

Mitochondria import most of their outer and inner membrane proteins from the cytosol, but also encode a small number of inner membrane proteins in their own genome. As noted above, for proteins that are translated in the mitochondrial matrix, insertion into the inner membrane has been found to depend on the YidC homologue Oxa1p (Hell *et al.,* 1997, 1998, 2001) and their topology seems to obey the positive inside rule (Rojo *et al.,* 1999).

Three different assembly pathways have been suggested for the imported inner membrane proteins. All three utilize the so-called TOM complex for translocation through the outer membrane. Two of the pathways further make use of the same TIM23 complex in the inner membrane that also handles soluble matrix proteins; the third involves both a distinct set of chaperones in the intermembrane space as well as a distinct inner membrane insertion machinery, the TIM22 complex.

The TIM22 pathway is used by, for example, proteins that belong to the mitochondrial carrier protein family (Koehler, 2000; Pfanner and Geissler, 2001). After being chaperoned through the intermembrane space by the "Tiny TIM complexes" TIM9/10 and TIM8/13, proteins on this pathway are delivered to the inner membrane TIM22 complex and are directly inserted into the bilayer. It appears that positively charged loops are driven across the inner membrane by the membrane potential, but not much else is known about the insertion mechanism.

The two TOM-TIM23 pathways—the "conservative sorting" and "stop-transfer" pathways—differ in that the former posits a process in which the inner membrane protein is first fully imported into the matrix space and then inserted into the inner membrane from the matrix side using the Oxa1p complex (Hell *et al.,* 2001; Stuart and Neupert, 1996). This process is "conservative" in an evolutionary sense, since the prokaryotic ancestor of present-day mitochondria presumably inserted all their inner membrane proteins from the cytoplasmic (i.e., matrix) side of the membrane.

In contrast, in the "stop-transfer" model, hydrophobic segments in the imported inner membrane protein get stuck in the TIM23 channel and leave the translocon laterally, in much the same way as is envisaged for the Sec61 and SecYEG translocons (Tokatlidis *et al.,* 1996). Thus, there is no matrix-localized intermediate in this case. Possibly, this may explain an early observation (Gavel and von Heijne, 1992) that some imported inner membrane proteins follow the positive inside rule (presumably those that use the "conservative sorting" or the TIM22 pathways) whereas others do not (presumably those that are inserted by the stop-transfer mechanism).

Whether the different mitochondrial inner membrane assembly pathways impose different constraints on the final structure of the inner membrane proteins is not known.

VI. Membrane Protein Assembly in Chloroplasts

Chloroplasts in higher plants have three membranes: the outer and inner envelope membranes and the thylakoid membrane. Very little is known about membrane protein assembly into the two envelope membranes (Soll and Tien, 1998). The thylakoid has been better studied and in fact appears to use mechanisms very similar to those found in *E. coli* for membrane protein insertion (Dalbey and Robinson, 1999). Thus, SRP, SecA, SecYEG, YidC, and Tat homologues are all present in the thylakoid membrane or in the stroma (the Tat system was first identified in thylakoids, in fact). In contrast to *E. coli*, however, there are thylakoid proteins that appear to insert "spontaneously" into the membrane, insofar as no requirement for any of the known translocation machineries has been detected (Mant *et al.*, 2001).

VII. Membrane Protein Assembly in Peroxisomes

So far, studies of protein import into peroxisomes have focused largely on soluble matrix proteins (Fujiki, 2000; Sacksteder and Gould, 2000). The few membrane proteins studied to date appear to have targeting signals with no obvious consensus sequence and that may be located in either matrix or cytoplasmic domains (Honsho and Fujiki, 2001). No targeting signal receptor has been identified so far. A requirement for ATP and (unknown) cytosolic factors has been demonstrated (Just and Diestelkotter, 1996; Pause *et al.*, 1997), but this is where the story ends for the time being.

VIII. Conclusions

In what sense is what we know of membrane protein assembly *in vivo* relevant for our general understanding of membrane protein structure? Except for the rather trivial fact that most membrane proteins have a unique orientation relative to the membrane and cannot reorient on the time scale of their typical lifetime—which suggests a nonequilibrium distribution resulting from the assembly process—it is unlikely that a membrane protein is not at thermodynamic equilibrium with the surrounding bilayer. From this point of view the membrane insertion process may be viewed as one part of the folding process: interesting from a basic chemistry point of view but not necessarily very important for understanding or predicting the 3D structure.

However, it is already clear that the mechanisms responsible for membrane protein integration into the ER membrane and the inner bacterial membrane do place certain constraints on the "allowable" structures and in some (rare?) instances even prevent thermodynamic equilibrium from being reached. Thus, it is hard to imagine that proteins with "frustrated" topologies (i.e., with one or more strongly hydrophobic segments prevented from forming transmembrane helices by strongly charged flanking regions) represent equilibrium structures; rather, they most likely represent kinetically trapped states.

Whether natural proteins with "frustrated" topologies exist is an open question, but it is likely that another constraint discussed above—the differences in the requirements for inducing a helical hairpin with a lumenal versus a cytoplasmic loop—has helped shape natural membrane proteins. This does not appear to be a very strong constraint, however, and will only become apparent when large numbers of 3D structures have been solved.

Whether the transmembrane helices themselves are constrained by the assembly process is less clear. It is not unlikely that there are fine differences in the sequence requirements for insertion of a transmembrane helix via a translocon compared with direct partitioning into a lipid bilayer, although no such comparative studies have been made (and they may be difficult to conduct). For instance, one can well imagine a situation where a marginally hydrophobic segment in a protein may be sufficiently shielded from the surrounding lipid to simply pass through a translocon even though it could in principle form a stable transmembrane helix if given the opportunity. From this point of view, it would be interesting to compare the final structures of a given membrane protein when targeted either to the SecYEG-YidC or the Tat translocon in the *E. coli* inner membrane; they may well be different.

To conclude, membrane protein assembly processes are both diverse and highly evolved, attesting to the fact that strongly hydrophobic segments in proteins need to be guarded against and contained by tightly regulated processes in order not to wreak havoc on the cell. The lipid membrane may be a perfect shield against the environment, but "functionalizing" it with proteins is not an easy task!

Acknowledgments

This work was supported by grants from the Swedish Cancer Foundation and the Swedish Research Council.

References

Andersson, H., and von Heijne, G. (1993). *Sec*-dependent and *Sec*-independent assembly of *E. coli* inner membrane proteins—the topological rules depend on chain length. *EMBO J.* **12,** 683–691.

Andersson, H., and von Heijne, G. (1994). Membrane protein topology: effects of $\Delta\mu H^+$ on the translocation of charged residues explain the "positive inside" rule. *EMBO J.* **13,** 2267–2272.

Beck, K., Eisner, G., Trescher, D., Dalbey, R. E., Brunner, J., and Müller, M. (2001). YidC, an assembly site for polytopic *Escherichia coli* membrane proteins located in immediate proximity to the SecYE translocon and lipids. *EMBO Rep.* **2,** 709–714.

Beckmann, R., Spahn, C. M. T., Eswar, N., Helmers, J., Penczek, P. A., Sali, A., Frank, J., and Blobel, G. (2001). Architecture of the protein-conducting channel associated with the translating 80S ribosome. *Cell* **107,** 361–372.

Berks, B. C., Sargent, F., and Palmer, T. (2000). The Tat protein export pathway. *Mol. Microbiol.* **35,** 260–274.

Borel, A. C., and Simon, S. M. (1996a). Biogenesis of polytopic membrane proteins: membrane segments assemble with translocation channels prior to membrane integration. *Cell* **85,** 379–389.

Borel, A. C., and Simon, S. M. (1996b). Biogenesis of polytopic membrane proteins: membrane segments of P-glycoprotein sequentially translocate to span the ER membrane. *Biochemistry* **35,** 10587–10594.

Dalbey, R., and Robinson, C. (1999). Protein translocation into and across the bacterial plasma membrane and the plant thylakoid membrane. *Trends Biochem. Sci.* **24,** 17–22.

Dalbey, R., and von Heijne, G., Eds. (2002). *Protein Targeting, Transport and Translocation.* Academic Press, London.

de Cock, H., Schäfer, U., Potgeter, M., Demel, R., Müller, M., and Tommassen, J. (1999). Affinity of the periplasmic chaperone Skp of *Escherichia coli* for phospholipids, lipopolysaccharides and non-native outer membrane proteins—role of Skp in the biogenesis of outer membrane protein. *Eur. J. Biochem.* **259,** 96–103.

de Gier, J. W., and Luirink, J. (2001). Biogenesis of inner membrane proteins in *Escherichia coli. Mol. Microbiol.* **40,** 314–322.

Denzer, A. J., Nabholz, C. E., and Spiess, M. (1995). Transmembrane orientation of signal-anchor proteins is affected by the folding state but not the size of the N-terminal domain. *EMBO J.* **14,** 6311–6317.

Do, H., Falcone, D., Lin, J., Andrews, D. W., and Johnson, A. E. (1996). The cotranslational integration of membrane proteins into the phospholipid bilayer is a multistep process. *Cell* **85,** 369–378.

Engelman, D. M., and Steitz, T. A. (1981). The spontaneous insertion of proteins into and across membranes: the helical hairpin hypothesis. *Cell* **23,** 411–422.

Fujiki, Y. (2000). Peroxisome biogenesis and peroxisome biogenesis disorders. *FEBS Lett.* **476,** 42–46.

Gafvelin, G., and von Heijne, G. (1994). Topological "frustration" in multi-spanning *E. coli* inner membrane proteins. *Cell* **77,** 401–412.

Gafvelin, G., Sakaguchi, M., Andersson, H., and von Heijne, G. (1997). Topological rules for membrane protein assembly in eukaryotic cells. *J. Biol. Chem.* **272,** 6119–6127.

Gavel, Y., and von Heijne, G. (1992). The distribution of charged amino acids in mitochondrial inner membrane proteins suggests different modes of membrane integration for nuclearly and mitochondrially encoded proteins. *Eur. J. Biochem.* **205,** 1207–1215.

Gavel, Y., Steppuhn, J., Herrmann, R., and von Heijne, G. (1991). The positive-inside rule applies to thylakoid membrane proteins. *FEBS Lett.* **282,** 41–46.

Glick, B. S., and von Heijne, G. (1996). *Saccharomyces cerevisiae* mitochondria lack a bacterial-type Sec machinery. *Protein Sci.* **5,** 2651–2652.

Goder, V., and Spiess, M. (2001). Topogenesis of membrane proteins: determinants and dynamics. *FEBS Lett.* **504,** 87–93.

Hamman, B. D., Hendershot, L. M., and Johnson, A. E. (1998). BiP maintains the permeability barrier of the ER membrane by sealing the lumenal end of the translocon pore before and early in translocation. *Cell* **92,** 747–758.

Harms, N., Koningstein, G., Dontje, W., Müller, M., Oudega, B., Luirink, J., and de Cock, H. (2001). The early interaction of the outer membrane protein PhoE with the periplasmic chaperone Skp occurs at the cytoplasmic membrane. *J. Biol. Chem.* **276,** 18804–18811.

Heinrich, S., Mothes, W., Brunner, J., and Rapoport, T. (2000). The Sec61p complex mediates the integration of a membrane protein by allowing lipid partitioning of the transmembrane domain. *Cell* **102,** 233–244.

Hell, K., Herrmann, J., Pratje, E., Neupert, W., and Stuart, R. (1997). Oxa1p mediates the export of the N- and C-termini of pCoxII from the mitochondrial matrix to the intermembrane space. *FEBS Lett.* **418,** 367–370.

Hell, K., Herrmann, J. M., Pratje, E., Neupert, W., and Stuart, R. A. (1998). Oxa1p, an essential component of the N-tail protein export machinery in mitochondria. *Proc. Natl. Acad. Sci. USA* **95,** 2250–2255.

Hell, K., Neupert, W., and Stuart, R. A. (2001). Oxa1p acts as a general membrane insertion machinery for proteins encoded by mitochondrial DNA. *EMBO J.* **20,** 1281–1288.

Hermansson, M., Monné, M., and von Heijne, G. (2001). Formation of "helical hairpins" during membrane protein integration into the ER membrane. Role of the N- and C-terminal flanking regions. *J. Mol. Biol.* **313,** 1171–1179.

Honsho, M., and Fujiki, Y. (2001). Topogenesis of peroxisomal membrane protein requires a short, positively charged intervening-loop sequence and flanking hydrophobic segments—study using human membrane protein PMP34. *J. Biol. Chem.* **276,** 9375–9382.

Jayasinghe, S., Hristova, K., and White, S. H. (2001). Energetics, stability, and prediction of transmembrane helices. *J. Mol. Biol.* **312,** 927–934.

Johnson, A. E., and van Waes, M. A. (1999). The translocon: a dynamic gateway at the ER membrane. *Annu. Rev. Cell Dev. Biol.* **15,** 799–842.

Just, W. W., and Diestelkotter, P. (1996). Protein insertion into the peroxisomal membrane. *Ann. N.Y. Acad. Sci.* **804,** 60–75.

Koehler, C. M. (2000). Protein translocation pathways of the mitochondrion. *FEBS Lett.* **476,** 27–31.

Kuroiwa, T., Sakaguchi, M., Mihara, K., and Omura, T. (1991). Systematic analysis of stop-transfer sequence for microsomal membrane. *J. Biol. Chem.* **266,** 9251–9255.

Kutay, U., Ahnert-Hilger, G., Hartmann, E., Wiedenmann, B., and Rapoport, T. (1995). Transport route for synaptobrevin via a novel pathway for insertion into the endoplasmic reticulum. *EMBO J.* **14,** 217–223.

Liao, S., Lin, J., Do, H., and Johnson, A. (1997). Both lumenal and cytosolic gating of the aqueous ER translocon pore are regulated from inside the ribosome during membrane protein integration. *Cell* **90,** 31–41.

Luirink, J., Samuelsson, T., and de Gier, J. W. (2001). YidC/Oxa1p/Alb3: evolutionarily conserved mediators of membrane protein assembly. *FEBS Lett.* **501,** 1–5.

Mant, A., Woolhead, C. A., Moore, M., Henry, R., and Robinson, C. (2001). Insertion of PsaK into the thylakoid membrane in a "horseshoe" conformation occurs in the absence of signal recognition particle, nucleoside triphosphates, or functional Albino3. *J. Biol. Chem.* **276,** 36200–36206.

Molik, S., Karnauchov, I., Weidlich, C., Herrmann, R. G., and Klösgen, R. B. (2001). The Rieske Fe/S protein of the cytochrome b_6/f complex in chloroplasts: missing link in the evolution of protein transport pathways in chloroplasts? *J. Biol. Chem.* **276,** 42761–42766.

Monné, M., Gafvelin, G., Nilsson, R., and von Heijne, G. (1999a). N-tail translocation in a eukaryotic polytopic membrane protein—synergy between neighboring transmembrane segments. *Eur. J. Biochem.* **263,** 264–269.

Monné, M., Hermansson, M., and von Heijne, G. (1999b). A turn propensity scale for transmembrane helices. *J. Mol. Biol.* **288,** 141–145.

Monné, M., Nilsson, I., Elofsson, A., and von Heijne, G. (1999c). Turns in transmembrane helices: determination of the minimal length of a "helical hairpin" and derivation of a fine-grained turn propensity scale. *J. Mol. Biol.* **293,** 807–814.

Nilsson, I., and von Heijne, G. (1998). Breaking the camel's back: proline-induced turns in a model transmembrane helix. *J. Mol. Biol.* **284,** 1185–1189.

Nilsson, I., Witt, S., Kiefer, H., Mingarro, I., and von Heijne, G. (2000). Distant downstream sequence determinants can control N-tail translocation during protein insertion into the endoplasmic reticulum membrane. *J. Biol. Chem.* **275,** 6207–6213.

Ota, K., Sakaguchi, M., von Heijne, G., Hamasaki, N., and Mihara, K. (1998). Forced transmembrane orientation of hydrophilic polypeptide segments in multispanning membrane proteins. *Mol. Cell* **2,** 495–503.

Pause, B., Diestelkotter, P., Heid, H., and Just, W. W. (1997). Cytosolic factors mediate protein insertion into the peroxisomal membrane. *FEBS Lett.* **414,** 95–98.

Pfanner, N., and Geissler, A. (2001). Versatility of the mitochondrial protein import machinery. *Nature Rev. Mol. Cell Biol.* **2,** 339–349.

Rojo, E. E., Guiard, B., Neupert, W., and Stuart, R. A. (1999). N-terminal tail export from the mitochondrial matrix—adherence to the prokaryotic "positive-inside" rule of membrane protein topology. *J. Biol. Chem.* **274,** 19617–19622.

Sääf, A., Andersson, H., Gafvelin, G., and von Heijne, G. (1995). SecA-dependence of the translocation of a large periplasmic loop in the *E. coli* MalF inner membrane protein is a function of sequence context. *Mol. Membr. Biol.* **12,** 209–215.

Sääf, A., Wallin, E., and von Heijne, G. (1998). Stop-transfer function of pseudo-random amino acid segments during translocation across prokaryotic and eukaryotic membranes. *Eur. J. Biochem.* **251,** 821–829.

Sääf, A., Johansson, M., Wallin, E., and von Heijne, G. (1999). Divergent evolution of membrane protein topology: the *Escherichia coli* RnfA and RnfE homologues. *Proc. Natl. Acad. Sci. USA* **96,** 8540–8544.

Sääf, A., Hermansson, M., and von Heijne, G. (2000). Formation of cytoplasmic turns between two closely spaced transmembrane helices during membrane protein integration into the ER membrane. *J. Mol. Biol.* **301,** 191–197.

Sääf, A., Baars, L., and von Heijne, G. (2001). The internal repeats in the Na^+/Ca^{2+} exchanger–related *Escherichia coli* protein YrbG have opposite membrane topologies. *J. Biol. Chem.* **276,** 18905–18907.

Sacksteder, K. A., and Gould, S. J. (2000). The genetics of peroxisome biogenesis. *Annu. Rev. Genet.* **34,** 623–652.

Samuelson, J. C., Chen, M. Y., Jiang, F. L., Moller, I., Wiedmann, M., Kuhn, A., Phillips, G. J., and Dalbey, R. E. (2000). YidC mediates membrane protein insertion in bacteria. *Nature* **406,** 637–641.

Samuelson, J. C., Jiang, F. L., Yi, L., Chen, M. Y., de Gier, J. W., Kuhn, A., and Dalbey,

R. E. (2001). Function of YidC for the insertion of M13 procoat protein in *Escherichia coli*—translocation of mutants that show differences in their membrane potential dependence and Sec requirement. *J. Biol. Chem.* **276,** 34847–34852.

Schäfer, U., Beck, K., and Müller, M. (1999). Skp, a molecular chaperone of gram-negative bacteria, is required for the formation of soluble periplasmic intermediates of outer membrane proteins. *J. Biol. Chem.* **274,** 24567–24574.

Schulz, G. E. (2000). β-Barrel membrane proteins. *Curr. Opin. Struct. Biol.* **10,** 443–447.

Scotti, P. A., Urbanus, M. L., Brunner, J., de Gier, J. W. L., von Heijne, G., van der Does, C., Driessen, A. J. M., Oudega, B., and Luirink, J. (2000). YidC, the *Escherichia coli* homologue of mitochondrial Oxa1p, is a component of the Sec translocase. *EMBO J.* **19,** 542–549.

Sipos, L., and von Heijne, G. (1993). Predicting the topology of eukaryotic membrane proteins. *Eur. J. Biochem.* **213,** 1333–1340.

Soll, J., and Tien, R. (1998). Protein translocation into and across the chloroplastic envelope membranes. *Plant Mol. Biol.* **38,** 191–207.

Stuart, R. A., and Neupert, W. (1996). Topogenesis of inner membrane proteins of mitochondria. *Trends Biochem. Sci.* **21,** 261–267.

Sundberg, E., Slagter, J., Fridborg, I., Cleary, S., Robinson, C., and Coupland, G. (1997). Albino3, an *Arabidopsis* nuclear gene essential for chloroplast differentiation, encodes a chloroplast protein that shows homology to proteins present in bacterial membranes and yeast mitochondria. *Plant Cell* **9,** 717–730.

Tamm, L. K., Arora, A., and Kleinschmidt, J. H. (2001). Structure and assembly of beta-barrel membrane proteins. *J. Biol. Chem.* **276,** 32399–32402.

Tokatlidis, K., Junne, T., Moes, S., Schatz, G., Glick, B. S., and Kronidou, N. (1996). Translocation arrest of an intramitochondrial sorting signal next to TIM11 at the inner-membrane import site. *Nature* **384,** 585–589.

Tusnady, G. E., and Simon, I. (1998). Principles governing amino acid composition of integral membrane proteins: application to topology prediction. *J. Mol. Biol.* **283,** 489–506.

Urbanus, M. L., Scotti, P. A., Froderberg, L., Sääf, A., de Gier, J. W. L., Brunner, J., Samuelson, J. C., Dalbey, R. E., Oudega, B., and Luirink, J. (2001). Sec-dependent membrane protein insertion: sequential interaction of nascent FtsQ with SecY and YidC. *EMBO Rep.* **2,** 524–529.

van der Laan, M., Houben, E. N. G., Nouwen, N., Luirink, J., and Driessen, A. J. M. (2001). Reconstitution of Sec-dependent membrane protein insertion: nascent FtsQ interacts with YidC in a SecYEG-dependent manner. *EMBO Rep.* **2,** 519–523.

van de Vossenberg, J. L., Albers, S. V., van der Does, C., Driessen, A. J., and van Klompenburg, W. (1998). The positive inside rule is not determined by the polarity of the delta psi (transmembrane electrical potential) [letter]. *Mol. Microbiol.* **29,** 1125–1127.

van Klompenburg, W., Nilsson, I. M., von Heijne, G., and de Kruijff, B. (1997). Anionic phospholipids are determinants of membrane protein topology. *EMBO J.* **16,** 4261–4266.

von Heijne, G. (1986). The distribution of positively charged residues in bacterial inner membrane proteins correlates with the trans-membrane topology. *EMBO J.* **5,** 3021–3027.

Wahlberg, J. M., and Spiess, M. (1997). Multiple determinants direct the orientation of signal-anchor proteins: the topogenic role of the hydrophobic signal domain. *J. Cell Biol.* **137,** 555–562.

Wallin, E., and von Heijne, G. (1995). Properties of N-terminal tails in G-protein coupled receptors—a statistical study. *Protein Eng.* **8,** 693–698.

Wallin, E., and von Heijne, G. (1998). Genome-wide analysis of integral membrane proteins from eubacterial, archaean, and eukaryotic organisms. *Protein Sci.* **7,** 1029–1038.

Whitley, P., Grahn, E., Kutay, U., Rapoport, T., and von Heijne, G. (1996). A 12 residues long poly-leucine tail is sufficient to anchor synaptobrevin to the ER membrane. *J. Biol. Chem.* **271,** 7583–7586.

CONSTRUCTION OF HELIX-BUNDLE MEMBRANE PROTEINS

By AARON K. CHAMBERLAIN, SALEM FAHAM, SARAH YOHANNAN, and JAMES U. BOWIE

Department of Chemistry and Biochemistry, University of California, Los Angeles, California 90095

I. Introduction

Integral membrane proteins comprise roughly 30% of all proteins and are involved in many vital cellular processes such as environmental perception, cellular communication, and nutrient metabolism. Consequently, our understanding of cell biology will never be complete without a detailed understanding of membrane protein function and the intermolecular forces that act in the bilayer. Although progress on membrane proteins has been slow because of many technical challenges, reports of new structures and investigations of folding and stability are becoming more frequent as we develop more experimentally accessible systems. In this review we present some of our current understanding of the structure of α-helical membrane proteins and the forces that stabilize their structures. We focus on the folding and stability of membrane proteins within the bilayer rather than on the thermodyamics of helix insertion into the membrane.

ADVANCES IN PROTEIN CHEMISTRY, Vol. 63

0065-3233/03 $35.00

II. Transmembrane Helix Structure

Because transmembrane helices must span the membrane, they are longer on average than helices in soluble proteins, averaging about 26 residues (Bowie, 1997b). The membrane-spanning requirement also limits the orientation of the helices in the bilayer. The greater the tilt angle (the angle between the helix axis and the membrane normal) the longer the helix needs to be to span the bilayer. Tilt angles range from 0° to 40° in a fairly even distribution according to a survey of transmembrane helices (Bowie, 1997b). When normalized for the probability of an observation (0° angles are much less probable than 40° angles in three dimensions), it is clear that transmembrane helices strongly prefer smaller angles. Indeed, single TM helices in the absence of tertiary interactions tend to align with the bilayer normal (Huschilt *et al.,* 1989). These tilt angle restrictions have a significant effect on helix packing interactions (see below).

Many TM helices are not ideal, straight, and regular, as is illustrated by the seven helices of rhodopsin (Palczewski *et al.,* 2000) shown in Fig. 1. TM helices exhibit frequent distortions from ideality, including kinks and regions of π and 3_{10} helix. These distortions can cause the axis of

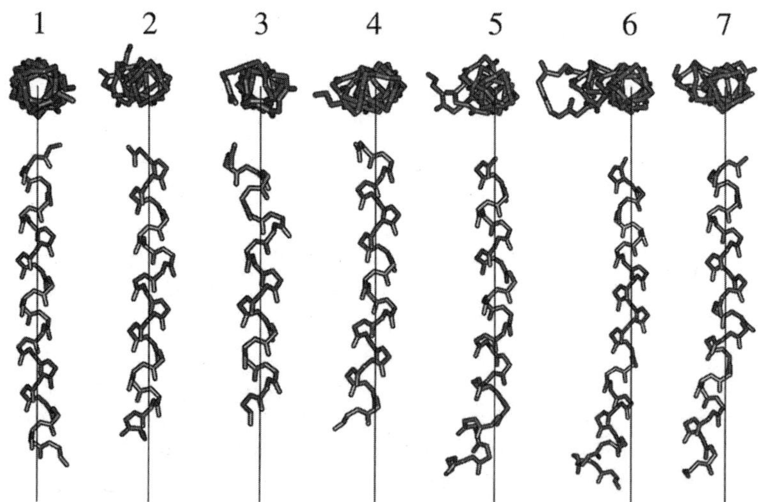

Fig. 1. Curvature or kinks in rhodopsin helices. Individual helices from rhodopsin (1JGJ) (Palczewski *et al.,* 2000) are shown in two perspectives in their order in the amino acid sequence. Lines are drawn approximately through the helix axes for reference. The amino acids shown are 1–26 (helix 1), 34–57 (helix 2), 71–90 (helix 3), 95–117 (helix 4), 122–149 (helix 5), 155–185 (helix 6), and 190–217 (helix 7).

the helix to bend as much as 30° and lead to irregularities in side-chain placement (Riek *et al.,* 2001). Oftentimes, these distortions are associated with Pro residues. Pro is strongly destabilizing in the middle of a helix because of steric clashes and the loss of a backbone hydrogen bond. Nevertheless, Pro is relatively common in TM helices, probably because of the high stability of helices in membranes (White and Wimley, 1999). In addition, weak C–H · · · O hydrogen bonds have been noted between the C_δ atom of the Pro side chain and the otherwise unsatisfied carbonyl oxygen of the preceding residue, which may help to minimize helix destabilization (Chakrabarti and Chakrabarti, 1998). The presence of a Pro residue does not always lead to significant helix distortions, however, and about one-third of the observed helical distortions do not involve Pro (Riek *et al.,* 2001). The cause of these nonproline distortions is not obvious and the extent to which the local sequence versus tertiary structure defines the helical distortions is still not clear. These noncanonical helical regions may be functionally important, providing weak points in the helix that allow for conformational changes in the membrane (Jacob *et al.,* 1999; Sansom and Weinstein, 2000). For example, Jacob *et al.* (1999) found that a GXXP motif in alamethecin was responsible for bending motions in the peptide. Interestingly, both the Pro and the preceding Gly were required for this flexibility, suggesting that it may be possible to define local sequence motifs that facilitate motion in transmembrane domains.

A. *Transmembrane Helix Packing*

The range of helix packing angles in membrane proteins is more restricted than in soluble proteins and the distribution of helix packing angles shows a strong peak at around +20° (Fig. 2) (Bowie, 1997b). In fact, 62% of helix packings fall in the range from 0° to +40°. Part of the preference for these packing angles is due to the restrictions on tilt angles (noted above) that lead to a high probability of observing small packing angles. It is not the only factor, however, since small positive packing angles are greatly favored over small negative angles. Small positive packing angles are likely to be favored because of side-chain packing constraints. Packing angles around +20° allow side-chain interdigitation without steric conflicts according to both the ridges-into-grooves or knobs-into-holes helix packing models (Crick, 1953; Chothia *et al.,* 1981; Walther *et al.,* 1996). Thus, the +20° angle allows packing of long helices over much of their length without steric conflicts. Langosch and Heringa pointed out that contacting TM helices can exhibit a superhelical twist commonly observed in coiled–coil

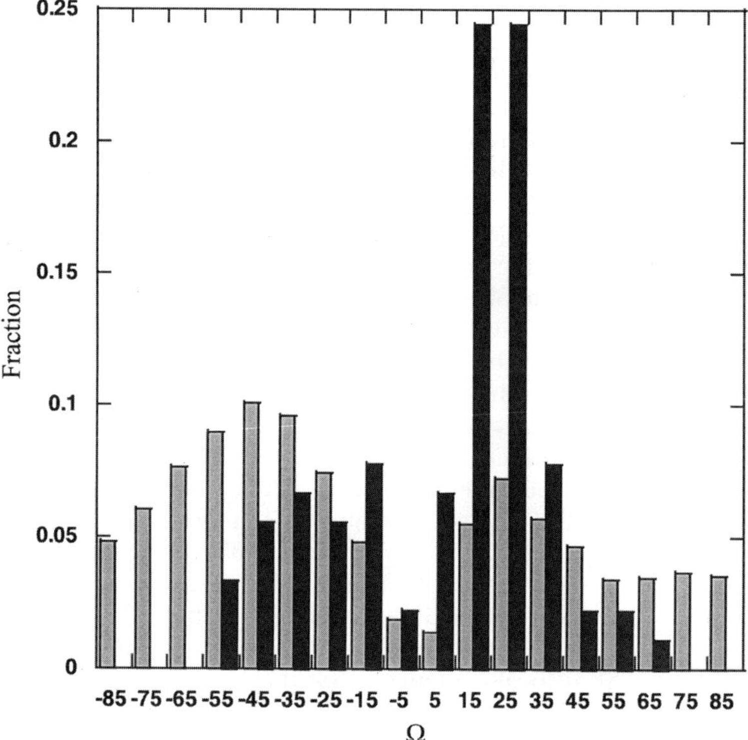

FIG. 2. Helix packing angle (Ω) distributions. Gray bars: The distribution of angles seen in 3498 helix packing interactions from soluble proteins. Black bars: The distribution seen in 88 helix interactions from the membrane proteins, bacteriorhodopsin, bovine cytochrome *c* oxidase, and photosynthetic reaction center (Bowie, 1997a).

proteins (Langosch and Heringa, 1998). This superhelical bending allows helices to extend their area of contact where straight helices would diverge.

The majority of helix packing interactions in membrane proteins are antiparallel, but this preference is greatly influenced by topological constraints. Transmembrane helices that are adjacent in sequence virtually always pack against each other and topological restrictions dictate that these be antiparallel. If one examines helix packings between protein subunits, which are largely free of topological constraints, the preference for antiparallel packing largely disappears (Bowie, 1997b). Although helix dipole interactions might be expected to favor antiparallel packings, it must not be a strong factor. Calculations by Ben-Tal and Honig suggest that helix dipole interactions largely disappear due to solvation once the helix emerges from the membrane (Ben-Tal and Honig, 1996).

B. *Residue Environment Preferences*

When the first soluble protein structures were revealed, it was immediately obvious that there was a striking preference for hydrophobic side chains in the interior and polar side chains on the surface (Perutz *et al.,* 1965). This simple observation has had a powerful impact on our ability to predict structural features of soluble proteins (Eisenberg and McLachlan, 1986; Bowie *et al.,* 1990, 1991). An examination of the first membrane protein structures found a much more subtle contrast between interior and surface residues. Rees *et al.* (1989a) found that the interiors of membrane proteins are about as hydrophobic as the interiors of soluble proteins, but the lipid facing residues were somewhat more apolar. Thus, residue preferences for buried or surface environments are much weaker for TM helices than for soluble proteins. A number of more recent surveys have noted a preference for small residues at helix interfaces (Javadpour *et al.,* 1999; Jiang and Vakser, 2000). Possible reasons for this preference include the smaller entropy cost for fixing small side chains on folding, or better packing by the smaller residues. Consistent with the latter possibility, several groups report better packing for small side chains than for large side chains in membrane proteins (Javadpour *et al.,* 1999; Adamian and Liang, 2001). The opposite is apparently true for soluble proteins (Adamian and Liang, 2001). The origin of this improved packing still remains to be explained.

One of the most conspicuous features of residue distributions in membrane proteins is an "aromatic belt," which sandwiches the apolar transmembrane domain. When aromatic residues are specifically displayed in membrane protein structures, a clustering in the interfacial region is often quite obvious. One example is shown in Fig. 3 (see color insert). Umschneider and Samsom (2001) have noted that this irregular distribution is restricted to Trp, His, and Tyr side chains, with Phe showing little preference for its location in the membrane. Thus, it appears that aromaticity and amphilicity are the preferred features. This aromatic belt reflects the favorable partitioning of aromatic amino acids into the interfacial region (Wimley and White, 1996; Yau *et al.,* 1998) and may act to stabilize the orientation of the entire membrane protein or individual helices in the bilayer.

III. Thermodynamic Studies

A. *Models*

Several models have emerged to describe the folding process of membrane proteins. In 1990, Engelman and co-workers proposed a two-stage folding model (Popot and Engelman, 1990) based on experimental results of bacteriorhodopsin fragments. [See Fig. 4A and Booth (2000).] In stage 1, the helices form and insert into the membrane. The unfavorable entropy of folding the backbone is overcome by the favorable partitioning of the hydrophobic side chains into the membrane. Deber and co-workers have defined a threshold hydrophobicity that, once surpassed, enables helix insertion into the bilayer (Liu *et al.*, 1996). Indeed, hydrophobicity is such a strong determinant of partitioning and consequent secondary structure formation that algorithms for predicting the TM segments are remarkably successful compared to secondary structure prediction algorithms in soluble proteins (Rost

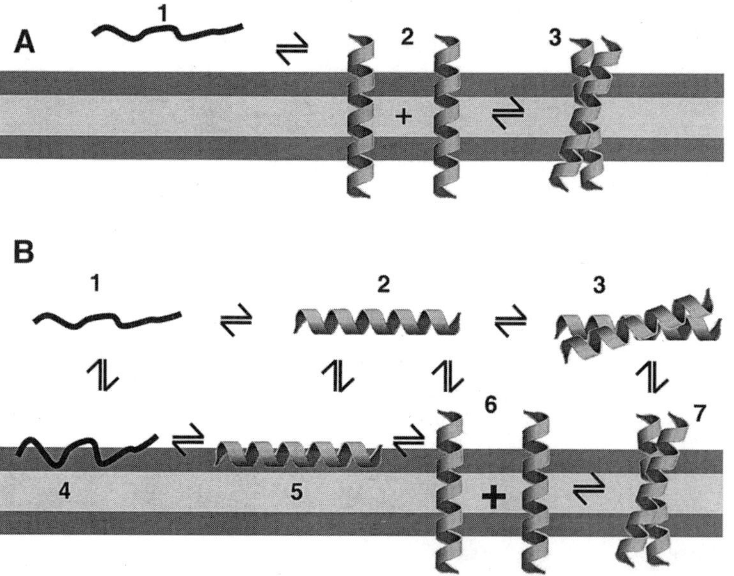

Fig. 4. Thermodynamic models of membrane protein folding. (A) In the two-step model of folding, the TM helices fold and insert into the membrane and subsequently move laterally to associate (Popot and Engelman, 1990). (B) White and co-workers constructed an expanded model, which includes the protein residing in water (states 1–3), the interfacial region (states 4 and 5), and the membrane core (states 6 and 7) (White and Wimley, 1999).

et al., 1996; Deber *et al.*, 2001). In stage 2, the individual helices associate within the membrane to form the folded structure. Consistent with this model, some individual helices of bacteriorhodopsin were shown to insert into bilayers as isolated helices. The tertiary structure reformed on mixing the fragments with the complementary fragments containing the other helices (Popot *et al.*, 1987; Kahn *et al.*, 1992; Hunt *et al.*, 1997; Marti, 1998; Luneberg *et al.*, 1998). Thus, many TM segments appear to be independently stable and follow this two-stage folding model.

The two-stage model is conceptually very useful because it separates the thermodynamics of insertion from helix assembly. When examining the stability of membrane proteins in the bilayer, the second stage is presumably the most relevant since the extrusion of the protein back into the aqueous environment should be extremely unfavorable. This idea is supported by the experiments discussed below. The two-stage model may not contain sufficient detail to describe the folding process in all cases, however. For example, the F and G helices of bacteriorhodopsin do not spontaneously form helices in vesicles, indicating that these peptides require assistance from the remainder of the protein to fold properly (Hunt *et al.*, 1997).

To develop a more complete thermodynamic framework for membrane protein folding, White and co-workers have created a three-stage and then a four-stage model of folding (White and Wimley, 1999). As shown in Fig. 4B, the model includes various folding and partitioning steps. The protein may be unfolded (states 1 and 4), folded into noninteracting helices (states 2, 5, and 6), or fully folded with the TM helices packed together (states 3 and 7). These different conformations can exist in water (states 1–3), the interfacial region (states 4 and 5), or the hydrocarbon region of the membrane (states 6 and 7). Cofactor binding to the packed helices and loop folding would be additional states not shown. Certainly, this model is a more complete formalism for discussing membrane protein stability. The experimental determination of the free energies involved is a Herculean task, however. Many of the states are not significantly populated or are prone to aggregation. Nevertheless, a quantitative description of all the folding equilibria is the ultimate prize.

B. *Experiments*

Thermodynamic measurements will, in the end, provide the quantitative descriptions of the atomic interactions needed for computational predictions of membrane proteins. With membrane proteins, of course, these interactions must include all the interactions among the water, its solutes, the membrane, and the protein itself. Unfortunately, true thermodynamic measurements are difficult with membrane proteins, because it is often difficult to find solution conditions that yield reversible transitions. Usually we must rely on nonequilibrium measurements, such as the time required for inactivation at a given temperature, to judge the relative stability of different proteins. Furthermore, the systems used are often complex and the nature of the denatured state is still poorly understood, limiting our ability to interpret the free energies found in thermodynamic folding studies. Thus, many of our ideas concerning the forces that stabilize membrane proteins are based on theoretical arguments or qualitative comparisons. Ultimately, equilibrium measurements of stability will be essential for validating theories and for developing a better understanding of the dominant forces in membrane protein stabilization.

Measurement of unfolding free energies requires the ability to measure the relative populations of folded and unfolded proteins. This measurement is usually difficult under native conditions because only a tiny fraction of the protein is unfolded and this fraction is difficult to detect. Consequently, unfolding free energies are usually measured under conditions that produce marginal stability and the results are extrapolated back to conditions without denaturant. In membrane protein experiments, the protein stability has been modulated by urea, guanidine hydrochloride, pH, sodium dodecyl sulfate (SDS), and temperature. Reversible unfolding has been achieved in a few cases with guanidine hydrochloride, urea, and SDS, but not with temperature (Surrey and Jahnig, 1992; Klug *et al.*, 1995; Lau and Bowie, 1997; Klug and Feix, 1998; Chen and Gouaux, 1999).

Thermal unfolding has been examined for a number of membrane proteins, including bacteriorhodopsin, cytochrome *c* oxidase, band 3, and photosystem II. These studies have been thoroughly reviewed by Haltia and Freire (1995), so we will simply highlight some of the general conclusions that seem to emerge: (1) Complete thermal denaturation of the membrane domains is invariably irreversible. In the case of bacteriorhodopsin, several studies reported that the unfolding transition was scan rate independent, suggesting that thermodynamic analysis could still be applied (Galisteo and Sanchez-Ruiz, 1993; Shnyrov *et al.*, 1994). This conclusion was subsequently disputed, however (Galisteo and Sanchez-Ruiz, 1993; Shnyrov *et al.*, 1994). (2) Unfolding of the transmembrane domains involves minimal loss of helical secondary structure. For example, complete denaturation of bacteriorhodopsin leads to loss of only about 15% of the helical content (Kahn *et al.*, 1992) and in bacterial cytochrome *c* oxidase, a protein with large extramembrane domains, roughly two-thirds of the helical content remains after complete denaturation (Haltia *et al.*, 1994). (3) The denaturation enthalpies are generally much lower than for soluble proteins. In fact, the denaturation enthalpies of membrane proteins can be largely accounted for by denaturation of the water-soluble portions only (Haltia and Freire, 1995). Overall, the results are consistent with the idea that the membrane-embedded portions of transmembrane proteins are more stable than the soluble portions. At high temperatures much of the transmembrane helical structure, and perhaps some of the tertiary interactions, remains intact.

Equilibrium unfolding experiments using SDS as a denaturant have been performed on bacteriorhodopsin and diacylglycerol kinase (DGK) (Lau and Bowie, 1997; Chen and Gouaux, 1999). In these experiments, increasing concentrations of SDS are added to a protein solubilized in a nondenaturing detergent or lipid/detergent mixture. For both systems the unfolding curves were largely reversible. In the case of DGK, the unfolding occurred within the detergent micelles since the unfolding curves depended on the mole fraction of SDS rather than its bulk concentration (Lau and Bowie, 1997). For bacteriorhodopsin, SDS denaturation exhibited a single unfolding transition, whereas for DGK two unfolding transitions were observed. The soluble domain of DGK unfolded at lower SDS concentrations than the transmembrane portions, suggesting that the transmembrane domain has higher overall stability—a result consistent with the thermal denaturation studies. Extrapolations of the unfolding free energies to zero denaturant, with admittedly dubious theoretical justification, indicated stabilities of 6 kcal/mol for the soluble domain and 16 kcal/mol for the membrane embedded domain at 25°C. Thus, the membrane-embedded domain appears to be at the high end of the typical stability range of soluble proteins, even in detergent solution. Presumably, the stability would be even higher in a bilayer environment since detergent solubilization is invariably destabilizing (Brouillette *et al.*, 1989; Bowie, 2001).

The nature of the unfolded state in denaturant and how it relates to the denatured state under native conditions in the bilayer is a major issue in all denaturation experiments. Thermodynamic arguments from the two-stage model suggest that the relevant denatured state has lost its tertiary structure and maintained the transmembrane helix secondary structure. As noted above, CD spectra on thermally denatured bacteriorhodopsin suggest that the denatured protein maintains most of its helical secondary structure. The extent to which tertiary structure is disrupted is unclear, however. It is possible that some stable interhelical interactions are maintained even at high temperature. The helical secondary structure content is also maintained in SDS micelles, and near-UV circular dichroism (CD) spectra suggest substantial loss or rearrangement of the tertiary structure in diacylglycerol kinase (Zhou and Bowie, unpublished). Thus, the SDS unfolding may reasonably mimic the second step of the two-stage model and has the advantage of being reversible. Clearly, however, the complexities of detergent mixing in the protein–detergent complex may be difficult to separate from the protein stability.

Measurement of dissociation constants of TM helix peptides is perhaps the most promising approach for evaluating the strengths of specific interactions. For example, equilibrium sedimentation and fluorescence quenching have been used to obtain dissociation constants for various TM peptides in detergent solutions (Fleming *et al.*, 1997; Fisher *et al.*, 1999; Choma *et al.*, 2000; Zhou *et al.*, 2000, 2001; Fleming and Engelman, 2001; Gratkowski *et al.*, 2001). These systems offer the opportunity to compare the free energies of association for TM helix variants. The fact that the measurements must be made in detergents rather than in a lipid bilayer is a drawback. Fleming and Engelman (2001) showed, however, that the stabilities of a series of glycophorin A transmembrane helix mutants follow the same order in different environments, suggesting that relative stabilities may be environment independent. Thus, it may be possible to obtain a quantitative picture of the contributions of particular interactions in the bilayer from measurements in detergent. Naturally it would be ideal to obtain equilibrium measurements in a bilayer environment, but this is extremely challenging. In some cases with weak association, it has been possible to use fluorescence quenching to measure dissociation constants (Mall *et al.*, 2001). Additionally, Isenbarger and Krebs (2001) have developed a method to measure the equilibrium stability of the bacteriorhodopsin lattice and have used it to measure the lattice stability of various mutants. These sorts of experiments, which quantitatively measure the association or folding reactions in different conditions, will allow us to understand the energetic contributions of the different forces that stabilize membrane proteins.

IV. The Contribution of Loops versus Transmembrane Helices

Is the manner in which transmembrane (TM) helices pack together dictated by the sequences contained within the membrane or by the extramembranous sequences, or "loops"? There is now considerable evidence that most loops are not essential in specifying the fold of membrane proteins. First, in many cases, the TM helices can encode considerable information for specifying the fold. Many single TM helices, such as the TM helix from glycophorin A, self-associate in the absence of their extramembranous domains (Lemmon *et al.*, 1992a,b). Second, there are many examples in which the loops between TM helices in membrane proteins can be clipped without losing the ability to fold and function. For example, Lac permease, when expressed as two complementary halves in *E. coli,* reassembles into a functional permease (Bibi and Kaback, 1990). Two of the three loops between TM helices could be cleaved in the anion channel, CLC-1, without loss of activity (Schmidt-Rose and Jentsch, 1997). The band 3 anion exchanger in red blood cells can even be cleaved up to 4 times in the loops and still retain function (Groves *et al.,* 1998). Moreover, a peptide of the first two helices of bacteriorhodopsin can associate specifically with a fragment containing the remaining helices to reconstitute activity (Marti, 1998). Finally, the loops of bacteriorhodopsin can be shortened (Gilles-Gonzalez *et al.,* 1991) or replaced with heterologous sequence (Allen *et al.,* 2001; Kim *et al.,* 2001) without the loss of function.

Despite the fact that many loops are not critical for the organization of some TM helices, loops still play a stabilizing role. In the case of bacteriorhodopsin, clipping of the loops or replacing them with heterologous sequences leads to a significant loss of stability. Katragadda *et al.* (2000) showed that the solution structure of three loop peptides from bacteriorhodopsin resemble the structure of the same sequence in the full-length structure. This finding suggests the sequences may have been optimized for a particular geometry. Moreover, like loops, some TM helices can be replaced with heterologous sequences without loss of function (Pohlschroder *et al.,* 1996; Guzman *et al.,* 1997; Zhou *et al.,* 1997). In diacylglycerol kinase, a variant in which the entire first TM helix is replaced with polyalanine still maintains activity within twofold of the wild-type protein (Zhou *et al.,* 1997). This helix is nevertheless an integral part of the structure since sequence changes within this helix both stabilize (Zhou and Bowie, 2000) and destabilize the structure (Zhou *et al.,* 1997) and key active site residues occur both N-terminal and C-terminal to it (Wen *et al.,* 1996). This transmembrane helix must be largely passive in the sense that the sequence does not play an active role in defining the structure of the protein. Thus, while some TM helices encode structural specificity, other helices do not and the loop regions contribute significantly to the overall stability of membrane proteins.

V. FORCES THAT STABILIZE TRANSMEMBRANE HELIX INTERACTIONS

A. *Lipid Interactions with Transmembrane Proteins*

Lipid bilayers are complex environments and the activity of membrane proteins can be modulated by the overall properties of the lipid bilayer, as well as by specific interactions with individual lipids. Overall bilayer properties that can influence membrane protein stability include thickness, curvature, and rigidity. These properties can differ in the various cell compartments. Moreover, within a given compartment, different bilayer environments can exist. Although the ideal fluid mosaic model of biological membranes includes the free diffusion of lipids and complete mixing of all the membrane components, the heterogenous mixing of certain lipids and the formation of lipid microdomains, or "rafts," occur in the bilayer (Kurzchalia and Parton, 1999). Rafts rich in sphingolipids and cholesterol have been implicated in important biological functions such as protein sorting and cell signaling (Simons and Ikonen, 1997; Brown and London, 1998; Ikonen, 2001). Rafts can selectively include or exclude various proteins. For example, certain rafts attract glucosylphosphatidylinositol (GPI) anchored proteins (Hooper, 2001). The targeting of the raft-attached proteins could be directed by the destination of the raft. A growing body of literature supports the involvement of rafts in vesicle formation and protein sorting, such as the apical targeting of proteins in epithelial cells (Ikonen and Simons, 1998) and the sorting of prohormone convertase 2 (Blazquez *et al.*, 2000). Preferential sorting of proteins into microdomains could have important concentration effects in organizing membrane protein complexes (Mall *et al.*, 2001). Receptor clustering, an important step in signal transduction, sometimes occurs in lipid microdomains. The actual clustering of the receptor and the recruitment of other required proteins can enhance the signal processing. Rafts have been implicated in the signaling of various receptors, such as the FcεRI receptor (Sheets *et al.*, 1999), the T-cell receptor (Langlet *et al.*, 2000), the EGF receptor (Waugh *et al.*, 1999), and the ephrin-B1 receptor (Brueckner *et al.*, 1999). The different bilayer environments could also play important roles in modulating membrane protein structure and stability. In this section, we describe some of the ways different bilayer properties may influence membrane proteins.

1. *Hydrophobic Matching*

Hydrophobic matching occurs when the thickness of the hydrophobic surface of the lipid bilayer equals the hydrophobic thickness of a membrane protein. The stability of membrane proteins may be enhanced by hydrophobic matching or reduced if the hydrophobic region is either too long or too short, creating a "hydrophobic mismatch." For example, the activity of diacylglycerol kinase (DGK) is dependent on the length of the hydrophobic tails of the bilayer (Pilot *et al.*, 2001). Lipids with tail lengths of 18 carbons were found to enhance DGK activity over the activity seen in lipids of tail lengths 16 and 20. Maneri and Low (1988) found that the stability of Band3 increases with increasing hydrocarbon

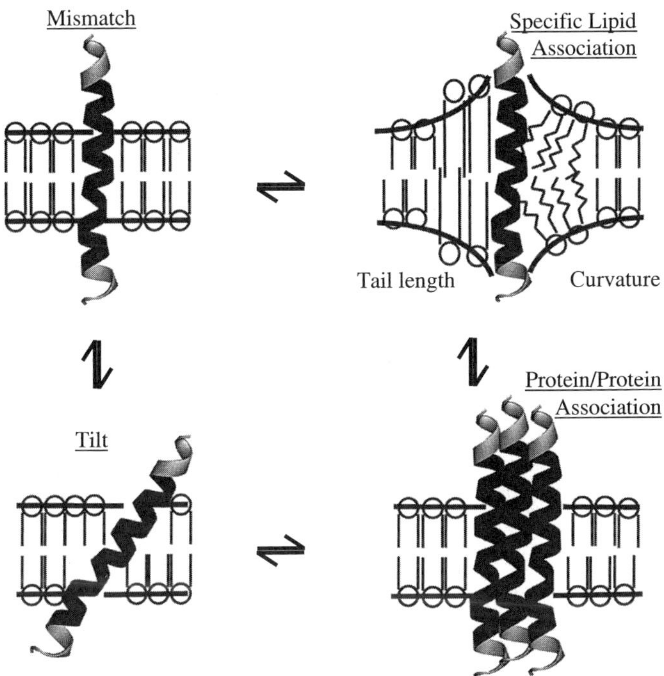

FIG. 5. Responses of a TM protein to hydrophobic mismatch. The hydrophobic regions of a TM protein (black regions) may be too long for the lipid core, creating a mismatch. To help reduce this stress, the protein may change its tilt angle or undergo more favorable associations. The protein may associate with a specific lipid, with a different tail length or curvature, or with another protein to reduce the lipid-facing surface area.

chain length. Hydrophobic matching could also play a role in determining the tilt angle of transmembrane helices as shown in Fig. 5. Indeed, Williamson *et al.* (2002) found that the structure of KCSA adjusts to match the thickness of the bilayer in which it is placed.

The hydrophobic matching principle can also lead to heterogenous mixing of lipids in biological membranes. The length of the hydrocarbon tails found in membrane lipids varies between 12 and 24 carbons, potentially leading to a sizable ~100% change in the thickness of the hydrophobic region of the corresponding bilayers. If different lipid species cannot match their hydrophobic regions, a lipid might preferentially associate with its own type, forcing a partitioning into microdomains (Lehtonen *et al.*, 1996). Furthermore, membrane proteins may force a nonrandom distribution of lipids in the bilayer. When bacteriorhodopsin is in mixed lipid vesicles of dilauroylphosphatidylcholine

(DLPC) and distearoylphosphatidylcholine (DSPC), it preferentially associates with DLPC at low temperatures (Dumas *et al.,* 1997). Therefore, in the presence of bacteriorhodopsin, the mixing of lipids within the bilayer is not complete as would be expected in the ideal fluid mosaic model. This preferential association of the protein with a certain lipid has been referred to as "molecular sorting" (Dumas *et al.,* 1997).

Hydrophobic matching is just one example in which a region of a molecule, that is, the protein, has chemical properties that match the properties of its surroundings, the lipids. With this view in mind, we should also consider how well the surface of the membrane protein matches the interfacial and the water-exposed regions of the bilayer. For example, aromatic residues have an affinity for the interface region (Wimley and White, 1996; Yau *et al.,* 1998) and the positively charged residues have an affinity for the phosphate head groups.

Despite our simple representations of the membrane, the borders separating the water, the interfacial regions, and the core regions are not clearly defined and discrete. A better description might be to view the positions of the chemical groups of the bilayer in distributions that have a certain mean position and a sizable width (Nagle and Tristram-Nagle, 2000). These "fuzzy" boundaries between regions in the bilayers should give the bilayer some flexibility in trying to relieve the stress caused from mismatching chemical properties. Figure 5 illustrates how some membranes may adjust to relax the strain of a hydrophobic mismatch. The strain may lead to the tilting of the protein or the formation of protein–protein or lipid–protein associations.

2. *Lateral Capillary Forces*

A hydrophobic mismatch between a membrane protein and the surrounding lipids may create a lateral force that would pull membrane proteins together. A general theoretical description of this force, referred to as a "lateral capillary force," has been presented by Kralchevsky and co-workers (Kralchevsky, 1997; Kralchevsky and Nagayama, 2000). Although experimental verification of this force for membrane proteins in a bilayer has not been demonstrated, the force can be observed in larger systems, such as 1.7 μm latex beads at an air/water interface, and would be expected to operate on membrane proteins (Kralchevsky, 1997).

Basically, the lateral force originates from the deformation of the bilayer in the presence of a protein with a hydrophobic region of different thickness than the bilayer. An example in which the hydrophobic region of the protein is too long for the bilayer tails is illustrated in Fig. 6. (Similar arguments apply if the protein is too short for the lipid tails.)

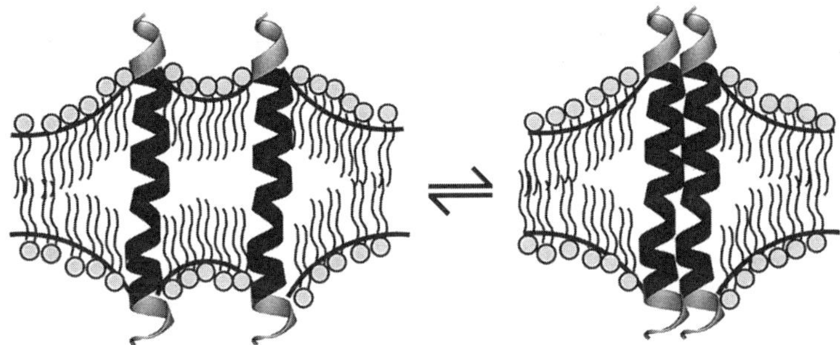

FIG. 6. Lateral capillary forces can assist protein–protein associations. A meniscus is formed around each TM protein when the hydrophobic regions of the protein (black regions) are not the same length as the hydrophobic regions of the membrane. The unfavorable deformation can, in part, be relieved by the lateral association of two proteins, which reduces the surface area the proteins expose to lipids.

The deformation in the bilayer, the meniscus, forms around the protein to reduce the exposure of the hydrophobic region. If two protein molecules contact each other, the number of deformed lipid molecules would be reduced because of the reduction in the exposed surface area of the protein.

The magnitude of this force for membrane proteins is unknown and is certain to be affected by various properties of the bilayer. First, the lateral capillary force diminishes as the mismatch becomes smaller. Second, the diagram in Fig. 6 presumes that the lipids do not have an inherent curvature. Many lipids have a tendency to curve around their head groups, however (see below). In the mismatch shown, the meniscus could actually relieve the curvature stress created from the lipids tendency to curve. If the shape of the meniscus matched the curvature of the lipids, no lateral capillary force would be produced. On the other hand, the opposite hydrophobic mismatch, with the lipid tail being longer than the protein, would exacerbate the curvature stress and produce a larger lateral capillary force. In this manner, the lipid curvature and the hydrophobic length both interact to determine the magnitude of the lateral capillary force for a given membrane protein.

3. Curvature

Bilayers are ideal for cylindrically shaped lipids, where the hydrocarbon tails are the same size as the head groups (see Fig. 7). The tails of many lipids prefer to occupy a larger area, however, giving these lipids an inherent tendency to curve. The flattening of these lipids into a planar bilayer causes a curvature stress, in which the center of the bilayer

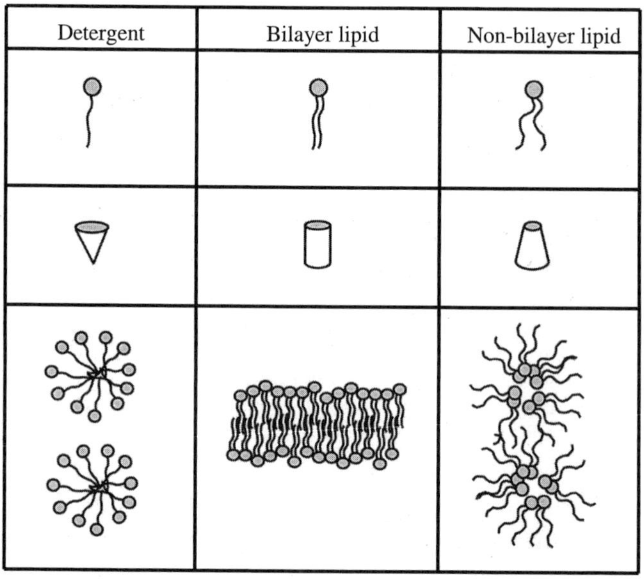

Detergent	Bilayer lipid	Non-bilayer lipid

FIG. 7. Schematic diagrams of different phases of amphiphiles. Amphiphiles, such as detergents with large head groups (gray) containing only one small hydrophobic chain (tail), can be thought of as cones that pack together to form spherical micelles. Bilayer-forming lipids are more cylindrical in shape and pack together into planes. The head groups of nonbilayer lipids occupy less area than the tails and therefore the lipids form inverted phases.

is at higher pressure than the head groups. In some solutions containing a single lipid component, such as cardiolipin or monogalactosyldiacylglycerol (MGDG), the inherent curvature leads to the formation of other lipid phases, such as the inverted hexagonal H_{II} phase. These non-bilayer-forming lipids are present in biological membranes, but in these cases, the formation of the bilayer is maintained by the other lipids or membrane proteins. For example, the inverted hexagonal phase of MGDG, a thylakoid membrane lipid, is inhibited by the addition of the light-harvesting complex II, a major protein of the thylakoid membrane (Simidjiev *et al.*, 2000). It was also shown that cytochrome *c* oxidase can help cardiolipin into a bilayer structure (Rietveld *et al.*, 1987). On the other hand, the formation of nonlamellar phases was induced by the addition of a designed helical TM peptide (van der Wel *et al.*, 2000). In this way, the interaction of the lipids and the TM proteins determines the phase behavior of the entire system.

Curvature stress could play an important role not only in the lipid phase but also in membrane protein stability. A membrane protein might relieve the curvature stress if it is shaped like an hourglass, that is, its cross section in the center of the membrane is small compared to its cross section at the edges. Alternatively, an egg-shaped protein would exacerbate the stress. The importance of curvature was demonstrated in refolding experiments with bacteriorhodopsin, where changes in curvature stress influenced the protein yield on refolding (Curran *et al.*, 1999).

A different type of curvature in bilayers can also result from the unequal distribution of lipids between the two monolayers, with one monolayer having more lipid molecules than the other. This curvature is different from the curvature stress described above in that it originates from the total number of lipids on each side of the membrane and not from an inherent curvature in each lipid molecule. This curvature could be important in the budding of lipid vesicles (Huttner and Zimmerberg, 2001) and would be expected to influence and be influenced by the curvature stress described above.

4. *Specific Interactions*

Membrane proteins can also have specific interactions with individual lipids. Bound lipids have now been observed in various high-resolution crystal structures of membrane proteins. A particularly striking example is the 1.55 Å structure of bacteriorhodopsin (PDB code 1C3W) with 14 lipids bound per monomer (Luecke *et al.*, 1999). Specific interactions have been found both with the head groups and with the hydrocarbon tails. For example, in the bacteriorhodopsin and photosynthetic reaction center structures, the hydrocarbon chains can be seen to follow grooves on the surface of the proteins (Luecke *et al.*, 1999; McAuley *et al.*, 1999). In the bacterial cytochrome *c* oxidase, interactions with the head groups of two phosphatidylcholine lipids are stabilized by salt bridges to Arg side chains (Harrenga and Michel, 1999). The conformation of the bound lipids does not always conform to bilayer geometry, underscoring the influence of membrane proteins on their surrounding lipids.

Some of these bound lipids may play significant roles in stabilizing the folded structure of membrane proteins. Indeed, removal of cardiolipin from cytochrome *c* oxidase inactivates the enzyme (Abramovitch *et al.*, 1990). Specific lipid interactions may also play a role in the folding process. Dowhan and co-workers have shown that Lac permease will not fold properly in the absence of phosphatidylethanolamine (PE), but once folded, it does not require PE (Bogdanov and Dowhan, 1999). Thus, PE appears to possess the properties of a chaperone, perhaps stabilizing an obligatory folding intermediate. With the high-resolution structures in hand, it should now be possible to probe the contributions of some of these lipid interactions using a combination of mutagenesis and lipid analogues.

The entire thermodynamic system of the membrane and TM protein must be considered to understand how the protein and bilayer achieve their native state. We have summarized four of the mechanisms, hydrophobic matching, tilt angles, and specific protein/lipid and protein/protein interactions that are important in determining the stability (Fig. 5). Other important factors, such as the stability of lipid/lipid interactions, have been left out of our protein-centric view. We describe a hydrophobic mismatch as an unfavorable interaction that can be relieved by the other three processes, but we would expect all these properties of the system to interact. We could easily describe the same equilibria by saying that a strain in curvature is relieved by a hydrophobic mismatch or that strong protein/protein packing interactions might help relieve the hydrophobic mismatch or curvature stress. The complex interplay between all these interactions is at the heart of what determines membrane protein stability and will no doubt be difficult to quantify.

B. van der Waals Interactions

Any cursory look at the structures of membrane proteins will reveal that the side chains between neighboring helices make close contacts and interdigitate, implying that these van der Waals contacts stabilize the folded structure. The solution structure of the TM portions of glycophorin A (GpATM) suggested that van der Waals interactions may be a dominant force in membrane protein stabilization (MacKenzie *et al.*, 1997), because the structure revealed no traditional hydrogen bonds or charge/charge interactions. (See Section V, C on hydrogen bonding for an alternative possibility.) Glycophorin A dimerization occurs through a critical GxxxG motif in which the two Gly residues lie on the same side of the TM helix and allow the close approach and packing of the two helices. The GxxxG motif can lead to TM helix dimerization in a variety of sequence contexts. For example, the GxxxG motif is sufficient to induce dimerization of polyvaline and polymethione helices (Brosig and Langosch, 1998). Furthermore, in a library of random TM sequences that oligomerize, GxxxG was commonly observed (Russ and Engelman, 1999, 2000). Interestingly, GxxxG was found to be statistically overrepresented in a database of TM segment sequences (Arkin and Brunger, 1998; Senes *et al.*, 2000). All of these lines of evidence demonstrate that the GxxxG motif is sufficient to cause TM helix association, even without the use of charge/charge interactions or traditional hydrogen bonds.

The interest in van der Waals forces as a primary determinant of TM helix association prompted studies on the efficiency of packing of membrane proteins. If van der Waals interactions dominate membrane protein structures, it might be expected that membrane proteins would be very well packed together. Are membrane proteins better packed than soluble proteins, in which the hydrophobic effect can operate? At the moment the answer is not clear. In the earliest examination of membrane protein packing, Rees *et al.* (1989b) found the photosynthetic reaction center was about as well packed as soluble proteins. In a more recent examination, Adamian and Liang (2001) found that residues in TM helices are more likely to contact an empty space or "pocket" than helical residues in soluble proteins, suggesting that membrane proteins may be less well packed. In contrast, Eilers *et al.* (2000) compared the packing of helical residues in seven different helical membrane proteins to those of 37 soluble proteins and concluded that the TM helices were generally more tightly packed. The membrane proteins examined had an average packing value of 0.431 compared to 0.405 for the soluble proteins. The packing values of individual membrane proteins span a broad range, however, from 0.389 to 0.469, compared to 0.333 to 0.456 for soluble proteins. Thus, if membrane proteins are better packed than soluble proteins, it appears the difference is modest.

If membrane proteins are not dramatically better packed that soluble proteins, does this suggest that van der Waals interactions are not important? Certainly not. As noted early on in the study of proteins (Richards, 1977), soluble proteins are very well packed, so there is perhaps little room for improvement. There is clearly a strong driving force in both environments to pack well. In both cases for van der Waals interactions to favor the folded state over an unfolded state, the protein must pack better with itself than with the solvent. Thus, well-packed proteins, making extensive van der Waals contacts, occur within and outside membrane bilayers. Even if TM helices are not more tightly packed than soluble helices, van der Waals forces probably play a stronger role in specifying TM helix association because of the reduced role of the hydrophobic effect and the paucity of charged residues.

C. *Hydrogen Bonding*

1. *Traditional Hydrogen Bonds Involving Nitrogen and Oxygen*

Hydrogen bonds can be extremely stable in the membrane where there is a low dielectric and limited competition from water. The energetic cost of moving a non-hydrogen-bonded peptide unit into a hydrophobic environment from water is much higher than the cost of burying the same peptide unit when it forms a hydrogen bond. For example, the transfer free energy of a non-hydrogen-bonded peptide unit from water into carbon tetrachloride is ~6 kcal/mol, whereas a hydrogen-bonded peptide unit costs only ~0.6 kcal/mol (see White and Wimley, 1998; White *et al.*, 1998; White and Wimley, 1999 and references therein). These values imply that, within the membrane, the hydrogen-bonded peptide unit is more stable by ~5 kcal/mol over the non-hydrogen-bonded peptide unit. Thus, the cost of placing an unsatisfied hydrogen bond in the bilayer is enormous, implying that essentially all polar groups in the membrane must be hydrogen bonded. The importance of satisfying potential hydrogen bonds is illustrated by Smith and co-workers. Their magic-angle spinning NMR experiments suggest the structure of the glycophorin A dimer in bilayers is slightly different than the micellar structure, allowing two threonine residues to hydrogen bond in the dimeric interface (Smith *et al.*, 2001).

A striking demonstration of the strength of hydrogen bonding in membranes came from two separate efforts to design TM peptide oligomers based on the soluble GCN4 leucine zipper peptide (Choma *et al.*, 2000; Zhou *et al.*, 2000). In both studies, hydrophobic peptides were designed that preserved the residues buried in the GCN4 dimer, including a single Asn residue which hydrogen bonds across the GCN4 dimer interface. The designed peptides were found to oligomerize into dimers and trimers, but surprisingly, Asn was the only residue critical for oligomerization. In fact, a single Asn in a polyleucine sequence is sufficient to drive oligomerization. Subsequently both groups examined the oligomerization of TM helices with a series of amino acid substitutions at a single position. Residues with two potential hydrogen bonding atoms (Asn, Asp, Gln, and Glu) were very effective in driving oligomerization, whereas other residues were not (Gratkowski *et al.*, 2001; Zhou *et al.*, 2001).

Interhelical hydrogen bonds are fairly common in membrane protein structures. Adamian and Liang (2002) examined the known membrane protein structures and found that interacting helices that have interhelical hydrogen bonds make more intimate contacts than helices that interact without traditional hydrogen bonds. Serine was the most frequent amino acid in hydrogen bonds, followed by tyrosine, histidine, threonine, and arginine. Serine residues were frequently seen to be spaced by seven residues, creating "serine zippers" in analogy to the leucine zippers of coiled–coil peptides (O'Shea *et al.*, 1991).

The recent structure of the glycerol-conducting channel shows a unique feature that at first seems to undermine the importance of hydrogen bonding in TM regions: two helices transverse only halfway through the membrane (Fu *et al.*, 2000). A helix that ends in the membrane core would normally leave unsatisfied backbone hydrogen bonds. The N-termini of these half-helices, M3 and M7, have conserved NPA (Asn-Pro-Ala) motifs, however, and come into contact in the center of the membrane to form a network of hydrogen bonds. The backbone of three turn residues makes hydrogen bonds with the Asn side chains in the NPA motifs. The OD1 atom of Asn-68 hydrogen bonds with the NH of Ala-70 (N $\cdots$ O distance = 2.93 Å) as does the ND2 of Asn-203 with the backbone O of Leu-67 (N $\cdots$ O distance = 2.72 Å). Therefore, although the existence of helical termini in the TM region would not be expected from the energetics of hydrogen bond formation, the glycerol-conducting channel shows how a protein can use nonstandard conformations to satisfy the hydrogen-bonding potential.

As pointed out by Zhou *et al.* (2001), the strength of hydrogen bonding in the bilayer presents a significant danger. Mutations that introduce polar side chains in the bilayer could distort structures or lead to the formation of inappropriate complexes in the drive to satisfy hydrogen bonding potential. Thus, polar residues may need to be carefully protected in membrane protein structures. Indeed, the *neu* oncogene, a mutant of the neu/erbB-2 tyrosine kinase receptor, bears a Val-to-Glu mutation in the transmembrane helix. The mutation is thought to enhance dimerization of the receptor leading to cell transformation (Weiner *et al.*, 1989). Smith *et al.* (1996) have shown that this Glu does indeed form a hydrogen bond in a bilayer. A second example occurs in the cystic fibrosis transmembrane conductance regulator. Indirect evidence suggests that a point mutant causing a mild form of the disease, V232D, disrupts the structure through the formation of a hydrogen bond between the D232 and Q207 (Therien *et al.*, 2001). These examples illustrate how a fundamental understanding of physical forces can lead to a molecular understanding of disease.

2. *Nontraditional Hydrogen Bonds*

CH $\cdots$ O hydrogen bonds are weak hydrogen bonds that are known to influence small molecule energetics and are gaining acceptance as a force that stabilizes proteins. CH $\cdots$ O bonds form when an electronegative carbon atom (the donor) and an oxygen atom (the acceptor) share a hydrogen atom. Like all hydrogen bonds, they include dipole/dipole, monopole/monopole, and van der Waals interactions (Desiraju and Steiner, 1999). The strength of the hydrogen bond is proportional to the strength of the interacting dipoles making hydrogen bonds with oxygen and nitrogen atoms relatively strong. Weaker hydrogen bonds can be made with CH $\cdots$ O, CH $\cdots$ N, OH $\cdots$ π, or NH $\cdots$ π interactions

(π refers to a π electron cloud). In proteins, the strongest CH···O bonds occur with histidine Cε or glycine Cα atoms as hydrogen donors, because they are the most polar carbons. The dipole/dipole interactions make hydrogen bonds directional, as opposed to van der Waals interactions that are radially symmetric. CH···O hydrogen bonds, like all hydrogen bonds, can be identified by having two characteristics: (1) the proper geometric structure, and (2) an influence on the energetics, most often seen as a reduced C—H stretching frequency in the infrared or Raman spectra (Green, 1974). In proteins, hydrogen bonds must simply be identified by the geometry alone, because of the difficulty in isolating one specific donor-hydrogen resonance in IR spectroscopy.

CH···O bonds are well established in small molecule chemistry. In 1937, CH···O bonds were first proposed to contribute to intermolecular interactions [see Green (1974)]. Ramachandran used CH···O bonds to help explain the structures of polyglycine and collagen in the 1960s (Ramachandran and Sasisekharan, 1965; Ramachandran *et al.*, 1966). A survey of 113 small-molecule crystal structures in the Cambridge Structural Database (Taylor and Kennard, 1982) found various features that suggest the cohesiveness of CH···O interactions, including (1) a hydrogen bound to a carbon has a statistical preference to associate with O over another C or H, (2) the H···O distance is often less than the sum of their van der Waals radii, and (3) the CH···O geometry is similar to the typical OH···O hydrogen bond geometry. The enthalpy of formation of the CH···O bond in *N,N*-dimethylforamide dimers has been estimated with high-level, quantum-mechanical calculations to be a somewhat surprising -3.0 ± 0.5 kcal/mol, or about half the strength of a NH···O hydrogen bond (Vargas *et al.*, 2000). This suggests that these interactions could be quite significant in stabilizing protein structures.

Derewenda *et al.* (1995) analyzed the CH···O contacts in 13 protein structures and found that a large proportion of short CH···O contacts show geometric configurations typical of hydrogen bonds. Most of these close CH···O interactions involved Hα protons in β-sheet structures, while short CH···O contacts were less common in helices. The same group has identified a CH···O bond with the histidine Cε in the active site of serine hydrolases (Derewenda *et al.*, 1994). Meadows *et al.* (2000) describe a CH···O interaction in lysozyme with good geometry (distance H···O = 2.76 Å, angle CH···O = 158°). In addition, the Cδ atom of proline can form a CH···O bonds with a backbone oxygen to replace the NH···O bond in the backbone of α-helices (Chakrabarti and Chakrabarti, 1998).

Although considered as "weak" hydrogen bonds, CH···O bonds are more likely to have a significant energetic impact in the membrane with

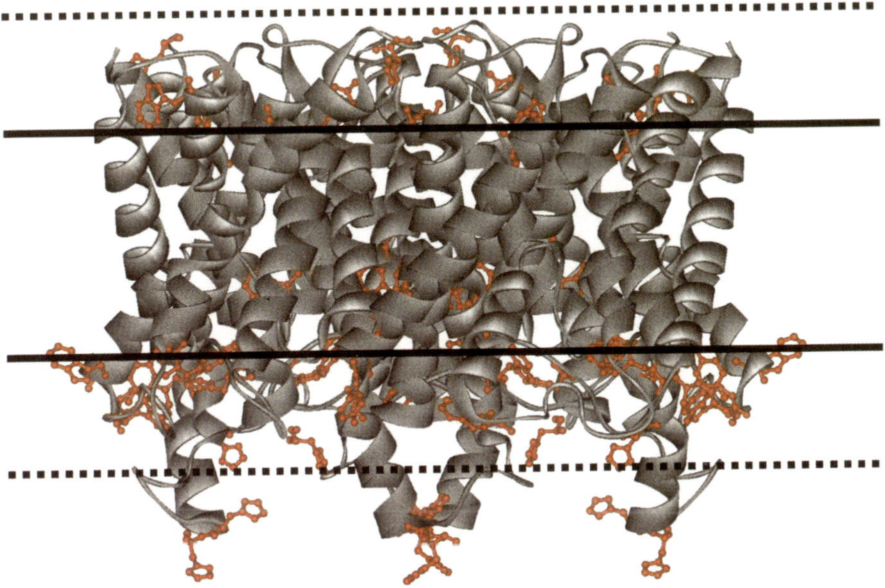

Chamberlain *et al.*, Fig. 3. An example of aromatic belts. A ribbon representation of the glycerol-conducting channel tetramer (1FX8) (Fu *et al.*, 2000) is shown with the amphiphilic aromatic residues (Trp, Tyr, His) highlighted in red ball-and-stick representations. The region between the two solid black bars roughly corresponds to the most hydrophobic 30 Å slice of the protein perpendicular to the *z*-axis of the tetramer. The dashed lines are spaced roughly 15 Å from the solid lines and are intended to mark the approximate extent of the interfacial region (White and Wimley, 1999).

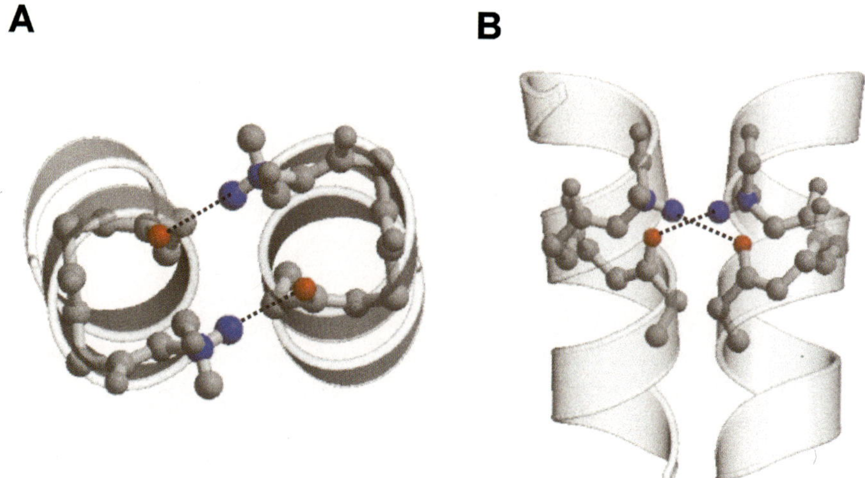

Chamberlain *et al.*, FIG. 8. Potential CH···O hydrogen bonds found in the glycophorin ATM dimer. The two CH···O contacts, one per monomer, are shown. (A) The second Hα of Gly-83 (blue) nearly makes the ideal 180° angle with the donor Cα (blue) and acceptor O of Val-80 (red). (B) The acceptor carbonyl is well positioned. The lone pair of electrons of the carbonyl O point toward the hydrogen atom. The angle from the carbonyl C (gray) to the carbonyl O (red) to the hydrogen (blue) should ideally be 120°. The coordinates were taken from the structure of glycophorin ATM (MacKenzie *et al.*, 1997) and the figure is made with Molscript (Kraulis, 1991).

its low dielectric constant. The prevalence of glycine in TM helix interfaces has been noted (Javadpour *et al.*, 1999; Russ and Engelman, 2000) and is often attributed to a lack of side-chain entropy costs upon folding or the ability to faciliate closer, more intimate helix associations. Javadpour *et al.* (1999) have suggested that CH···O bonds may help explain the predominance of glycine in helical interfaces, however, noting the polar interactions of Gly-420 and Gly-457 cytochrome *c* oxidase as an example. Senes *et al.* (2001) surveyed 11 structures of TM helix proteins specifically to assess the possible role of CH···O bonds in TM helix association. Their analysis identified potential CH···O bonds between Cα-H···O as well as Cβ-H···O and Cγ-H···O. The ratio of the number of CH···O interactions involving Cα to the number involving Cβ and Cγ increases when the H···O distance is less than 2.7 Å. This result is consistent with the stronger dipole moment of the Cα–H atoms exerting a larger attractive force on the acceptor dipole. They identified networks of the CH···O bonds reminiscent of the traditional hydrogen bonding networks described by Adamian and Liang (2002). In particular, as shown in Fig. 8 (see color insert), the GxxxG motif in glycophorin ATM forms three CH···O bonds, two involving the HA2 protons of glycine and one involving the HA proton of the valine residing in the first "x" position of the GxxxG motif. Of helical glycine residues, 23% were found to be CH···O bond donors and 10% were CH···O bond acceptors. Additionally, 24% of serine residues and 20% of threonine residues were involved in CH···O bonds. These results imply that the use of CH···O bonds strongly influences the association of TM helices, especially in the absence of the stronger, traditional hydrogen bonds with N and O donor atoms. Thus, it is possible that the relatively weak CH···O bonds play an important role, along with van der Waals and traditional hydrogen bonds, in determining in the stability and specificity of TM associations. An experimental demonstration of the strength of these interactions is still needed, however, before we can assess their influence on membrane protein structures.

D. *Salt Bridges*

Because of the extremely large energetic cost of moving a charge from the high dielectric environment of water to the low dielectric of the membrane, it is unlikely that isolated, charged side chains would exist in the membrane. It is much more likely that they would exist in a neutral form (Smith *et al.*, 1996) or be neutralized by interaction with an oppositely charged side chain, that is, form a salt bridge. To our knowledge there has not been a systematic study of salt bridges in membrane proteins, but we would expect them to be relatively rare in TM helices because of the low abundance of charged residues (Senes *et al.*, 2000). There is indirect evidence, however, for important salt-bridge interactions in membrane proteins. For example, an Asp and Lys in two TM helices of Lac permease appear to interact. The permease activity is retained if the positions of both the Asp and Lys are exchanged in the double mutant D237K/K358D, or if both charges are neutralized, as in the mutant D237C/K358C (King *et al.*, 1991; Sahin-Toth *et al.*, 1992). The individual point mutants, with only one charge neutralized, are inactive. Second, an Ala to Lys mutation, A19K, in the aspartate chemoreceptor, Tar, abolishes activity, while some second site suppressor mutations that could form an interhelical salt bridge restore activity (Unemura *et al.*, 1998). Other examples of possible TM salt bridges are found in the P_i-linked antiport carrier (Hall and Maloney, 2001), gastrin-releasing peptide receptor (Donohue *et al.*, 1999), α_{1b} adrenergic receptor (Porter and Perez, 1999), cystic fibrosis transmembrane conductance regulator (Cotten and Welsh, 1999), and T-cell receptor assembly (Cosson *et al.*, 1991). Salt bridges have also been documented in side-chain interactions with lipid head groups. For example, in the bacterial cytochrome *c* oxidase structure, Arg side chains interact with the phosphate of a phosphatidylcholine head group (Harrenga and Michel, 1999). To the extent that salt bridges exist in membrane proteins, it seems likely that they would predominate in the interfacial region where they would be more accessible to lipid head groups and water. Nevertheless, more work is needed to better assess the role of salt bridges in membrane protein structures.

VI. Conclusions

Unlike the case of soluble proteins in which the hydrophobic effect dominates folding, it is still not possible to point to a force that dominates in the construction of helix bundle membrane proteins. Clearly hydrogen bonds can be extremely strong, but are relatively sparse and are not critical for the development of stable helix–helix interactions. Van der Waals interactions are certainly important, but the fact that a well-packed leucine zipper interface is not sufficient to drive TM helix association suggests that packing alone is not enough. Although the complex interplay between lipid structure and protein structure must play an important role, the fact that many membrane proteins remain folded and functional in detergent micelles suggests that lipid structure alone cannot entirely explain membrane protein architecture. Thus, it would seem that membrane protein structure is defined by a subtle balancing of many small interactions. This is of course the worst possible situation for workers addressing the structure prediction problem as it may not be possible to focus on one aspect of stability. Certainly, we will need to develop a much better understanding of the molecular forces operating in the membrane environment. Perhaps our most acute need is a better understanding of lipid–protein interactions and weakly polar interactions in the membrane—a major challenge for the future.

References

Abramovitch, D. A., Marsh, D., and Powell, G. L. (1990). *Biochim. Biophys. Acta* **1020,** 34–42.

Adamian, L., and Liang, J. (2001). *J. Mol. Biol.* **311,** 891–907.

Adamian, L., and Liang, J. (2002). *Proteins* **47,** 209–218.

Allen, S. J., Kim, J. M., Khorana, H. G., Lu, H., and Booth, P. J. (2001). *J. Mol. Biol.* **308,** 423–435.

Arkin, I. T., and Brunger, A. T. (1998). *Biochim. Biophys. Acta* **1429,** 113–128.

Ben-Tal, N., and Honig, B. (1996). *Biophys. J.* **71,** 3046–3050.

Bibi, E., and Kaback, H. R. (1990). *Proc. Natl. Acad. Sci. USA* **87,** 4325–4329.

Blazquez, M., Thiele, C., Huttner, W. B., Docherty, K., and Shennan, K. I. J. (2000). *Biochem. J.* **349,** 843–852.

Bogdanov, M., and Dowhan, W. (1999). *J. Biol. Chem.* **274,** 36827–36830.

Booth, P. J. (2000). *Biochim. Biophys. Acta* **1460,** 4–14.

Bowie, J. U. (1997a). *J. Mol. Biol.* **272,** 780–789.

Bowie, J. U. (1997b). *Nat. Struct. Biol.* **4,** 915–917.

Bowie, J. U. (2001). *Curr. Opin. Struct. Biol.* **11,** 397–402.

Bowie, J. U., Reidhaar-Olson, J. F., Lim, W. A., and Sauer, R. T. (1990). *Science* **247,** 1306–1310.

Bowie, J. U., Lüthy, R., and Eisenberg, D. (1991). *Science* **253,** 164–170.

Brosig, B., and Langosch, D. (1998). *Protein Sci.* **7,** 1052–1056.

Brouillette, C., McMichens, R., Stern, L., and Khorana, H. (1989). *Proteins: Struct. Func. Genet.* **5,** 38.

Brown, D. A., and London, E. (1998). *Annu. Rev. Cell Dev. Biol.* **14,** 111–136.

Brueckner, K., Labrador, J. P., Scheiffele, P., Herb, A., Seeburg, P. H., and Klein, R. (1999). *Neuron* **22,** 511–524.

Chakrabarti, P., and Chakrabarti, S. (1998). *J. Mol. Biol.* **284,** 867–873.

Chen, G. Q., and Gouaux, E. (1999). *Biochemistry* **38,** 15380–15387.

Choma, C., Gratkowski, H., Lear, J. D., and DeGrado, W. F. (2000). *Nat. Struct. Biol.* **7,** 161–166.

Chothia, C., Levitt, M., and Richardson, D. (1981). *J. Mol. Biol.* **145,** 215–250.

Cosson, P., Lankford, S., Bonifacino, J., and Klausner, R. (1991). *Nature* **351,** 414.

Cotten, J. F., and Welsh, M. J. (1999). *J. Biol. Chem.* **274,** 5429–5435.

Crick, F. (1953). *Acta Crystallogr.* **6,** 689–697.

Curran, A. R., Templer, R. H., and Booth, P. J. (1999). *Biochemistry* **38,** 9328–9336.

Deber, C. M., Wang, C., Liu, L. P., Prior, A. S., Agrawal, S., Muskat, B. L., and Cuticchia, A. J. (2001). *Protein Sci.* **10,** 212–219.

Derewenda, Z. S., Derewenda, U., and Kobos, P. M. (1994). *J. Mol. Biol.* **241,** 83–93.

Derewenda, Z. S., Lee, L., and Derewenda, U. (1995). *J. Mol. Biol.* **252,** 248–262.

Desiraju, G. R., and Steiner, T. (1999). *The Weak Hydrogen Bond in Structural Chemistry and Biology.* Oxford Univ. Press, Oxford.
Donohue, P. J., Sainz, E., Akeson, M., Kroog, G. S., Mantey, S. A., Battey, J. F., Jensen, R. T., and Northup, J. K. (1999). *Biochemistry* **38,** 9366–9372.
Dumas, F., Sperotto, M. M., Lebrun, C., Tocanne, J. F., and Mouritsen, O. G. (1997). *Biophys. J.* **73,** 1940–1953.
Eilers, M., Shekar, S. C., Shieh, T., Smith, S. O., and Fleming, P. J. (2000). *Proc. Natl. Acad. Sci. USA* **97,** 5796–5801.
Eisenberg, D., and McLachlan, A. D. (1986). *Nature* **319,** 199–203.
Fisher, L. E., Engelman, D. M., and Sturgis, J. N. (1999). *J. Mol. Biol.* **293,** 639–651.
Fleming, K. G., and Engelman, D. M. (2001). *Proc. Natl. Acad. Sci. USA* **98,** 14340–14344.
Fleming, K. G., Ackerman, A. L., and Engelman, D. M. (1997). *J. Mol. Biol.* **272,** 266–275.
Fu, D., Libson, A., Miercke, L. J., Weitzman, C., Nollert, P., Krucinski, J., and Stroud, R. M. (2000). *Science* **290,** 481–486.
Galisteo, M., and Sanchez-Ruiz, J. (1993). *Eur. Biophys. J.* **22,** 25–30.
Gilles-Gonzalez, M. A., Engelman, D. M., and Khorana, H. G. (1991). *J. Biol. Chem.* **266,** 8545–8550.
Gratkowski, H., Lear, J. D., and DeGrado, W. F. (2001). *Proc. Natl. Acad. Sci. USA* **98,** 880–885.
Green, R. D. (1974). *Hydrogen Bonding by C—H Groups.* Wiley, New York/Toronto.
Groves, J. D., Wang, L., and Tanner, M. J. (1998). *FEBS Lett.* **433,** 223–227.
Guzman, L., Weiss, D., and Beckwith, J. (1997). *J. Bacteriol.* **179,** 5094–5103.
Hall, J. A., and Maloney, P. C. (2001). *J. Biol. Chem.* **276,** 25107–25113.
Haltia, T., and Freire, E. (1995). *Biochim. Biophys. Acta* **1228,** 1–27.
Haltia, T., Semo, N., Arrondo, J. L., Goni, F. M., and Freire, E. (1994). *Biochemistry* **33,** 9731–9740.
Harrenga, A., and Michel, H. (1999). *J. Biol. Chem.* **274,** 33296–33299.
Hooper, N. M. (2001). *Proteomics* **1,** 748–755.
Hunt, J. F., Earnest, T. N., Bousche, O., Kalghatgi, K., Reilly, K., Horvath, C., Rothschild, K. J., and Engelman, D. M. (1997). *Biochemistry* **36,** 15156–15176.
Huschilt, J. C., Millman, B. M., and Davis, J. H. (1989). *Biochim. Biophys. Acta* **979,** 139–141.
Huttner, W. B., and Zimmerberg, J. (2001). *Curr. Opin. Cell Biol.* **13,** 478–484.
Ikonen, E. (2001). *Curr. Opin. Cell Biol.* **13,** 470–477.
Ikonen, E., and Simons, K. (1998). *Semin. Cell Devel. Biol.* **9,** 503–509.
Isenbarger, T. A., and Krebs, M. P. (2001). *Biochemistry* **40,** 11923–11931.
Jacob, J., Duclohier, H., and Cafiso, D. S. (1999). *Biophys. J.* **76,** 1367–1376.
Javadpour, M. M., Eilers, M., Groesbeek, M., and Smith, S. O. (1999). *Biophys. J.* **77,** 1609–1618.
Jiang, S., and Vakser, I. A. (2000). *Proteins* **40,** 429–435.
Kahn, T. W., Sturtevant, J. M., and Engelman, D. M. (1992). *Biochemistry* **31,** 8829–8839.
Katragadda, M., Alderfer, J. L., and Yeagle, P. L. (2000). *Biochim. Biophys. Acta* **1466,** 1–6.
Kim, J. M., Booth, P. J., Allen, S. J., and Khorana, H. G. (2001). *J. Mol. Biol.* **308,** 409–422.
King, S. C., Hansen, C. L., and Wilson, T. H. (1991). *Biochim. Biophys. Acta* **1062,** 177–186.
Klug, C. S., and Feix, J. B. (1998). *Protein Sci.* **7,** 1469–1476.
Klug, C. S., Su, W., Liu, J., Klebba, P. E., and Feix, J. B. (1995). *Biochemistry* **34,** 14230–14236.
Kralchevsky, P. A. (1997). *Adv. Biophys.* **34,** 25–39.
Kralchevsky, P. A., and Nagayama, K. (2000). *Adv. Colloid Interface Sci.* **85,** 145–192.
Kraulis, P. J. (1991). *J. Appl. Crystallogr.* **24,** 946–950.
Kurzchalia, T. V., and Parton, R. G. (1999). *Curr. Opin. Cell Biol.* **11,** 424–431.

Langlet, C., Bernard, A.-M., Drevot, P., and He, H.-T. (2000). *Curr. Opin. Immunol.* **12,** 250–255.

Langosch, D., and Heringa, J. (1998). *Proteins* **31,** 150–159.

Lau, F., and Bowie, J. (1997). *Biochemistry* **36,** 5884–5892.

Lehtonen, J. Y. A., Holopainen, J. M., and Kinnunen, P. K. J. (1996). *Biophys. J.* **70,** 1753–1760.

Lemmon, M. A., Flanagan, J. M., Hunt, J. F., Adair, B. D., Bormann, B. J., Dempsey, C. E., and Engelman, D. M. (1992a). *J. Biol. Chem.* **267,** 7683–7689.

Lemmon, M. A., Flanagan, J. M., Treutlein, H. R., Zhang, J., and Engelman, D. M. (1992b). *Biochemistry* **31,** 12719–12725.

Liu, L. P., Li, S. C., Goto, N. K., and Deber, C. M. (1996). *Biopolymers* **39,** 465–470.

Luecke, H., Schobert, B., Richter, H. T., Cartailler, J. P., and Lanyi, J. K. (1999). *J. Mol. Biol.* **291,** 899–911.

Luneberg, J., Widmann, M., Dathe, M., and Marti, T. (1998). *J. Biol. Chem.* **273,** 28822–28830.

MacKenzie, K. R., Prestegard, J. H., and Engelman, D. M. (1997). *Science* **276,** 131–133.

Mall, S., Broadbridge, R., Sharma, R. P., East, J. M., and Lee, A. G. (2001). *Biochemistry* **40,** 12379–12386.

Maneri, L. R., and Low, P. S. (1988). *J. Biol. Chem.* **263,** 16170–16178.

Marti, T. (1998). *J. Biol. Chem.* **273,** 9312–9322.

McAuley, K. E., Fyfe, P. K., Ridge, J. P., Isaacs, N. W., Cogdell, R. J., and Jones, M. R. (1999). *Proc. Natl. Acad. Sci. USA* **96,** 14706–14711.

Meadows, E. S., De Wall, S. L., Barbour, L. J., Fronczek, F. R., Kim, M.-S., and Gokel, G. W. (2000). *J. Am. Chem. Soc.* **122,** 3325–3335.

Nagle, J. F., and Tristram-Nagle, S. (2000). *Biochim. Biophys. Acta* **1469,** 159–195.

O'Shea, E. K., Klemm, J. D., Kim, P. S., and Alber, T. (1991). *Science* **254,** 539–544.

Palczewski, K., Kumasaka, T., Hori, T., Behnke, C. A., Motoshima, H., Fox, B. A., Le Trong, I., Teller, D. C., Okada, T., Stenkamp, R. E., Yamamoto, M., and Miyano, M. (2000). *Science* **289,** 739–745.

Perutz, M. F., Kendrew, J. C., and Watson, H. C. (1965). *J. Mol. Biol.* **13,** 669–678.

Pilot, J. D., East, J. M., and Lee, A. G. (2001). *Biochemistry* **40,** 8188–8195.

Pohlschroder, M., Murphy, C., and Beckwith, J. (1996). *J. Biol. Chem.* **271,** 19908–19914.

Popot, J. L., and Engelman, D. M. (1990). *Biochemistry* **29,** 4031–4037.

Popot, J. L., Gerchman, S. E., and Engelman, D. M. (1987). *J. Mol. Biol.* **198,** 655–676.

Porter, J. E., and Perez, D. M. (1999). *J. Biol. Chem.* **274,** 34535–34538.

Ramachandran, G. N., and Sasisekharan, V. (1965). *Biochim. Biophys. Acta* **109,** 314–316.

Ramachandran, G. N., Sasisekharan, V., and Ramakrishnan, C. (1966). *Biochim. Biophys. Acta* **112,** 168–170.

Rees, D., DeAntonio, L., and Eisenberg, D. (1989a). *Science* **245,** 510.

Rees, D. C., Komiya, H., Yeates, T. O., Allen, J. P., and Feher, G. (1989b). *Annu. Rev. Biochem.* **58,** 607–633.

Richards, F. M. (1977). *Annu. Rev. Biophys. Bioeng.* **6,** 151–176.

Riek, R. P., Rigoutsos, I., Novotny, J., and Graham, R. M. (2001). *J. Mol. Biol.* **306,** 349–362.

Rietveld, A., Van Kemenade, T. J. J. M., Hak, T., Verkleij, A. J., and De Kruijff, B. (1987). *Eur. J. Biochem.* **164,** 137–140.

Rost, B., Fariselli, P., and Casadio, R. (1996). *Protein Sci.* **5,** 1704–1718.

Russ, W. P., and Engelman, D. M. (1999). *Proc. Natl. Acad. Sci. USA* **96,** 863–868.

Russ, W. P., and Engelman, D. M. (2000). *J. Mol. Biol.* **296,** 911–919.

Sahin-Toth, M. S., Dunten, R. L., Gonzalez, A., and Kaback, R. (1992). *Proc. Natl. Acad. Sci. USA* **89,** 10547–10551.

Sansom, M. S., and Weinstein, H. (2000). *Trends Pharmacol. Sci.* **21,** 445–451.
Schmidt-Rose, T., and Jentsch, T. J. (1997). *J. Biol. Chem.* **272,** 20515–20521.
Senes, A., Gerstein, M., and Engelman, D. M. (2000). *J. Mol. Biol.* **296,** 921–936.
Senes, A., Ubarretxena-Belandia, I., and Engelman, D. M. (2001). *Proc. Natl. Acad. Sci. USA* **98,** 9056–9061.
Sheets, E. D., Holowka, D., and Baird, B. (1999). *J. Cell Biol.* **145,** 877–887.
Shnyrov, V. L., Azuaga, A. I., and Mateo, P. L. (1994). *Biochem. Soc. Trans.* **22,** 367S.
Simidjiev, I., Stoylova, S., Amenitsch, H., Javorfi, T., Mustardy, L., Laggner, P., Holzenburg, A., and Garab, G. (2000). *Proc. Natl. Acad. Sci. USA* **97,** 1473–1476.
Simons, K., and Ikonen, E. (1997). *Nature* (London) **387,** 569–572.
Smith, S., Smith, C., and Bormann, B. (1996). *Nat. Struct. Biol.* **3,** 252–258.
Smith, S. O., Song, D., Shekar, S., Groesbeek, M., Ziliox, M., and Aimoto, S. (2001). *Biochemistry* **40,** 6553–6558.
Surrey, T., and Jahnig, F. (1992). *Proc. Natl. Acad. Sci. USA* **89,** 7457–7461.
Taylor, R., and Kennard, O. (1982). *J. Am. Chem. Soc.* **104,** 5063–5070.
Therien, A. G., Grant, F. E., and Deber, C. M. (2001). *Nat. Struct. Biol.* **8,** 597–601.
Ulmschneider, M. B., and Sansom, M. S. (2001). *Biochim. Biophys. Acta* **1512,** 1–14.
Unemura, T., Tatsuno, I., Shibasaki, M., Homma, M., and Kagawishi, I. (1998). *J. Biol. Chem.* **273,** 30110–30115.
van der Wel, P. C. A., Pott, T., Morein, S., Greathouse, D. V., Koeppe, R. E., II, and Killian, J. A. (2000). *Biochemistry* **39,** 3124–3133.
Vargas, R., Garza, J., Dixon, D. A., and Hay, B. P. (2000). *J. Am. Chem. Soc.* **122,** 4750–4755.
Walther, D., Eisenhaber, F., and Argos, P. (1996). *J. Mol. Biol.* **25,** 536–553.
Waugh, M. G., Lawson, D., and Hsuan, J. J. (1999). *Biochem. J.* **337,** 591–597.
Weiner, D., Liu, J., Cohen, J., Williams, W., and Greene, M. (1989). *Nature* **339,** 230–231.
Wen, J., Chen, X., and Bowie, J. (1996). *Nat. Struct. Biol.* **3,** 141–148.
White, S. H., and Wimley, W. C. (1998). *Biochim. Biophys. Acta* **1376,** 339–352.
White, S. H., and Wimley, W. C. (1999). *Annu. Rev. Biophys. Biomol. Struct.* **28,** 319–365.
White, S. H., Wimley, W. C., Ladokhin, A. S., and Hristova, K. (1998). *Methods Enzymol.* **295,** 62–87.
Williamson, I. M., Alvis, S. J., East, J. M., and Lee, A. G. (2002). *Biophys. J.* **83,** 2026–2038.
Wimley, W. C., and White, S. H. (1996). *Nat. Struct. Biol.* **3,** 842–848.
Yau, W. M., Wimley, W. C., Gawrisch, K., and White, S. H. (1998). *Biochemistry* **37,** 14713–14718.
Zhou, F. X., Cocco, M. J., Russ, W. P., Brunger, A. T., and Engelman, D. M. (2000). *Nat. Struct. Biol.* **7,** 154–160.
Zhou, F. X., Merianos, H. J., Brunger, A. T., and Engelman, D. M. (2001). *Proc. Natl. Acad. Sci. USA* **98,** 2250–2255.
Zhou, Y., and Bowie, J. U. (2000). *J. Biol. Chem.* **275,** 6975–6979.
Zhou, Y., Wen, J., and Bowie, J. U. (1997). *Nat. Struct. Biol.* **4,** 986–990.

TRANSMEMBRANE β-BARREL PROTEINS

By GEORG E. SCHULZ

Institute for Organic Chemistry and Biochemistry, Albert-Ludwigs-Universität, D-79104 Freiburg im Breisgau, Germany

I. INTRODUCTION

After numerous complete genomic DNA sequences had been established and after these had been interpreted to a great extent in terms of protein sequences, it became evident that about 20% of all proteins are located in membranes (Liu and Rost, 2001). This number was derived from a search for transmembrane α-helices with a computerized prediction system, the results of which are known to come with a high confidence level. Such helices can be recognized from a continuous stretch of 20 to 30 nonpolar residues with a predominance of aliphatic side chains at the center and aromatic residues at both ends (Sipos and von Heijne, 1993). The number of transmembrane α-helices per protein is broadly distributed and averages around six.

The main chain amides of these α-helices are all locally complemented, so that the interface to the nonpolar membrane interior is exclusively formed by the nonpolar side chains. This explains the usefulness of an α-helix as a membrane-crossing element. The helix orientation can be deduced from the charge patterns of the interhelical segments, positive charges being inside and negative charges outside the cell. As a consequence the coarse structures, usually called topologies, of all these α-helical membrane proteins can be assessed from the sequence. The presence of additional β-sheets in these proteins is discussed. However, these must be internal because it cannot be expected that the membrane is faced by a mixture of α-helices and β-sheets as the main chain hydrogen bond donors and acceptors at the sheet edges cannot be complemented by those of α-helices.

0065-3233/03 $35.00

Because of their construction, transmembrane β-sheets have an amide saturation problem at their edge strands. Fusing both edges to form a barrel, however, permits complementation of all amide hydrogen bond donors and acceptors. Like a transmembrane helix, a β-barrel can face the nonpolar interior of the membrane if its outer surface is coated with nonpolar side chains. Such barrels occur, indeed, in the outer membrane of gram-negative bacteria. They should be detectable in the sequence because every second residue is nonpolar. The significance of this information is low, however, because the individual β-strands are only slightly more than half a dozen residues long and the intermittent residues pointing to the barrel interior can be both polar and nonpolar.

From the start the β-barrels require cooperative folding of a polypeptide chain of 100 or more residues which constitute an entropic hurdle. In contrast, an α-helical transmembrane protein can traverse the membrane as soon as a local segment of about 20 residues becomes nonpolar. The remaining transmembrane part can then be added piece by piece, which is entropically much more favorable. Therefore the β-barrel membrane proteins arose probably rather late during protein structure evolution, constituting an addition to the much simpler α-helical membrane proteins.

At present, transmembrane β-barrel proteins have been found exclusively in the outer membrane of gram-negative prokaryotes; these membranes seem to lack α-helical proteins. Accordingly, a separation exists between α-proteins in all cytoplasmic membranes and β-proteins in the specialized outer membranes. Following the endosymbiotic hypothesis, β-proteins are also expected in the outer membranes of mitochondria and chloroplasts, but none of these proteins has yet been structurally elucidated. Given the limited abundance of such membranes, the β-proteins are likely to make up only a small, special class of membrane proteins.

In contrast to their limited importance in nature, the β-barrel proteins are most prominent in the list of established membrane protein structures. Moreover, they show a high degree of internal chain-fold symmetry and therefore convey the impression of beautiful proteins (Fig. 1). One should not forget that the first protein structure, myoglobin, caused some disappointment among those who solved it as it showed no symmetry whatsoever; even the α-helices were not whole-numbered but about 3.6 residues per turn. Accordingly, the symmetric transmembrane β-barrels stand out from the bulk of asymmetric chain folds of water-soluble proteins.

The number of distinct chain folds of integral membrane proteins is probably much smaller than the respective number for water-soluble

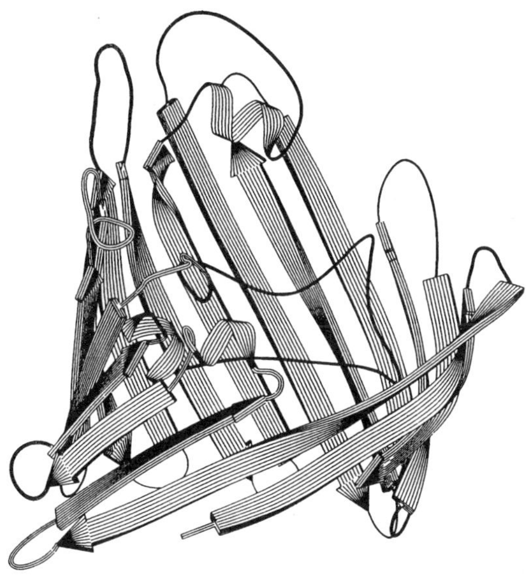

FIG. 1. Ribbon plot of the 16-stranded β-barrel of the general porin from *Rhodobacter capsulatus* viewed from the molecular threefold axis (Weiss and Schulz, 1992). Note the large variation of the inclination angle α and the difference between the high barrel wall facing the membrane and the low wall at the trimer interface.

proteins, which ranges around a thousand (Schulz, 1981; Brenner and Levitt, 2000). Proteins of the cytoplasmic membrane consist mostly of transmembrane α-helices, and the bacterial outer membrane proteins contain β-barrels. Both types show a high neighborhood correlation which limits the number of different topologies appreciably (Schulz and Schirmer, 1979). The α-helices run, in general, perpendicular to the membrane plane and connections are formed between neighboring helix ends (Bowie, 1999). All transmembrane β-barrels contain meandering all-next-neighbor antiparallel sheets, the topologies of which are completely described by the number of strands.

II. STRUCTURES

Here we discuss structures that have been established at the atomic level revealing the exact conformation of the polypeptide chain. All were determined by X-ray diffraction analysis of a *three*-dimensional protein crystal. Some α-helical membrane protein structures have been analyzed by electron diffraction of *two*-dimensional crystals, although generally with a lower accuracy. For a long time structural analyses by NMR

TABLE I
β-Barrels in Water-Soluble Proteins

	n	S	R(Å)	α(°)	Secondary structure
LβH proteins	1	−18	—	—	Left-handed β-helix
Metalloprotease	1	18	—	—	Right-handed β-helix
Pectate lyase	1	20	—	—	Right-handed β-helix
Chymotrypsin	6	8	6.5	45	Symmetric antiparallel β-sheet
TIM	8	8	7.2	37	All-next-neighbor parallel β-sheet
Streptavidin	8	10	7.9	43	All-next-neighbor antiparallel β-sheet
Lipocalins	8	12	8.7	48	All-next-neighbor antiparallel β-sheet

had been considered inapplicable to membrane proteins, because these proteins can only be solubilized in large micelles. However, two small membrane proteins have now been tackled with this method, and success has been reported (Arora *et al.,* 2001; Fernandez *et al.,* 2001a,b). Moreover, the native structure of gramicidin A in the membrane was deduced from solid-state NMR data (Ketchem *et al.,* 1993). Other methods providing important structural information on proteins are light spectroscopy, electron microscopy, and atomic force microscopy. However, none of them reaches atomic resolution.

The existence of β-barrels was established for chymotrypsin at a very early stage in the now common protein crystal structure analyses. This enzyme contains two distorted six-stranded β-barrels with identical topologies (Birktoft and Blow, 1972). A selection of β-barrels in water-soluble proteins is given in Table I. The very abundant TIM-barrel consisting of eight parallel β-strands was also detected rather early (Banner *et al.,* 1975). Additional eight-stranded β-barrels of this group are those of streptavidin (Hendrickson *et al.,* 1989) and of the lipocalins (Newcomer *et al.,* 1984).

The β-helices also belong to this group as they can be taken as single-stranded β-barrels ($n = 1$) with large shear numbers of $S = 18$ and more (Table I, Fig. 2). They were first detected with pectate lyase (Yoder *et al.,* 1993). The right-handed and left-handed versions have positive and negative S values, respectively. The cross-sections of these β-helical barrels deviate drastically from circles. The cross-section of the pectate lyase type resembles a boomerang and those of the metalloproteases a flat ellipse (Baumann *et al.,* 1993), whereas the LβH (left-handed β-helix) types of proteins form triangles (Raetz and Roderick, 1995).

X-Ray diffraction analysis is a suitable and convenient method for obtaining exact structures of membrane proteins, but it requires three-dimensional crystals. Membrane protein crystallization has always

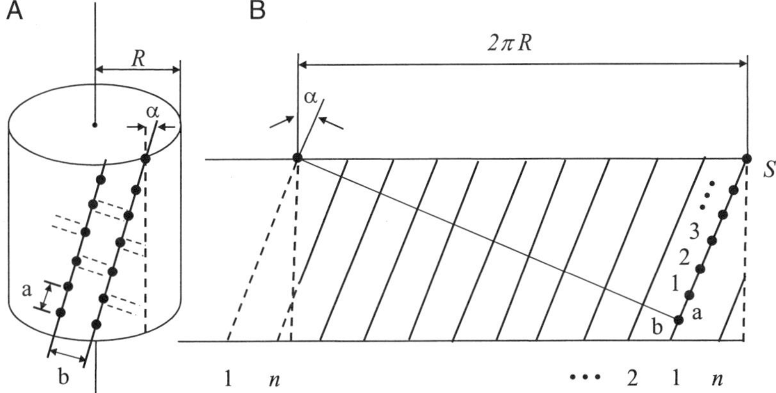

FIG. 2. General architecture of a β-barrel. The description depends neither on the sequence of the strands nor on their directions. Residues are represented by their Cα atoms. (A) Barrel geometry with the tilt angle α of the β-strands versus the barrel axis. The hydrogen bonding pattern is shown for antiparallel β-strands. (B) The barrel is cut where the first strand reaches the upper end, flattened out, and viewed from the outside. The sequence of the n strands along the circumference is arbitrary. The thin line follows a pleat of the sheet, that is, the hydrogen bonds. The shear number S is derived by running from a given strand to the left along the hydrogen bonds once around the barrel and counting the residue number S to the point of return to the same strand (Murzin *et al.*, 1994a,b; Liu, 1998). The depicted β-strand tilt corresponds to a positive S-value, and a tilt to the left to a negative S.

been a bottleneck. Part of this obstacle is the preparation of sufficient homogeneous membrane protein material, because the limited volume of the two-dimensional entity membrane cannot incorporate large amounts of a recombinant protein. Moreover, any tampering with the membrane is highly hazardous for the respective organism so that high expression levels are generally rare.

This problem was circumvented by expressing membrane proteins into the cytosol and (re)naturing it from there into micelles (Schmid *et al.*, 1996), which is possible for a number of β-barrel proteins. As a general observation the crystallization of the bacterial outer membrane proteins appears to be easier than that of the α-helical proteins from the plasma membrane. Accordingly, the list of the structurally established β-barrel membrane proteins is comparatively long. It is given in Table II. The resolution of the analyses ranges from 1.6 Å for OmpA (neglecting the nonnative gramicidin-A crystals) to 3.2 Å for the porin OmpC (OmpK36). The crystals are usually loosely packed, except for one crystal form of OmpA that reached 50% (v/v) protein in the crystal but diffracted merely to medium resolution (Pautsch and Schulz, 2000).

TABLE II
Transmembrane β-Barrel Proteins[a]

	n	S	R(Å)[b]	α(°)[b]	Oligomer	Reference
Gramicidin A (native)	1	6	3.4	77	Dimer	Ketchem *et al.*, 1993
Gramicidin A (CsCl, CH_3OH)[c]	2	−6	Cs^+ fits	—	Dimer	Wallace and Ravikumar, 1988
Gramicidin A (C_6H_6, ethanol)[c]	2	−4	narrow	—	Dimer	Langs, 1988
Nanotube	1	'8'	5.6	90	Stack	Ghadiri *et al.*, 1994
OmpX	8	8	7.2	37	Mono	Vogt and Schulz, 1999
OmpA	8	10	7.9	43	Mono	Pautsch and Schulz, 1998
OmpT	10	12	9.5	42	Mono	Vandeputte-Rutten *et al.*, 2001
OmpLA[d]	12	12	10.6	37	Mono	Snijder *et al.*, 1999
TolC[e]	12	'20'	13.6	51	"Trimer"	Koronakis *et al.*, 2000
α-Hemolysin[e]	14	'14'	12.3	37	"Hepta"	Song *et al.*, 1996
Porin *R. capsulatus*	16	20	15.5	43	Trimer	Weiss *et al.*, 1990
Porin OmpF (PhoE, OmpC)	16	20	15.5	43	Trimer	Cowan *et al.*, 1992
Porin *R. blasticus*	16	20	15.5	43	Trimer	Kreusch and Schulz, 1994
Porin *P. denitrificans*	16	20	15.5	43	Trimer	Hirsch *et al.*, 1997
Porin Omp32	16	20	15.5	43	Trimer	Zeth *et al.*, 2000
Maltoporin (two species)	18	20	17.1	40	Trimer	Schirmer *et al.*, 1995
Sucrose porin	18	20	17.1	40	Trimer	Forst *et al.*, 1998
FhuA	22	24	19.9	39	Mono	Locher *et al.*, 1998
FepA	22	24	19.9	39	Mono	Buchanan *et al.*, 1999

[a] All sheets are antiparallel except for native gramicidin A. The topologies are always all-next-neighbor.

[b] The radius is calculated for a circular cross section. The angle α can vary by ±15° around the barrel.

[c] Nonnative conformations of short peptides depend on the crystallization condition.

[d] This enzyme exists as a monomer in the membrane and becomes active on dimerization.

[e] The barrel itself consists of parts from several subunits, violating the rule of one barrel per subunit.

The first data showing a transmembrane barrel came from electron microscopy and electron diffraction of two-dimensional porin crystals (Jap, 1989). They followed spectroscopic studies of the amide bands in the infrared range indicating that the porin contains β-strands tilted at a 45° angle against the membrane plane (Nabredyk *et al.*, 1988). As can be visualized in Fig. 2A, a 45° tilt in a barrel can point either to the right or to the left, the two alternatives being mirror images. This ambiguity is resolved, however, because only one of them corresponds to the energetically favorable and therefore commonly observed β-sheet twist. In conjunction with the barrel diameter known from the electron diffraction data, these data would have sufficed for modeling the β-barrel of the analyzed porin in a rather detailed manner. Shortly thereafter an X-ray analysis of three-dimensional porin crystals clarified this structure in atomic detail (Weiss *et al.*, 1990).

The pentadecameric antibiotic peptide gramicidin A forms channels through membranes that allow the passage of alkali ions. It has been included in Table II because it forms β-barrels in two artificial conformations as well as in its native conformation. The artificial conformations were obtained by crystallizing from nonpolar solvents in the presence and absence of Cs^+ ions (Wallace and Ravikumar, 1988; Langs, 1988). They showed very narrow β-barrels, indicating that they present artifacts instead of channels. The real structure of gramicidin A consists of a somewhat wider barrel in the membrane, which could only be determined by solid-state NMR of membranes containing gramicidin A (Ketchem *et al.*, 1993). It revealed a helical $n = 1$ type of β-barrel, two of which associate head-to-head forming a channel through the membrane. Such narrow barrels can only be assumed if L-amino acid residues alternate with D-amino acids (or glycines) along the peptide chain, and this is here actually the case. The artificial nanotubes listed in Table II follow the design of the gramicidin A channel except that the chain is an eight-membered ring instead of a β-helix (Ghadiri *et al.*, 1994). The rings are stacked forming a channel.

The smallest established transmembrane β-barrel of the canonical type is that of OmpX (Vogt and Schulz, 1999). It contains eight strands with a shear number $S = 8$ and appears to represent the minimum construction for a transmembrane β-protein (Table II). In contrast to OmpX, the ubiquitous outer membrane protein A possesses an N-terminal 171-residue domain (here called OmpA) as a membrane anchor and a C-terminal periplasmic domain binding to the peptidoglycan cell wall. OmpA contains an eight-stranded β-barrel with a shear number $S = 10$, which is larger than that of OmpX, giving rise to a larger barrel cross-section (Pautsch and Schulz, 1998). The relationship between strand numbers, shear numbers, and barrel radii is illustrated in Fig. 3. Applying point mutations, the OmpA crystals could be improved to diffract to 1.6 Å resolution (Pautsch and Schulz, 2000). This allowed for an anisotropic structure refinement revealing the major mobility directions of loops and turns. It demonstrated that the turns and loops

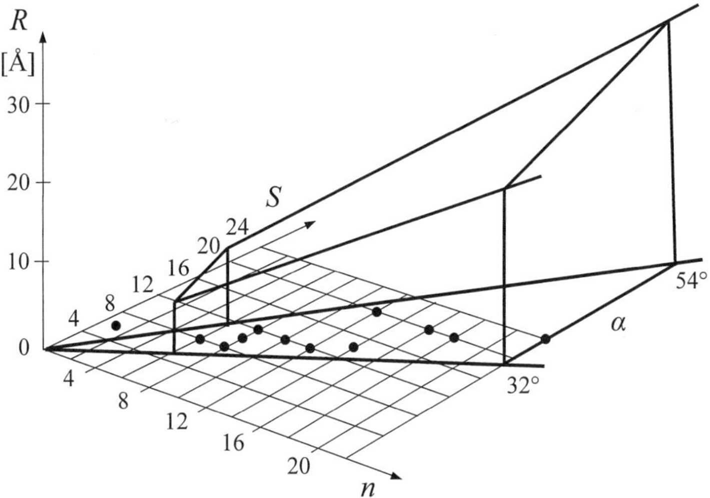

FIG. 3. The observed β-barrels (no β-helices included) concentrate at tilt angles α between 35° and 50°. Because of the two-residue repeat in the hydrogen bond pattern (Fig. 2A), the shear number S is always even. The radius R of the barrel increases with the number of strands n as well as with the shear number S. Up to OmpLA ($n = S = 12$) the interior of the barrel can be filled by side chains and fixed water molecules.

have asymmetric mobilities which correspond to those of a model in which the polypeptide is replaced by a resilient wire.

The bacterial outer membranes also contain some enzymes, among them the protease OmpT consisting of a 10-stranded β-barrel with a shear number $S = 12$ (Vandeputte-Rutten *et al.*, 2001). While the "lower" part of the β-barrel is immersed in the membrane as usual, its "upper" part protrudes into the external medium and contains the catalytic center where foreign proteins are split. OmpT is of medical interest because it contributes to the pathogenicity of bacteria. A further surfacial enzyme is the phospholipase A (OmpLA) that destroys lipopolysaccharides. It consists of a 12-stranded β-barrel with a shear number $S = 12$ (Snijder *et al.*, 1999). Its barrel contains a solid interior hydrogen bonding network without a pore, a nonpolar outer surface, and the catalytic center at the external end. OmpLA is active as a dimer accommodated in the membrane. Interestingly, the dimer interface is for the most part nonpolar.

The β-barrels of TolC (Koronakis *et al.*, 2000) and of α-hemolysin (Song *et al.*, 1996) are special because they are composed of several pieces coming from different subunits (Table II). The TolC barrel consists of three four-stranded all-next-neighbor antiparallel β-sheet pieces coming from the three subunits. The major parts of the subunits are α-helical and not in the outer membrane. The shear number S is as large as 20, giving rise to a wide channel along the barrel axis (see Fig. 3). This is in contrast with the 12-stranded β-barrel of OmpLA that has a smaller S and a solid core. The α-hemolysin barrel consists of seven β-hairpin loops coming from the seven subunits. Each subunit is water-soluble. The β-hairpin loop undergoes a large conformational change during the cooperative process of β-barrel formation, which is likely to occur on membrane insertion and after the large globular parts have formed an annular heptamer (Olson *et al.,* 1999).

The porins in the outer membrane of gram-negative bacteria are passive channels showing various grades of selectivity. The most common type consists of 16-stranded β-barrels with a shear number $S = 20$. The structure of the porin from *Rhodobacter capsulatus* illustrated in Fig. 1 was the first transmembrane β-barrel structure to be established at atomic resolution (Weiss *et al.,* 1990). It revealed the general construction principles (Schulz, 1992, 1994), which were subsequently also observed in other membrane proteins (e.g., the aromatic girdles), in other transmembrane β-barrels (e.g., the short periplasmic turns), and in other porins of this type (e.g., the transversal electric field for polarity separation). The structures of two channels that are highly selective for maltooligosaccharides and sucrose, respectively, showed 18-stranded barrels with kidney-shaped cross sections (Schirmer *et al.,* 1995; Meyer *et al.,* 1997; Forst *et al.,* 1999). These cross sections deviate strongly from circles and allow long narrow channels which are required for the selection process.

The largest β-barrels have been observed with the monomeric iron transporter proteins FhuA and FepA. The structure of FhuA was established independently by two groups (Locher *et al.,* 1998; Ferguson *et al.,* 1998). It is known with and without a ligated siderophore. The structure of the ferric enterobactin receptor FepA is homologous to that of FhuA showing identical topology and a similar transport mechanism (Buchanan *et al.,* 1999). In both cases there are more than 700 residues assembled in two domains: an N-terminal 150-residue domain is located inside a C-terminal 22-stranded β-barrel with a shear number $S = 24$.

III. Construction Principles

The general structural properties of β-barrels are illustrated in Fig. 2. The barrel is sketched as a cylinder with circular cross section and all β-strands are assumed to run at the same inclination angle α. The β-pleated sheet parameters $a = 3.3$ Å and $b = 4.4$ Å refer to all kinds of

β-sheets: parallel, antiparallel, or mixed. The hydrogen bonds in Fig. 2A are sketched for the predominant antiparallel sheet. When cutting the cylindrical barrel at the place where the first strand reaches the upper barrel end and flattening it out, the relationship between the number of strands n, the shear number S, and the tilt angle α becomes obvious:

$$R = [(Sa)^2 + (nb)^2]^{0.5}/2\pi$$
$$\tan\alpha = Sa/nb$$
$$R = nb/2\pi \cos\alpha$$

Negative values of the shear number S are observed in the extreme cases of left-handed β-helices (Table I) and gramicidin-A artifacts (Table II). In canonical β-barrels S is positive and ranges between n and $n + 4$, allowing for an optimum β-sheet twist. TolC with $S = n + 8$ is an exception. Furthermore, S is always an even number because after running around the barrel, ridges and valleys of the pleated sheet have to be joined to ridges and valleys again. In other words, the hydrogen bonds repeat only every second residue.

The relationship between n, S, α and R in β-barrels is shown graphically in Fig. 2B and in Fig. 3. The smallest barrel considered has six strands. Two strongly distorted copies of it were found in chymotrypsin (Table I). Eight-stranded β-barrels are more regular and much more common; the most prominent example is the TIM barrel. Among the water-soluble proteins a series of eight-stranded barrels exists with shear numbers ranging from 8 to 12 in TIM, streptavidin, and lipocalins, respectively. In all these cases the barrel interior contains a hydrophobic core. In TIM barrels the active centers are invariably found at the carboxy-terminal end of the barrel. Streptavidin binds biotin at one barrel end. The same applies for lipocalins where the bound large nonpolar compounds reach down to the barrel center. The increasing ligand sizes and binding site depths from TIM to the lipocalins correspond to the increasing barrel radii caused by larger shear numbers.

In contrast to the smaller β-barrels of water-soluble proteins, those of membrane-inserted β-barrels start at eight strands and run up to 22 (Table II). Presumably, the required tightly packed nonpolar barrel core of the water-soluble proteins limits the radius of circular barrels to small values. In contrast, transmembrane barrels form polar cores the stability of which depends on hydrogen bonds rather than on geometrically exact nonpolar packing contacts. Such polar cores can be constructed much more leisurely and may also include water molecules that increase the interior volume. Accordingly, the interiors of the barrels with up to 12 strands are polar and solid, except for TolC with its exceptionally large shear number of 20. The large shear number gives rise to a large barrel radius (Fig. 3) which causes TolC to form a channel. The same applies for α-hemolysin, which, however, has a smaller shear number but two more β-strands.

Channels are of course also formed by all porins. A general porin contains 16 β-strands and has a shear number of 20 and a nearly circular cross section (Table II). Three parallel barrels associate to form trimers. The type of residues outlining the channel determines the specificity of a porin which, however, is usually not very strict. The two 18-stranded porins are very specific. Their channel cross-sections are actually smaller than those of the general porins in agreement with their higher selectivity. The 22-stranded barrels of the iron transporter proteins have circular cross sections and would form a very wide channel if they were not filled with the globular N-terminal 150-residue domain.

Apart from the construction principles dictated by the β-barrel geometry and illustrated in Fig. 2, the transmembrane β-barrels follow additional rules that are probably dictated by factors other than the covalent peptide structure:

1. Both the N- and C- termini are at the periplasmic barrel end restricting the strand number n to even values.
2. All β-strands are antiparallel and locally connected to their next neighbors.
3. On trimerization a nonpolar core is formed at the molecular threefold axis of the porins so that the central part of the trimer resembles a water-soluble protein.
4. The external β-strand connections are long loops named L1, L2, etc., whereas the periplasmic strand connections are generally minimum length turns named T1, T2, etc.
5. Cutting the barrel as shown in Fig. 2 and placing the periplasmic end at the bottom, the chain runs from the right to the left.
6. In all porins the constriction at the barrel center is formed by an inserted long loop L3.
7. The β-barrel surface contacting the nonpolar membrane interior is coated with aliphatic side chains forming a nonpolar ribbon. The two rims of this ribbon are lined by girdles of aromatic side chains.
8. The sequence variability in transmembrane β-barrels is higher than in water-soluble proteins and exceptionally high in the external loops.

The first three rules are likely to reflect the folding process of the trimeric porins. Presumably, the central part folds in the periplasm like a water-soluble protein. The membrane-exposed parts of the barrels are then formed on insertion into the membrane. The short turns of rule 4 may facilitate barrel formation inside the membrane. Presumably rule 5 is a consequence of rule 4 because appropriate short turns can only be formed in one of the two possible chain directions. Rule 6 seems to reflect an early evolutionary event that has not yet been revised. The aromatic girdles of rule 7 were suggested to stabilize the β-barrel and its vertical position in the membrane (Schulz, 1992). The stabilization has been confirmed experimentally by demonstrating the preference of aromatic compounds for the two nonpolar–polar transition regions of the membrane (Wimley and White, 1996; Yau *et al.*, 1998). Rule 8 came as a surprise to those with a high respect for membrane proteins which, of course, is mainly caused by our difficulties in solving membrane protein structures. Rule 8 explains these difficulties because it indicates that membrane proteins are subjected to fewer structural restraints than water-soluble ones and for this reason in general are more mobile and thus less crystallizable.

Given the drastic mobility differences between the external loops and the membraneous and periplasmic moieties of the barrels, it was suggested that these proteins can be crystallized by creating suitable packing contacts through semirandom mutagenesis at loops and turns (Pautsch *et al.*, 1999). Without structural knowledge, these loops and turns can often be predicted from the sequence and from interaction studies. For OmpA and OmpX this approach resulted in surprising successes. The procedure is also applicable to α-helical membrane proteins inasmuch as their crystallization problem is governed by their small polar surfaces. It is quite possible, however, that they hesitate to crystallize because their transmembrane α-helices are stabilized by the native laminar membrane environment and become flexible in the detergent/lipid micelles used for crystallization.

With so many rules, the prediction of transmembrane β-barrels from the sequence should be achievable at a high confidence level. However, the simple approach of looking for alternating polar and nonpolar residues inside and outside the barrel is not very helpful because this pattern is frequently broken by nonpolar residues on the inside. Moreover, the β-strands are merely slightly more than half a dozen residues long which limits their information content appreciably. These problems have been tackled in several prediction programs (Welte *et al.*, 1991; Schirmer and Cowan, 1993; Gromiha *et al.*, 1997; Seshadri *et al.*, 1998; Diederichs *et al.*, 1998; Jacoboni *et al.*, 2001) but cannot be considered solved. For some time to come, the safety of β-barrel prediction will remain well below that of transmembrane α-helices with their nonpolar 25-residue segments.

IV. Functions

OmpX is synthesized in large amounts in stress situations and it is probably used as a defensive weapon binding to and thus interfering with foreign proteins. Half of the OmpX β-barrel protrudes into the external medium, presenting an inclined β-sheet edge that binds to any foreign protein with a β-strand in its surface layer (Vogt and Schulz, 1999). Such proteins are ubiquitous, an example being the large group of proteins with central parallel β-sheets surrounded by α-helices. In X-ray analysis, all loop residues of OmpX were located in the electron density distribution indicating that the β-sheet edge presented to the foreign proteins is rigid, as it is required for tight binding.

In contrast to OmpX the long external loops of OmpA are highly mobile and, on the whole, not visible in the respective electron density map. *In vivo,* the mobile loops are rather resistant to proteolytic attack, presumably because they bind to the surrounding lipopolysaccharides. Obviously, the mobile loops fulfill essential functions in bacterial life (Morona *et al.,* 1984). They are also known as docking points for bacteriophages.

Among the outer membrane enzymes, OmpT is a special protease that has been implicated in the pathogenicity of bacteria. It is monomeric with the active center pointing to the outside (Vandeputte-Rutten *et al.,* 2001). Another enzyme, the phospholipase A OmpLA, produces holes in the outer membrane when it is activated. The activation process has not yet been clarified, but it is known to require a dimerization of OmpLA in the membrane. The activation by dimer formation has been verified by a crystal structure analysis of an OmpLA dimer that was produced by a reaction with an inhibitor (Snijder *et al.,* 1999). It showed that each active center contained a catalytic triad Ser-His-Asn on one subunit and an oxanion hole formed by an amide together with a hydrated Ca^{2+} ion on the other. The active centers are well placed for deacylating lipopolysaccharides of the external leaflet of the outer bacterial membrane. OmpLA functions in the secretion of colicins and virulence factors.

All general porins contain pores with sizes allowing the permeation of molecules up to molecular masses of about 600 Da (Nikaido, 1994). The pores come with various selectivities. The porin from *Rhodobacter capsulatus,* for instance, contains a rather nonpolar binding site near the external end of the pore eyelet, indicating that it may pick up molecules such as adenosine at very low concentrations. The structure also revealed a transversal electric field across the pore eyelet that acts as a polarity separator, excluding the unwanted nonpolar compounds (Schulz, 1992).

Porin OmpF from *Escherichia coli* has been thoroughly analyzed by numerous groups and became the first membrane protein to form X-ray grade crystals (Garavito and Rosenbusch, 1980). It is closely homologous to the porins PhoE (Cowan *et al.,* 1992) and OmpC. These three porins show permeation properties adjusted to different environmental conditions. PhoE allows an efficient uptake of phosphate. OmpC from *E. coli* and its homologue OmpK36 from *Klebsiella pneumoniae* (Dutzler *et al.,* 1999) are osmoporins expressed at high osmotic pressures usually caused by high salt concentrations. The high salt OmpC differs from the low salt OmpF mainly by an increased number of charged residues pointing into the pore lumen. Presumably, the charge increase counteracts Debye–Hückel shielding at high ionic strength so that OmpF and OmpC (OmpK36) have comparable permeation properties in differing environments. Additional structures were established for the main porin from *Paracoccus denitrificans* (Hirsch *et al.,* 1997) and for Omp32 from *Comomonas acidovorans* (Zeth *et al.,* 2000). They confirmed the established general features of their homologues.

All structurally established porins are aggregates of three parallel β-barrels, each of which contains a single polypeptide chain. The β-barrels contain either 16 or 18 strands. The interfaces are usually large and tightly packed. Therefore, it is not conceivable that the subunits are stable as monomers, either in the membrane or in the periplasm. However, the existence of a functional monomeric porin has been reported (Conlan *et al.,* 2000), but its structure has not yet been elucidated.

With maltoporin the bacteria developed a particular small pore that is adapted to the amylose helix and accepts only glucose units (Schirmer *et al.,* 1995; Meyer *et al.,* 1997). The energetics of a maltooligosaccharide diffusing through such a pore has been examined in detail, revealing a combination of nonpolar and optimally spaced polar interactions that result in smooth gliding (Meyer and Schulz, 1997). This energy profile has been confirmed in a more detailed molecular dynamics study (Schirmer and Phale, 1999) and in an experimental study based on mutants (Dumas *et al.,* 2000). The sucrose porin is a homologue of the maltoporin and very specific for the small molecule sucrose, in contrast with the maltoporin specific for larger oligomers (Forst *et al.,* 1998).

Besides porin trimers with 16- and 18-stranded β-barrels, even larger 22-stranded β-barrel proteins were found in the outer membrane, namely the monomeric active iron transporters FhuA and FepA. The lack of ATP or an equivalent energy carrier in the periplasm restricts the outer membrane, in the first place, to more or less specific but passive pores that are not able to transport any solute against a concentration gradient. Bacteria overcame this problem by inventing plugged β-barrels and the TonB apparatus.

The structure of the two evolutionarily related plugged pores FhuA and FepA are known. Their β-barrels have diameters of about 40 Å (Table II). They bind siderophore-encapsulated iron in the external half of the barrel and are obstructed by an N-terminal 150-residue domain in the periplasmic barrel half. Mutational studies revealed their TonB binding sites. The directed iron transport through the outer membrane is energized by an interaction with TonB of the inner membrane that can draw energy from the cytosolic ATP pool (Rutz *et al.*, 1992; Larsen *et al.*, 1999). The plug formed by the 150-residue domain is removed after binding to TonB, making the siderophore available for internalization.

A completely different principle has been followed by the heptameric α-hemolysines (Song *et al.*, 1996; Olson *et al.*, 1999). These proteins associate with their extramembrane domains. Subsequently, each subunit donates a β-hairpin to form a common 14-stranded β-barrel through the membrane. In a similar manner TolC is assembled from three α-helical subunits (Koronakis *et al.*, 2000). The subunits form a long, wide channel that spans the periplasm in their α-helical part and that is prolonged through the outer membrane by the β-barrel. The channel is used for the export of xenobiotics.

According to the endosymbiotic theory, the outer membrane of gram-negative bacteria corresponds to the outer membranes of mitochondria and chloroplasts (Reumann *et al.*, 1999). All of them are porous and cannot hold an electric potential difference. One long-term candidate for a porin homologue is the voltage-dependent anion channel (VDAC) of the outer mitochondrial membrane (Mannella, 1997). A further candidate for a β-barrel channel in the outer mitochondrial membrane is Tom40, which contains β-structure and forms a pore (Hill *et al.*, 1998). Its molecular mass would suggest a β-barrel of the size of general porins. Unfortunately, none of these proteins has as yet yielded crystals suitable for structure analysis. Presumably, they are particularly difficult to crystallize because they face the soft cytosol, which does not require tough structures, in contrast to their bacterial counterparts that face the external medium demanding much higher stability.

V. Folding and Stability

On first sight the folding process of β-proteins appears to be much more complex than that of α-proteins. This does not apply, however, for the all-next-neighbor antiparallel transmembrane β-barrels discussed here. For the porins it was suggested that the central part of the homotrimer including all N- and C-termini folds in the periplasm like a water-soluble protein so that the membrane-facing parts of the β-barrels dangle as 200-residue loops into the solvent (Schulz, 1992). On membrane insertion, these loops can then easily meander forming the special β-sheet topology. The simplicity of the folding process is corroborated by the fact that porins and other transmembrane β-barrels such as OmpA (Pautsch and Schulz, 1998) can be (re)natured from inclusion bodies. This production method worked even for the large monomeric β-barrel of FepA.

In vivo, the folding process may be supported by a periplasmic chaperone called Skp. Skp is a 17 kDa protein associated with the plasma membrane that, together with peptidyl prolyl isomerases and disulfide-exchanging enzymes, helps folding freshly synthesized proteins in the periplasm (Schäfer *et al.,* 1999). Skp binds to partially unfolded polypeptides. Depending on the presence of phospholipids, lipopolysaccharides, and bivalent cations, Skp exists in two conformations, one of which is protease-sensitive (DeCock *et al.,* 1999). Moreover, it was shown that Skp binds to unfolded periplasmic proteins and inserts into phospholipid monolayers, corroborating its putative role as helper in folding and membrane insertion.

The stability of the β-barrel itself was demonstrated in engineering experiments with OmpA. The four external loops of OmpA were replaced by shortcuts in all possible combinations (Koebnik, 1999). The resulting deletion mutants lost their biological functions in bacterial F-conjugation and as bacteriophage receptors, but kept the transmembrane β-barrel as demonstrated by their resistance to proteolysis and thermal denaturation. The results confirm the expectation that the large external loops do not contribute to β-barrel folding and stability.

Less dramatic changes were applied to α-hemolysin where the β-strand sequence was altered by reversing the sequence within the β-hairpin contributed by each subunit to the β-barrel (Cheley *et al.,* 1999). With respect to the β-barrel this changed only the hydrogen bonding pattern, but it reversed the sequence in the β-turn at the end of the hairpin which should, therefore, have local conformational consequences. It turned out that the "retro"-barrel formed a channel but failed to invade erythrocytes. A high activity could be obtained, however, when the β-turn was left in its original amino acid sequence, demonstrating that the tight β-turns at one end of these barrels are important for the stability of the whole barrel. Unfortunately, the detailed structures of the "retro"-barrels remained unknown.

The interiors of the three small β-barrels of OmpA, OmpX, and OmpLA are filled with polar residues forming a hydrogen bonding network. A number of separated cavities mostly filled with water have been reported for OmpA and OmpX. Accordingly, these β-proteins are rigid inverse micelles and unlikely to form pores. For OmpA there was a lengthy discussion on the question of pore formation because small pores had actually been detected by ion fluxes (Sugawara and Nikaido, 1994; Arora *et al.*, 2000). It seems likely that these data were obtained with OmpA preparations that had lost the internal barrel structure during the protein purification process permitting channel formation. Such conformations should differ appreciably from the crystalline structure.

Using an atomic force microscope (AFM) operated in the tapping mode, the surface structure of OmpF was analyzed under several conditions. For this purpose, two-dimensional crystals of the porin OmpF from *E. coli* were mounted on graphite or mica and exposed to large voltages and low pH as well as high ionic strength environments (Müller and Engel, 1999). When operating in the contact mode at minimum force the AFM showed a large conformational change in the external loops reminiscent of a structural collapse. The authors relate this to voltage closure. Unfortunately, the experimental setup is somewhat artifical and the actual voltage across the porin cannot be measured. However, the effects are obvious and certainly report a structural weakness in the protrusion formed by the external loops. Since the AFM works with living material and is able to detect vertical differences in the range of atomic bond lengths, it is likely to yield most interesting results in the future.

A pore feature attracting continuous interest is voltage gating. This effect remains to be explained in detail. It is observed at comparatively high voltages across the membrane. Since the corresponding electric field strength is so high that it may disrupt hydrogen bonds, it seems likely that most of the *in vitro* voltage gating studies (Liu and Delcour, 1998; Saxena *et al.*, 1999) report nothing else than a structural breakdown inside a pore. This does not apply to the voltage-dependent anion channel of the outer mitochondrial membrane (VDAC), however, which shows channel closure *in vivo* (Mannella, 1998). Presumably, VDAC contains a solid but separate mobile and charged domain that can be driven by the electric field onto the pore so that it prevents any further ion flow.

VI. Channel Engineering

The diffusion of small solutes through porins is passive. The diameters of the pore eyelets range from 10 Å for the general porins to 6 Å for the highly selective porins. Larger pores are usually decorated with oppositely charged residues at opposite sides that form a local transversal electric field at the pore eyelet. This field constitutes an energy barrier for low-polarity solutes (Schulz, 1992) so that the bacterium can exclude the unwanted nonpolar molecules such as antibiotics despite presenting a spacious eyelet for picking up large polar molecules such as sugars.

The engineering of porins became popular after it had been demonstrated that a mass-produced porin (re)natured from inclusion bodies had assumed exactly the native conformation (Schmid *et al.*, 1996). A systematic study changing the pore properties by mutations showed a strong correlation between eyelet cross section and diffusion rate (Schmid *et al.*, 1998). Furthermore, a series of nine porins with mutations at the eyelet was analyzed with respect to ion conductance, ion selectivity, and voltage gating (Saxena *et al.*, 1999). It was shown that charge reversals affect selectivity and voltage gating. Similar results were obtained with mutation at loop L3 of PhoE (VanGelder *et al.*, 1997). In contrast to modifications at loop L3 inside the β-barrel, mutations at barrel wall residues lining the eyelet had only minor effects on voltage gating. This corroborates the suggestion that voltage gating reflects a structural breakdown in the pore.

Sucrose porin has a somewhat larger pore than its homologue maltoporin, the pore eyelet of which is closely adjusted to $\alpha(1 \rightarrow 4)$-bound glucose units (Meyer and Schulz, 1997). With mutations at loop L3 the specificity of the sucrose porin was changed toward that of a maltoporin (Ulmke *et al.*, 1999). The specificity change was achieved by introducing three eyelet-defining residues of the maltoporin and by removing the additional N-terminal 70-residue domain of the sucrose porin, the structure of which is not yet known.

The width of a passive channel can be monitored by ionic currents through a black-lipid membrane harboring the respective membrane protein. The channel may be clogged by organic molecules reducing the current over the residence time of such a molecule. This time as well as the amount of the current reduction is characteristic for the applied compound. Gu *et al.* (1999) used this principle by placing a cyclodextrin as an adapter into the 14-stranded β-barrel of an engineered α-hemolysin. Using this device they determined the current reductions and the residence times caused by a number of modified adamantans, demonstrating that they can measure these compounds in concentrations of around 10 μM and distinguish between various types. In a further study Braha *et al.* (1997) introduced a zinc ion binding site into the β-barrel of α-hemolysin. When occupied, this site changed the channel conductivity, so that this barrel could be used as a zinc ion sensor. These are marvelous engineering examples, converting a β-barrel pore to a molecular sensor with an electric output signal.

In addition to the manifold possibilities of engineering on native β-barrels, these can also be designed from scratch. In an *ab initio* approach Ghadiri *et al.* (1994) designed cyclic octapeptides and demonstrated that these assemble to socalled "nanotubes" forming channels through a membrane. The octapeptides consisted of alternating D- and L-amino acid residues and thus followed closely the construction principle of gramicidin A. In its native conformation gramicidin A forms a single strand β-barrel ($n = 1$) with a shear number $S = 6$ that may also be called a β-helix. The nanotubes are close to the same construction with $n = 1$ and a shear number $S =$ '8', but forming a ring instead of a β-helix. Stacking these rings across the membrane in a β-barrel with an inclination angle $\alpha = 90°$ (Fig. 2A) then produces the channel.

VII. Conclusions

Although the major fraction of transmembrane proteins is α-helical, their β-barrel counterparts have become popular because many of them could be analyzed in atomic detail and at high resolution. The transmembrane β-barrel proteins occur only in the outer membranes of bacteria, mitochondria, and chloroplasts. They assume astonishingly regular conformations giving rise to numerous rules for their construction. It has been suggested that these regularities are required by the folding process. They are likely to permit the detection of transmembrane β-barrels from the sequence at a reasonable confidence level in the future.

Whereas water-soluble proteins contain regular β-barrels with up to eight strands and nonpolar interiors, transmembrane β-barrels contain eight or more strands and have polar interiors. The smaller transmembrane β-barrels have solid cores partially filled with water. Accordingly, they can be considered inverse micelles. The larger barrels of this type have channels along their axis that allow for the permeation of various types of solutes. The largest known 22-stranded transmembrane β-barrels are used for the active transport of rare commodities through the bacterial outer membrane. Their interior contains a globular protein domain which functions as a plug that can be removed for transport.

The success rate for engineering transmembrane β-barrels appears to be superior to that for soluble proteins. On one hand, these barrels can be mass produced into inclusion bodies and (re)natured therefrom *in vitro.* On the other hand, the interiors of the β-barrels housing the permeation channels can be mutated with little effect on the barrel construction. The instabilities usually introduced with amino acid residue changes can be compensated for to a large degree by the surrounding β-barrel. This allows the application of a rather broad spectrum of mutations without destroying the protein construction as with water-soluble proteins in which mutations are frequently punished by deterioration. Taking advantage of this situation, the β-barrels have already been engineered to sensors for organic molecules and metal ions at micromolar concentrations, which, in the future, may lead to useful technical devices.

References

Arora, A., Rinehart, D., Szabo, G., and Tamm, L. K. (2000). Refolded outer membrane protein A of *Escherichia coli* forms ion channels with two conductance states in planar lipid bilayers. *J. Biol. Chem.* **275,** 1594–1600.

Arora, A., Abildgaard, F., Bushweller, J. H., and Tamm, L. K. (2001). Structure of outer membrane protein A transmembrane domain by NMR spectroscopy. *Nat. Struct. Biol.* **8,** 334–338.

Banner, D. W., Bloomer, A. C., Petsko, G. A., Phillips, D. C., Pogson, C. I., Wilson, I. A., Corran, P. H., Furth, A. J., Milman, J. D., Offord, R. E., Priddle, J. D., and Waley, S. G. (1975). Structure of chicken muscle triose phosphate isomerase determined crystallographically at 2.5 Å resolution using amino acid sequence data. *Nature* **255,** 609–614.

Baumann, U., Wu, S., Flaherty, K. M., and McKay, D. B. (1993). Three-dimensional structure of the alkaline protease of *Pseudomonas aeruginosa:* a two-domain protein with a calcium binding parallel beta roll motif. *EMBO J.* **12,** 3357–3364.

Birktoft, J. J., and Blow, D. M. (1972). Structure of crystalline α-chymotrypsin. *J. Mol. Biol.* **68,** 187–240.

Bowie, J. U. (1999). Helix-bundle membrane protein fold templates. *Protein Sci.* **8,** 2711–2719.

Braha, O., Walker, B., Cheley, S., Kasianowicz, J. J., Song, L., Gouaux, J. E., and Bayley, H. (1997). Designed protein pores as components for biosensors. *Chem. Biol.* **4,** 497–505.

Brenner, S. E., and Levitt, M. (2000). Expectations from structural genomics. *Protein Sci.* **9,** 197–200.

Buchanan, S. K., Smith, B. S., Venkatramani, L., Xia, D., Esser, L., Palnitkar, M., Chakraborty, R., van der Helm, D., and Deisenhofer, J. (1999). Crystal structure of the outer membrane active transporter FepA from *Escherichia coli. Nat. Struct. Biol.* **6,** 56–63.

Cheley, S., Braha, O., Lu, X., Conlan, S., and Bayley, H. (1999). A functional protein pore with a "retro" transmembrane domain. *Protein Sci.* **8,** 1257–1267.

Conlan, S., Zhang, Y., Cheley, S., and Bayley, H. (2000). Biochemical and biophysical characterization of OmpG: a monomeric porin. *Biochemistry* **39,** 11845–11854.

Cowan, S. W., Schirmer, T., Rummel, G., Steiert, M., Ghosh, R., Pauptit, R. A., Jansonius, J. N., and Rosenbusch, J. P. (1992). Crystal structures explain functional properties of two *E. coli* porins. *Nature* **358,** 727–733.

DeCock, H., Schäfer, U., Potgeter, M., Demel, R., Müller, M., and Tommassen, J. (1999). Affinity of the periplasmic chaperone Skp of *Escherichia coli* for phospholipids, lipopolysaccharides and non-native outer membrane proteins. *Eur. J. Biochem.* **259,** 96–103.

Diederichs, K., Freigang, J., Umhau, S., Zeth, K., and Breed, J. (1998). Prediction by a neural network of outer membrane β-strand protein topology. *Protein Sci.* **7,** 2413–2420.

Dumas, F., Koebnik, R., Winterhalter, M., and Van Gelder, P. (2000). Sugar transport through maltoporin of *Escherichia coli. J. Biol. Chem.* **275,** 19747–19751.

Dutzler, R., Rummel, G., Alberti, S., Hernández-Allés, S., Phale, P. S., Rosenbusch, J. P., Benedi, V. J., and Schirmer, T. (1999). Crystal structure and functional characterization of OmpK36, the osmoporin of *Klebsiella pneumoniae. Structure* **7,** 425–434.

Ferguson, A. D., Hofmann, E., Coulton, J. W., Diederichs, K., and Welte, W. (1998). Siderophore-mediated iron transport: crystal structure of FhuA with bound lipopolysaccharide. *Science* **282,** 2215–2220.

Fernández, C., Adeishvili, K., and Wüthrich, K. (2001a). Transverse relaxation-optimized NMR spectroscopy with the outer membrane protein OmpX in dihexanoyl phosphatidylcholine micelles. *Proc. Natl. Acad. Sci. USA* **98,** 2358–2363.

Fernández, C., Hilty, C., Bonjour, S., Adeishvili, K., Pervushin, K., and Wüthrich, K. (2001b). Solution NMR studies of the integral membrane proteins OmpX and OmpA from *Escherichia coli. FEBS Lett.* **504,** 173–178.

Forst, D., Welte, W., Wacker, T., and Diederichs, K. (1998). Structure of the sucrose-specific porin ScrY from *Salmonella typhimurium* and its complex with sucrose. *Nat. Struct. Biol.* **5,** 37–46.

Garavito, R. M., and Rosenbusch, J. P. (1980). Three-dimensional crystals of an integral membrane protein: an initial X-ray analysis. *J. Cell Biol.* **86,** 327–329.

Ghadiri, M. R., Granja, J. R., and Buehler, L. K. (1994). Artificial transmembrane ion channels from self-assembling peptide nanotubes. *Nature* **369,** 301–304.

Gromiha, M. M., Majumdar, R., and Ponnuswamy, P. K. (1997). Identification of membrane spanning β-strands in bacterial porins. *Protein Eng.* **10,** 497–500.

Gu, L. Q., Braha, O., Conlan, S., Cheley, S., and Bayley, H. (1999). Stochastic sensing of organic analytes by a pore-forming protein containing a molecular adapter. *Nature* **398,** 686–690.

Hendrickson, W. A., Pähler, A., Smith, J. L., Satow, Y., Merritt, E. A., and Phizackerley, R. P. (1989). Crystal structure of core streptavidin determined from multi-wavelength anomalous diffraction of synchrotron radiation. *Proc. Natl. Acad. Sci. USA* **86,** 2190–2194.

Hill, K., Model, K., Ryan, M. T., Dietmeier, K., Martin, F., Wagner, R., and Pfanner, N. (1998). Tom40 forms the hydrophilic channel of the mitochondrial import pore for preproteins. *Nature* **395,** 516–521.

Hirsch, A., Breed, J., Saxena, K., Richter, O. M. H., Ludwig, B., Diederichs, K., and Welte, W. (1997). The structure of porin from *Paracoccus denitrificans* at 3.1 Å resolution. *FEBS Lett.* **404,** 208–210.

Jacoboni, I., Martelli, P. L., Fariselli, P., de Pinto, V., and Casadio, R. (2001). Prediction of the transmembrane regions of β-barrel membrane proteins with a neural network-based predictor. *Protein Sci.* **10,** 779–787.

Jap, B. K. (1989). Molecular design of PhoE porin and its functional consequences. *J. Mol. Biol.* **205,** 407–419.

Ketchem, R. R., Hu, W., and Cross, T. A. (1993). High-resolution conformation of gramicidin A in a lipid bilayer by solid-state NMR. *Science* **261,** 1457–1460.

Koebnik, R. (1999). Structural and functional roles of the surface-exposed loops of the β-barrel membrane protein OmpA from *Escherichia coli. J. Bacteriol.* **181,** 3688–3694.

Koronakis, V., Sharff, A., Koronakis, E., Luisi, B., and Hughes, C. (2000). Crystal structure of the bacterial membrane protein TolC central to multidrug efflux and protein export. *Nature* **405,** 914–919.

Kreusch, A., and Schulz, G. E. (1994). Refined structure of the porin from *Rhodopseudomonas blastica:* comparison with the porin from *Rhodobacter capsulatus. J. Mol. Biol.* **243,** 891–905.

Langs, D. A. (1988). Three-dimensional structure at 0.86 Å of the uncomplexed form of the transmembrane ion channel peptide gramicidin A. *Science* **241,** 188–191.

Larsen, R. A., Thomas, M. G., and Postle, K. (1999). Protonmotive force, ExbB and ligand-bound FepA drive conformational changes in TonB. *Mol. Microbiol.* **31,** 1809–1824.

Liu, J., and Rost, B. (2001). Comparing function and structure between entire proteomes. *Protein Sci.* **10,** 1970–1979.

Liu, N., and Delcour, A. H. (1998). The spontaneous gating activity of OmpC porin is affected by mutations of a putative hydrogen bond network or of a salt bridge between the L3 loop and the barrel. *Protein Eng.* **11,** 797–802.

Liu, W. M. (1998). Shear numbers of protein β-barrels: definition, refinements and statistics. *J. Mol. Biol.* **275,** 541–545.

Locher, K. P., Rees, B., Koebnik, R., Mitschler, A., Moulinier, L., Rosenbusch, J. P., and Moras, D. (1998). Transmembrane signaling across the ligand-gated FhuA receptor: crystal structures of free and ferrichrome-bound states reveal allosteric changes. *Cell* **95,** 771–778.

Mannella, C. A. (1997). On the structure and gating mechanism of the mitochondrial channel, VDAC. *J. Bioenerg. Biomembr.* **29,** 525–531.

Mannella, C. A. (1998). Conformational changes in the mitochondrial channel protein, VDAC, and their functional implications. *J. Struct. Biol.* **121,** 207–218.

Meyer, J. E. W., and Schulz, G. E. (1997). Energy profile of maltooligosaccharide permeation through maltoporin as derived from the structure and from a statistical analysis of saccharide–protein interactions. *Protein Sci.* **6,** 1084–1091.

Meyer, J. E. W., Hofnung, M., and Schulz, G. E. (1997). Structure of maltoporin from *Salmonella typhimurium* ligated with a nitrophenyl-maltotrioside. *J. Mol. Biol.* **266,** 761–775.

Morona, R., Klose, M., and Henning, U. (1984). *Escherichia coli* K-12 outer membrane protein (OmpA) as a bacteriophage receptor: analysis of mutant genes expressing altered proteins. *J. Bacteriol.* **159,** 570–578.

Müller, D. J., and Engel, A. (1999). Voltage and pH-induced channel closure of porin OmpF visualized by atomic force microscopy. *J. Mol. Biol.* **285,** 1347–1351.

Murzin, A. G., Lesk, A. M., and Chothia, C. (1994a). Principles determining the structure of β-sheet barrels in proteins. I. A theoretical analysis. *J. Mol. Biol.* **236,** 1369–1381.

Murzin, A. G., Lesk, A. M., and Chothia, C. (1994b). Principles determining the structure of β-sheet barrels in proteins. II. The observed structures. *J. Mol. Biol.* **236,** 1382–1400.

Nabedryk, E., Garavito, R. M., and Breton, J. (1988). The orientation of β-sheets in porin. A polarized Fourier transform infrared spectroscopic investigation. *Biophys. J.* **53,** 671–676.

Newcomer, M. E., Jones, T. A., Åqvist, J., Sundelin, J., Eriksson, U., Rask, L., and Peterson, P. A. (1984). The three-dimensional structure of retinol-binding protein. *EMBO J.* **3,** 1451–1454.

Nikaido, H. (1994). Prevention of drug access to bacterial targets: permeability barriers and active efflux. *Science* **264,** 382–388.

Olson, R., Nariya, H., Yokota, K., Kamio, Y., and Gouaux, E. (1999). Crystal structure of staphylococcal LukF delineates conformational changes accompanying formation of a transmembrane channel. *Nat. Struct. Biol.* **6,** 134–140.

Pautsch, A., and Schulz, G. E. (1998). Structure of the outer membrane protein A transmembrane domain. *Nat. Struct. Biol.* **5,** 1013–1017.

Pautsch, A., and Schulz, G. E. (2000). High resolution structure of the OmpA membrane domain. *J. Mol. Biol.* **298,** 273–282.

Pautsch, A., Vogt, J., Model, K., Siebold, C., and Schulz, G. E. (1999). Strategy for membrane protein crystallization exemplified with OmpA and OmpX. *Proteins: Struct. Funct. Genet.* **34,** 167–172.

Raetz, C. R. H., and Roderick, S. L. (1995). A left-handed parallel β-helix in the structure of UDP-*N*-acetylglucosamine acyltransferase. *Science* **270,** 997–1000.

Reumann, S., Davila-Aponte, J., and Keegstra, K. (1999). The evolutionary origin of the protein-translocating channel of chloroplastic envelope membranes: identification of a cyanobacterial homolog. *Proc. Natl. Acad. Sci. USA* **96,** 784–789.

Rutz, J. M., Liu, J., Lyons, A. J., Goranson, J., Armstrong, S. K., McIntosh, M. A., Feix, J. B., and Klebba, P. E. (1992). Formation of a gated channel by a ligand-specific transport protein in the bacterial outer membrane. *Science* **258,** 471–475.

Saxena, K., Drosou, V., Maier, E., Benz, R., and Ludwig, B. (1999). Ion selectivity reversal and induction of voltage-gating by site-directed mutations in the *Paracoccus denitrificans* porin. *Biochemistry* **38,** 2206–2212.

Schäfer, U., Beck, K., and Müller, M. (1999). Skp, a molecular chaperone of gram-negative bacteria, is required for the formation of soluble periplasmic intermediates of outer membrane proteins. *J. Biol. Chem.* **274,** 24567–24574.

Schirmer, T., and Cowan, S. W. (1993). Prediction of membrane-spanning β-strands and its application to maltoporin. *Protein Sci.* **2,** 1361–1363.

Schirmer, T., and Phale, P. S. (1999). Brownian dynamics simulation of ion flow through porin channels. *J. Mol. Biol.* **294,** 1159–1167.

Schirmer, T., Keller, T. A., Wang, Y. F., and Rosenbusch, J. P. (1995). Structural basis for sugar translocation through maltoporin channels at 3.1 Å resolution. *Science* **267,** 512–514.

Schmid, B., Krömer, M., and Schulz, G. E. (1996). Expression of porin from *Rhodopseudomonas blastica* in *Escherichia coli* inclusion bodies and folding into exact native structure. *FEBS Lett.* **381,** 111–114.

Schmid, B., Maveyraud, L., Krömer, M., and Schulz, G. E. (1998). Porin mutants with new channel properties. *Protein Sci.* **7,** 1603–1611.

Schulz, G. E. (1981). Protein differentiation: emergence of novel proteins during evolution. *Angew. Chem. Int. Ed. Engl.* **20,** 143–151.

Schulz, G. E. (1992). Structure–function relationships in the membrane channel porin as based on a 1.8 Å resolution crystal structure. In: *Membrane Proteins: Structures, Interactions and Models* (A. Pullman, J. Jortner, and B. Pullman, Eds.), pp. 403–412. Kluwer Academic, Dordrecht.

Schulz, G. E. (1994). Structure–function relationships in porins as derived from a 1.8 Å resolution crystal structure. In: *New Comprehensive Biochemistry: Bacterial Cell Wall* (J. M. Ghuysen, and R. Hakenbeck, Eds.), vol. 27, pp. 343–352. Elsevier, Amsterdam.

Schulz, G. E., and Schirmer, R. H. (1979). *Principles of Protein Structure,* pp. 84–87. Springer-Verlag, New York.

Seshadri, K., Garemyr, R., Wallin, E., von Heijne, G., and Elofsson, A. (1998). Architecture of β-barrel membrane proteins: analysis of trimeric porins. *Protein Sci.* **7,** 2026–2032.

Sipos, L., and von Heijne, G. (1993). Predicting the topology of eukaryotic membrane proteins. *Eur. J. Biochem.* **213,** 1333–1340.

Snijder, H. J., Ubarretxena-Belandia, I., Blaauw, M., Kalk, K. H., Verhij, H. M., Egmond, M. R., Dekker, N., and Dijkstra, B. W. (1999). Structural evidence for dimerization-regulated activation of an integral membrane phospholipase. *Nature* **401,** 717–721.

Song, L., Hobaugh, M. R., Shustak, C., Cheley, S., Bayley, H., and Gouaux, J. E. (1996). Structure of staphylococcal α-hemolysin, a heptameric transmembrane pore. *Science* **274,** 1859–1865.

Sugawara, E., and Nikaido, H. (1994). OmpA protein of *Escherichia coli* outer membrane occurs in open and closed channel forms. *J. Biol. Chem.* **269,** 17981–17987.

Ulmke, D., Kreth, J., Lengeler, J. W., Welte, W., and Schmid, T. (1999). Site-directed mutagenesis of loop L3 of sucrose porin ScrY leads to changes in substrate selectivity. *J. Bacteriol.* **181,** 1920–1923.

Vandeputte-Rutten, L., Kramer, R. A., Kroon, J., Dekker, N., Egmond, M. R., and Gros, P. (2001). Crystal structure of the outer membrane protease OmpT from *Escherichia coli* suggests a novel catalytic site. *EMBO J.* **20,** 5033–5039.

VanGelder, P., Saint, N., van Boxtel, R., Rosenbusch, J. P., and Tommassen, J. (1997). Pore functioning of outer membrane protein PhoE of *Escherichia coli:* mutagenesis of the constriction loop L3. *Protein Eng.* **10,** 699–706.

Vogt, J., and Schulz, G. E. (1999). The structure of the outer membrane protein OmpX from *Escherichia coli* reveals possible mechanisms of virulence. *Structure* **7,** 1301–1309.

Wallace, B. A., and Ravikumar, K. (1988). The gramicidin pore: crystal structure of a cesium complex. *Science* **241,** 182–187.

Weiss, M. S., and Schulz, G. E. (1992). Structure of porin refined at 1.8 Å resolution. *J. Mol. Biol.* **227,** 493–509.

Weiss, M. S., Wacker, T., Weckesser, J., Welte, W., and Schulz, G. E. (1990). The three-dimensional structure of porin from *Rhodobacter capsulatus* at 3 Å resolution. *FEBS Lett.* **267,** 268–272.

Welte, W., Weiss, M. S., Nestel, U., Weckesser, J., Schiltz, E., and Schulz, G. E. (1991). Prediction of the general structure of OmpF and PhoE from the sequence and structure of porin from *Rhodobacter capsulatus.* Orientation of porin in the membrane. *Biochim. Biophys. Acta* **1080,** 271–274.

Wimley, W. C., and White, S. H. (1996). Experimentally determined hydrophobicity scale for proteins at membrane interfaces. *Nat. Struct. Biol.* **3,** 842–848.

Yau, W. M., Wimley, W. C., Gawrisch, K., and White, S. H. (1998). The preference of tryptophan for membrane interfaces. *Biochemistry* **37,** 14713–14718.

Yoder, M. D., Keen, N. T., and Jurnak, F. (1993). New domain motif: the structure of pectate lyase C, a secreted plant virulence factor. *Science* **260,** 1503–1507.

Zeth, K., Diederichs, K., Welte, W., and Engelhardt, H. (2000). Crystal structure of Omp32, the anion-selective porin from *Comamonas acidovorans,* in complex with a periplasmic peptide at 2.1 Å resolution. *Structure* **8,** 981–992.

LENGTH, TIME, AND ENERGY SCALES OF PHOTOSYSTEMS

By CHRISTOPHER C. MOSER,* CHRISTOPHER C. PAGE,*
RICHARD J. COGDELL,† JAMES BARBER,‡ COLIN A. WRAIGHT,§
and P. LESLIE DUTTON*

*Johnson Research Foundation and Department of Biochemistry and Biophysics, School of Medicine, University of Pennsylvania, Philadelphia, Pennsylvania 19104;
†Division of Biochemistry and Molecular Biology, Glasgow University, Lanark, Scotland G12 8QQ, United Kingdom;
‡Department of Biological Sciences, Imperial College of Science, Technology and Medicine, London SW7 2AY, United Kingdom; and
§Department of Biochemistry and Center for Biophysics and Computational Biology, University of Illinois at Urbana-Champaign, Illinois 61801

I. Introduction

Here we reflect on the functional design, engineering, and construction of the proteins engaged in the harvesting of light energy and transforming it into membrane electrochemical gradients and the oxidants and reductants that provide energy and material substrates for the cell. We focus our attention on light harvesting (LH) proteins and on reaction center (RC) proteins of photosynthetic bacteria, cyanobacteria, algae, and higher plants. Useful perspective comes from comparing photosystem design and engineering with the analogous membrane redox proteins of respiration.

ADVANCES IN PROTEIN CHEMISTRY, Vol. 63

Figure 1 (see color insert) describes a relay of membrane proteins that first transfers light energy captured by the network of the LH proteins to an RC protein, which then catalyzes electron tunneling–mediated charge separation and transmembrane electric potential generation and substrate oxidation and reduction. Redox connection of the RC with neighboring intrinsic and peripheral membrane proteins is then mediated through molecular diffusion of the substrates and products. These in turn provide the energy for the oxidoreductases cytochrome bc_1 or b_6f, which, like the RC, promote transmembrane charge separation and the buildup of an electrochemical potential gradient of protons. This electrochemical potential gradient ultimately provides the driving force for the majority of biochemical reactions in the cell, including the dark reactions of photosynthesis.

Although the proteins of photosynthetic machinery are large and complex, it appears that the engineering that has been favored by blind natural selection is comparatively simple and resilient and does not require an atom-by-atom examination to appreciate its design. Instead we can begin by considering how nature has worked within the time constraints imposed by various decay processes by exploiting the distinctive length scales that are associated with each stage of the relay (Fig. 2, see color insert). We hypothesize that the membrane proteins of the relay evolved with a strong selection on the simple distance between the cofactors. We contend that distance selection is dominant in providing photosynthetic and respiratory energy conversion with robust foundations that accommodate extensive structural and energetic tolerances. We further expect these tolerances within simple, robust frameworks have fostered the broad palette of evolutionary variance evident in the proteins of microorganisms, animals, and plants and have enabled them to generally operate well away from the boundaries of failure and pathogenesis. This relative simplicity and robustness also promises to be translatable to, and exploitable in, synthetic systems of energy conversion.

II. Overview of Length Scales in Bioenergetic Membranes

The physical process of energy conversion describes a few basic length scales that determine the general design of photosynthetic proteins. Electron tunneling defines the first and most conspicuous length scale in redox proteins that couple the early photosynthetic events of light energy transfer to the later proton transfer, redox catalysis, and diffusive motion reactions (Fig. 2). Indeed, RCs provided the first demonstration of electron tunneling in biology in the early 1960s (Devault and Chance, 1966) and have continued to provide much of the now extensive biological electron tunneling information in the years since. There is a tremendous experimental advantage in the natural ability to activate LH and RC photosystems with single-turnover flashes and time-resolve both donor oxidation and acceptor reduction from tens of seconds to tens of femtoseconds at temperatures from 300 K to 1 K. Furthermore, RCs have demonstrated the stability and the molecular biological and biochemical flexibility needed to alter structure and free energies of individual tunneling reactions for systematic studies, backed by the vital high-resolution structures. As a result, simple guidelines have emerged from the work on RCs that are now proving useful in the understanding of tunneling processes in other redox proteins, extending our insights to the tunneling engineering guidelines of electron transfer proteins in general.

The free energy (ΔG) dependence of the electron-tunneling rate gives rise to a second length scale, this time an energy/length scale. As shown in Fig. 2B, for any characteristic time and free energy, there is a maximum distance over which tunneling can extend. This is true for both the traditional exergonic electron transfers and the endergonic or uphill electron transfers (Page *et al.,* 1999). Electron transfer to an energetically unfavorable redox intermediate can take place within a given time as long the uphill electron transfer is followed eventually by a favorable downhill electron transfer. Indeed, endergonic electron transfers through intermediates hundreds of meV uphill can take place faster than typical enzymatic turnover times provided the redox centers are closer than 12 Å. The smaller the distance, the more energetically unfavorable the electron transfer intermediate can be.

Transmembrane electric field strengths provide a related length scale in these membrane-bound proteins, since the energetics of electron transfer are dependent on the length of the electron transfer along the membrane normal. Light-driven electron tunneling in RCs is coupled to electronic charge separation across the ~35 Å of the membrane low dielectric profile. The initial picosecond charge separation in RCs occurs within the several chlorophylls (Chl) or bacteriochlorophylls (BChl) of the RCs and is mostly parallel to the membrane. In the subsequent electron transfers, the electron and hole move in opposite directions to span the supporting membranes and generate a transmembrane electric potential ($\Delta\Psi$) that together with the transmembrane pH gradient (ΔpH) is an essential component of the electrochemical proton gradient ($\Delta\mu H^+$). A similar pattern is followed in many of the companion oxidoreductases in which the principal catalytic event involves mostly lateral electron transfer, succeeded by transmembrane electric field generating and proton moving steps. In the photosynthetic and respiratory machinery the longest transmembrane transfers contribute the most to

$\Delta\mu H^+$ and are also the most susceptible to inhibition by preexisting electric fields. In contrast, the lateral catalytic reactions are relatively unaffected by transmembrane fields.

The magnitude of the prevailing $\Delta\mu H^+$, as well as the proportions of the ΔpH and $\Delta\Psi$ to $\Delta\mu H^+$, varies from species to species (Dutton *et al.*, 2000). Across the 35 Å low dielectric region, bacteria develop more than 250 mV of electric field and 0.5 units of ΔpH. Mitochondrial membrane fields are a bit less at ~170 mV, with a similar pH drop, while chloroplasts may have little transmembrane field, and a much larger proton concentration gradient, up to 4 pH units. Thus bacterial reaction centers and oxidoreductases have to face an electric field of nearly 10^8 V/m when moving charges across the membrane. This can be enough to partially inhibit the bc_1 complex (Shinkarev *et al.*, 2001). On the other hand, the minor electric fields sensed by plant PSI, PSII, and b_6f means that the transmembrane electron transfers themselves are not inhibited. However, the large ΔpH will eventually inhibit the proton-exchanging quinone and water/oxygen redox reactions and slow the turnover of the system.

A third length scale is defined by the dependence of energy transfer upon the distance between chromophores. As shown in Fig. 1, light capture occurs predominantly by chlorophylls, carotenoids, and other pigments associated with LH proteins that transfer their energy to a reaction center (Cogdell *et al.*, 1999). Light is also captured directly by the pigments of the RC protein itself. However, RCs differ sharply from LHs by being equipped and engineered to couple absorbed light energy to electron tunneling and for the production of strong but stable oxidants and reductants. The length scales for energy transfer efficiency and electron tunneling in Fig. 2 are sharply different. The $1/r^3$ or $1/r^6$ decay of energy transfer efficiency, where r is the distance between cofactors, is less severe than the exponential decay of the tunneling rate with distance (Sundstrom *et al.*, 1999). However, the ultimate engineering limit of light energy transfer is the decay process of fluorescence, shown as the purple band in Fig. 2A. The need to complete energy transfer before fluorescence decay is one of two principal time constraints on the range of permissible intercofactor distances in these bioenergetic proteins.

Diffusion provides a fourth distance scale. Each of the protein complexes generates diffusible oxidants and reductants in the different compartments defined by the bioenergetic membranes. Thus, the PSI RC of cyanobacteria, algae, and plants deliver the strong stable reductant NADPH to the cytoplasm or chloroplast stroma, where it is used for biosynthesis. The corresponding strong oxidant dioxygen is released from the luminal side of the PSII RC and diffuses away to be consumed in respiratory processes within the cell or expelled from the organism as a by-product. Other oxidants and reductants produced by the different RCs are used locally as substrates with a redox potential difference (ΔEh) that drives further, respiratory-type electron transfer in cytochromes bc_1 and b_6f. Thus, RCs from purple bacteria produce hydroquinones in the membrane and ferricytochrome *c* in the periplasm that diffuse away to become substrates to the cytochrome bc_1 which regenerates ferrocytochrome *c* and quinone to complete a cycle. In plants, PSII and PSI produce the analogous hydroquinone and oxidized plastocyanin or cytochrome c_6 as substrates for the cytochromes b_6f in an overall linear electron transfer chain.

The distances that are covered by substrates and products to reach the next membrane protein are considerable. Ubiquinone diffuses tens to hundreds of angstroms to connect RC and bc_1 in bacteria, while plastoquinone diffuses even further between PSII and b_6f in chloroplasts. Figure 1 shows some approximations of diffusion of small organic oxidants and reductants such as NAD^+/NADH and redox proteins such as cytochrome *c*, plastocyanin, or ferredoxin. The diffusion of small molecules is sufficiently fast that distances greater than 5 Å can be covered in less time than a typical electron tunneling reaction with relatively small driving force. For larger molecules, such as cytochrome *c* and ubiquinone, diffusion is faster than typical electron tunneling reactions over distances greater than about 14 Å. Although there is little control over which direction the molecules diffuse in order to encounter their reaction partner, this difficulty is largely overcome by using pools of diffusing agents, such as the ubiquinone and plastoquinones in the membrane quinone pools, and the smaller pools of water-soluble cytochrome *c* or plastocyanin attracted to the membrane surface. At any rate, diffusion is fast enough to keep up with the rates of catalytic turnover (k_{cat}) of oxidoreductases of roughly 10^3 s^{-1} (Gupte *et al.*, 1984).

III. Managing Lengths in Natural Redox Protein Design

The time constraints required for competent bioenergetic function have influenced the geometry of cofactor assembly in photosynthetic proteins. Where time constraints are severe, distances will be confined. On the other hand, where time constraints are relaxed and there are conspicuous gaps between the characteristic times of different physical processes (Fig. 2), distances and other parameters such as free energy can be more relaxed. This relaxed situation allows robust operation of the redox protein, one well away from the limits of failure rather than a highly optimized but fragile design. This condition has apparently been favored repeatedly by natural selection. Thus light energy transfer is generally faster than fluorescence decay times for distances smaller than about 40 Å, and is also comfortably faster than electron tunneling at any distance. Electron tunneling in turn outstrips diffusion of modest-sized redox agents for distances smaller than about 12 Å. However, beyond these limits, diffusional considerations dominate the movement of oxidizing and reducing equivalents. Even so all of these processes are generally much faster than the natural rate-limiting steps of the enzymatic activities that terminate all photosynthetic and respiratory electron transfer chains (Suarez *et al.*, 2000). This fundamental time constraint apparently reflects the typical time and energy required to arrange the catalytic site into a catalytic intermediate state, as described by Michaelis (1951). By examining the structures of many natural electron-transfer proteins, it seems clear that nature has ensured that electron tunneling is faster than k_{cat} by keeping the tunneling distances between cofactors shorter than about 14–15 Å (Page *et al.*, 1999).

Electron tunneling over transmembrane dimensions is prohibitively slow relative to typical k_{cat} times. Instead, membrane proteins use the simple device of chains of redox cofactors. Figure 1 shows that each of the RC and bc_1 or b_6f membrane proteins accommodates a large number of redox cofactors obviously lined up in chains. This is true even for other membrane oxidoreductases with "surplus" cofactors that for many years were considered supernumerary. Consecutive tunneling over several short steps replaces the exponentially slower tunneling rates at long distances with an approximate linear dependence of rate decrease on distance tunneled. We note that the chains are largely arranged to promote tunneling across the membrane and beyond into the aqueous phase for distances up to 100 Å, while diffusive connections are largely lateral to the membrane.

Figure 1 illustrates that the overall design of the different transmembrane RCs and cytochrome bc_1 structures are remarkably similar. In each case the site of primary energy conversion, be it the array of redox (B)Chls in the RCs or the hydroquinone oxidizing Q_0 site of the cytochromes bc_1, comprises a cluster of redox centers separated by relatively short 4.5 ± 1 Å gaps and is located close to one membrane aqueous interface. In each case the electron transfer integral to the primary energy conversion yields a charge-separated state oriented almost parallel with the membrane plane and is itself not significantly electrogenic. In each case the initial charge separation is then propagated across the membrane profile through a series of electron transfers over somewhat larger distances of 8 ± 4 Å. The job is distributed between four or five redox cofactors that extend from one side of the membrane to the other.

Because nuclei are so much more massive than electrons, nuclear tunneling is much slower than electron tunneling at any distance. Cofactors must be in near contact for direct hydrogen or hydride tunneling over the few tenths of an angstrom that has been invoked in two and multiple-electron transfer in the catalytic mechanisms of flavin, quinone, and NAD(P), and in water oxidation and oxygen reduction. On the other hand, at the shortest distances even highly endergonic electron tunneling might be a faster alternative than direct hydride tunneling provided there are other proton donors and acceptors in the vicinity. For example, two one-electron transfers through a 0.6 eV (~14 kcal/mol) endergonic radical intermediate state can be as fast as a millisecond reaction even when the overall driving force for the two-electron reaction is zero (Page *et al.,* 1999).

Advances in structural work now permit us to consider the natural distribution of distances between cofactors in photosynthetic proteins. The X-ray crystal structures of RCs from *Rhodopseudomonas viridis* (Michel *et al.,* 1986) and *Rhodobacter sphaeroides* (Ermler *et al.,* 1994), species that had already been the subjects of 20 years of intensive investigation, provided the first precise tunneling distances between the variety of photosynthetic redox cofactors and details of the surrounding polypeptides. *Thermochromatium tepidum* added a third species to the bacterial reaction center structures determined to high resolution (Nogi *et al.,* 2000). More recently crystals of the RCs of PSI from cyanobacter *Synechococcus elongatus* yielded high-resolution structures (2.5 Å) of the polypeptide and the chlorophyll, pheophytin, phylloquinone, and iron sulfur cofactors (Jordan *et al.,* 2001). This was followed by 3.8 Å structures of the RC of PSII of *Syn. elongatus* (Zouni *et al.,* 2001) with enough secondary structural features to allow spectroscopy and biochemistry to help resolve the individual chlorophylls, pheophytins, quinone, tyrosines, and Mn. Structures of the intrinsic membrane LH protein partners to the RCs have appeared in the past few years: bacterial proteins LH1 and LH2 from certain species have been modeled to the RCs from *Rps. viridis* and *Rba. sphaeroides* (Cogdell *et al.,* 1999 and references therein). In the case of the PSI and PSII structures, extensive and quite different LH equipment is resolved in the many subunits surrounding the RCs. The structure of cytochrome bc_1, a respiratory companion to these photosynthetic proteins, has also provided us with the tunneling distances of the heme and iron sulfur cofactor constituents and one clearly located ubiquinone, probably a bound form of the substrate.

The distances for the physiologically productive electron transfers of reaction centers with crystal structures (some illustrated in Fig. 3; see color insert) can be divided into basic functional groups, as indicated by the different colors in the histogram of Fig. 4 (see color insert). There are obvious similarities and differences between the various reaction center structures that reflect the different physiologies of the reaction centers. The initial charge separation between chlorins takes place at the shortest distances. In many cases the central pair of these chlorophylls is noticeably closer together than the other chlorins, between 3 and 4 Å; however, the central pair in PSII has a spacing that falls in the average chlorin spacing of 4 to 6 Å. Even at these slightly longer spacings, electron tunneling will still be more than rapid enough to ensure efficient charge separation. Instead, the wider central pair in PSII may help to assure the relatively high oxidation potential of the ground state, reflecting its role in water oxidation (Barber and Archer, 2001).

The coupling into and out of the chlorin charge separation center takes place over longer distances of 6 to 11 Å. The various redox chain separations that move charges out to the extremities of the protein for delivery to other redox protein partners in diffusive reactions occur over still longer distances of 8 to 13 Å. Distances longer than 14 to 15 Å are generally not found for physiologically productive reactions, because such reactions would not be significantly faster than the typical times for catalytic turnover of the enzyme. Longer electron-transfer distances are covered by cofactor chains, which so obviously extend above or below the chlorin centers of the photosystems. An interesting variation on the redox chain theme is provided by the FeS center of the cytochrome bc_1 complex, which appears poised in evolutionary terms between a freely diffusive redox component and a fixed redox chain component. The double-headed arrow of Fig. 3 shows the FeS movement between Q_0 and heme c_1 implied by different crystal structures. There may be some advantages in robustness for maintaining a limited FeS diffusibility that leads to an escapement device that helps to ensure that the two-electron oxidation of Q_0 proceeds by two one-electron transfers in opposite directions. These distances in this histogram have a comparable distribution to the set of productive electron transfer reactions between all the redox centers in multiredox-center electron transfer proteins deposited in the PDB database.

IV. Managing Length and Size in Natural Light-Harvesting Design

There are now sufficient high-resolution structures of antenna complexes available to make the detailed comparison of their functional design worthwhile (Cogdell *et al.*, 1998). All light-harvesting systems by design must absorb light in the region of the spectrum that is actually available in the ecological niche in which that organism lives, and they must deliver the excitation energy efficiently to its reaction center. The first of these imperatives principally directs the choice of pigment, while the second directs the overall architectural design of the antenna complex. Can we see any common features that would allow us to think of a basic prototypical antenna complex on which all the present-day varieties have been based? The only feature that easily stands out when these antenna structures are compared is that there are no common structural features. There are clearly many ways to design an antenna complex that is sufficiently energy efficient to allow that organism to be successful on an evolutionary time scale.

Given that the structure of antenna complexes seems to be so variable, does this mean that the physics of the energy transfer process is rather nonstringent so that it will tolerate many design solutions to the problem of constructing an efficient light-harvesting complex? If a donor pigment is going to efficiently transfer excitation energy to an acceptor molecule then it must do it much faster then the excited state of the donor is lost by other competing processes. So what factors affect the rate of this energy transfer process? In general terms the rate of excitation energy transfer depends on energy terms (spectral overlaps in the Förster theory), an orientation factor (the relative disposition of the transition dipole movements of the donor and acceptor in the Förster theory) and distance (the precise distance dependence varies with the type of energy transfer) (Sundstrom *et al.*, 1999). Of these terms distance is the strongest determinate of the overall rate of energy transfer. How far apart the donor and acceptor molecules in a light-harvesting complex can be, if the energy term and the orientation factor are held constant, will depend on the fluorescence lifetime of the donor's excited singlet state. If, as is the case of carotenoids, this is very short, e.g., a few hundred femtoseconds, then the donor and acceptor molecules must be very close together. In all antenna complexes that contain chlorophylls and carotenoids, the carotenoids are always in van der Waals contact with at least one chlorophyll molecule. If the excited singlet state lifetime of the donor molecule is rather long, e.g., a few nanoseconds, then the distance between donor and acceptor can be much longer. In the LH1/RC "core" complex from purple bacteria the distance between an antenna BChl*a* molecule and the RC primary donor BChl*b* has been estimated to be ~43 Å, but the rate of energy transfer of 35–50 ps is still fast enough to maintain very high quantum efficiency.

In the classical Förster treatment of energy transfer between two well-separated donor/acceptor molecules the rate of energy transfer depends on $1/r^6$. Though there is a strong distance (r) dependence it still allows a relatively large leeway in the special organization of chlorophyll molecules in an antenna complex, while still maintaining high overall efficiency of energy transfer. In the LH1 complex from purple bacteria the distance between a B800 BChl molecule (the donor) and the nearest B850 BChl*a* molecule (the acceptor) is ~18 Å. The rate of this energy transfer renders as ~1 ps. Since the excited state lifetime of the donor is ~1 ns the overall efficiency of this energy transfer is very nearly 100%. If the distance between the donor and acceptor was increased to 30 Å then the rate would decrease to ~23 ps, but the overall efficiency would still be ~99%. Similar arguments pertain to Chl-containing LH complexes of eukaryotes such as LHCII (Kuehlbrandt *et al.*, 1994) or the inner antenna systems of PSI (Jordan *et al.*, 2001; Zouni *et al.*, 2001).

It is clear that the basic physics of the energy transfer process allows for a large degree of tolerance in the design of a light-harvesting system before overall efficiency is compromised. The only cases where this is not true are where the singlet excited-state lifetime is so short, i.e., in the case of carotenoids, that donor and acceptor molecules must be in van der Waals contact to retain any efficiency at all. Interestingly, the essential function of the carotenoids is one of photoprotection (Frank and Cogdell, 1996). They rapidly quench the excited triplet states of chlorophylls before they can react with molecular oxygen and produce the lethal singlet oxygen. The carotenoids prevent this by a triplet–triplet exchange reaction, and the resultant carotenoid triplet decays harmlessly giving off the excess energy as heat. This triplet–triplet transfer reaction occurs by electron exchange, which requires van der Waals contact. This distance constraint, then also allows the carotenoids to act as accessory light-harvesting molecules (singlet–singlet energy transfer) because they have perforce been placed so close to the chlorophylls (Frank and Cogdell, 1996).

It appears that all "core" antenna structures have a significant separation between the closest antenna chlorophylls and the redox active chlorophylls. There is almost a "cordon sanitaire" around the heart of the reaction center. Why should this be? Oxidation of an antenna chlorophyll produces a very effective quenching center that will "trap" and dissipate the energy long before it can reach the RC, where it is used productively (Law and Cogdell, 1998). It is therefore vital that the antenna chlorophylls closest to the reaction center chlorophylls be placed at a distance where electron transfer between the RC and the antenna system never occurs. Excitation energy transfer is still efficient over this distance and so the light-harvesting function is not compromised. Interestingly, though, this often means that paradoxically the slowest step is the energy transfer chains in the final "hop" into the reaction center.

This potential danger of electron transfer out into the antenna is there because the midpoint potential of an antenna chlorophyll molecule is not that much higher than that of the previous donor in the RC, and in PSII it is clearly well below that of P680.

In every basic biochemical textbook the photosynthesis unit is described as a "funnel." The energy transfer reactions are depicted as going "downhill" from the periphery to the reaction center. However, the real picture is not always like this. As Fig. 5 (see color insert) shows, in most cases the final energy transfer step to the RC is "uphill," and in the "core" complexes of PSI and PSII the antenna chlorophylls are largely isoenergetic. This then prompts the questions of how deep the funnel must be or how big the LH system can get in the absence of a funnel. An ever increasing body of work on artificial antenna mimics has shown that in the absence of a tightly coupled antenna system, as seen in the chlorosomes of green sulfur bacteria, above a certain critical antenna size the efficiency of energy transfer to the "trap" drops off dramatically. In effect in the absence of an energy gradient to "guide" the excitation energy to the trap it gets "lost" and never makes it to the trap. In most purple bacteria the size of the photosynthetic unit depends on the incident light intensity: the lower the light intensity the larger the size (Cogdell *et al.*, 1999). This is accomplished by making more LH2 relative to LH1. The energy of the BChl in LH2 is higher than that of those in LH1 and so the funnel is established. Interestingly, some purple bacteria can control the steps of the funnel, too. If some strains of *Rps. acidophila* are grown at very low light intensity this induces a new type of LH2 (called B800-820) where the large ring of closely interacting BChl molecules now absorb at 820 nm rather than 850 nm. This increases the slope of the "funnel" and prevents back transfer which allows these bacteria to grow at even lower light intensity than species which cannot accomplish this. Similar adjustments in the size of the LH system while maintaining the funnel in response to low light intensities are observed in other organisms, ranging from cyanobacteria (increasing phycobilin content) to higher plants (increased LHCII content). PSI, on the other hand, has about 100 Chls per P700 and there appears to be no discernible energy funnel present. That being so, is there a maximum size for a "core" antenna in the absence of a "funnel"? It is tempting to propose that PSI is operating at or near this limit. Indeed, this idea is reinforced in the case of iron-limited growth of cyanobacteria or in deep-living strains of the marine oxyphotobacteria, *Prochlorococcus,* where any extra ring of antenna complexes supplement the antenna size of their PSI complexes. In the case of *Prochlorococcus,* these extra complexes have Chls *a* and *b* which absorb to the blue of the "core" PSI Chl*a*' molecules, thereby establishing the funnel and allowing a doubling of the size of the light-harvesting components while maintaining high efficiency.

As mentioned above, the chlorosome antenna system of green sulfur bacteria breaches this size limit for an isoenergetic antenna. However, this is possibly due to its unique structure (Blankenship *et al.*, 1995). The BChl*c* molecules are packaged into the chlorosome forming tightly coupled, strongly interacting helical arrays. Within these arrays the excited state is extensively delocalized and, in effect, the array acts a super molecule. This enhances the rate of energy migration to such an extent that a single chlorosome can operate as an efficient antenna system even when it contains up to 20,000 BChl*c* molecules. The green sulfur bacteria are strict anaerobes and therefore live in rather deep anaerobic layers. The chlorosomes have such a high cross-structural area for photon capture that the cells appear black! In contrast, cyanobacteria and red algae use a range of phycobilin pigments in their LH systems. They are spectrally distinct and arranged within the phycobilisome so that light energy is absorbed at the peripheral ends of the rods by short wavelength pigments and transferred by the "funnel effect" to the longer wavelength absorbing allophycobilin adjacent to the membrane surface and close to the chlorophylls of the RC.

The take-home message for understanding the management of intercofactor distances in natural antenna system design is that although the rate of excitation energy transfer is extremely sensitive to distance, the overall efficiency of energy transfer is, in general, not (see Fig. 2). This explains how it is possible to have such a variety of structures of antenna complexes and yet always retain efficient functioning. This freedom, then, has allowed Nature to have a very wide design brief in the construction of antenna complexes.

V. Managing Distance in Electron Transfer

A. *Predicting Electron Transfer Rates from Structure and Distance*

Like plant RC proteins, bacterial RC proteins each support 10 or more individual electron transfer reactions, and many have long been experimentally accessible for direct measurement of tunneling rate constants. The distances between the cofactors range from near van der Waals contact to as far away as 24 Å and include tunneling steps both physiologically productive and counterproductive. The early application of molecular biological methods, as well as the natural stability and facile biochemical cofactor replacements of the bacterial RC, made the purple bacterial reaction centers a test bed of models of biological electron tunneling, opening the door to the examination of the role played by the electronic structure of the polypeptide tunneling medium between the cofactors. On the nuclear side of the equation, a variety of methods to alter and measure ΔG^0 values *in situ* combined with flash-activated kinetic analysis has verified the Marcus-like (Marcus, 1956) parabolic log rate versus free energy relationship at ambient temperatures, as shown in Fig. 2B. In some key cases, the free energy versus rate relationships have been examined down to cryogenic temperatures (Gunner and Dutton, 1989; Gunner *et al.*, 1986), verifying the original suggestion of Devault and Chance (1966) that at low temperature nuclear motion obeys a quantum tunneling rather than classical behavior originally described by Marcus. The parabolas shown in Fig. 2B are indeed quantized which has the effect of flattening the curve and changing the prefactor on the nuclear term from 4.2 to 3.1 in the equation in the figure legend.

For many of the bacterial RC reactions, the rate constant was measured or reasonably extrapolated to $-\Delta G^0 = \lambda$ (Dutton and Moser, 1994). This provided critical k_{max} or k_{opt} values isolating the electronic contribution to the tunneling rate of each reaction (unmodulated and uncomplicated by the natural variances of the Franck–Condon factor) and allows comparison of the tunneling medium between the cofactors of different electron transfers throughout the RCs. Figure 6 (see color insert) shows the now 10-year-old, remarkable length scale for tunneling in two bacterial RCs already presented in the general Fig. 2A focusing on the contribution of the protein structure to the electronic term. The approximate exponential change of k_{opt} over 12 orders of magnitude with distance from near van der Waals contact to 23 Å reveals a straightforward engineering blueprint for electron tunneling in the bacterial RC that is dominated by distance. This work provided the perspective needed to verify that the rate at van der Waals contact is near 10^{13} s^{-1}, about the rate-limiting time of one molecular vibration, and to suggest that tunneling reactions within the RC protein adhere reasonably well to this fundamental constant.

The work also provided evidence that the initial charge separation from the light-activated $BChl_2$ to BPh was two successive generic tunneling reactions using the intervening BChl monomer as a real intermediate, as opposed to a special single-step coherent electron transfer with the BChl monomer playing a special role as superexchanging virtual intermediate. In the end a two-hop mechanism has prevailed, and the single-step rate now appears to be far too slow (see van Brederode and van Grondelle, 1999).

Figure 6 offers a seductive menu of motifs that in principle could have been selected to raise or lower the tunneling barrier and hence slow or speed the electron transfer rate at any one distance. However, the structurally disparate reactions in the RCs have no distinction between physiologically productive and counterproductive electron transfers and closely adhere to a single exponential pointing to 10^{13} s^{-1}. This suggests that motifs that can be constructed in synthetic systems are not a conspicuous feature in the natural proteins (Moser *et al.*, 1992). The line fitted to the points in Fig. 6 provides an apparent "β value" of 1.4 $Å^{-1}$, where β is the coefficient of exponential decay of the tunneling rate with distance. This β can be interpreted simply as the weighted average of β values of two compartments in the protein structural volume between the cofactors. Protein volume contained within the van der Waals radii of atoms and bonds is assigned a β of 0.9 $Å^{-1}$, a value slightly less than that obtained for the aliphatic bridge, and the remaining interstitial volume is assigned a β of 2.8 $Å^{-1}$, the lesser value for vacuum. The electronic term for optimal tunneling rate constant can be modified to introduce a packing density term, ρ, the fraction of the protein volume between the cofactors that is inside the van der Waals radii. This offers a simple way to calculate optimal rates directly from structures.

The simple packing insensitive equation of Fig. 2 has a standard deviation with measured rate of about 10-fold (an order of magnitude). When the packing density is included in the equation of Fig. 6, the variance falls within the level of the collected experimental uncertainties in the structural resolution, and the kinetic, ΔG^0, and λ measurements. The packing rate expression lends itself to application to all electron transfer proteins. Calculated optimal tunneling rate constants from the ρ values and distances between the cofactors in other nonphotosynthetic proteins are also shown in Fig. 6 (pale points). This broader sample of proteins that perform a wide variety of oxidoreductase functions emphasizes the natural heterogeneity of the internal protein structure between reacting cofactors; again, there appears no statistical difference in ρ for physiologically productive and unproductive reactions. Thus, it is clear that generally nature has not evolved specific structural motifs in the protein to enhance productive and suppress unproductive electron tunneling.

The recurrent expectation over the past decade that aromatic amino acids commonly positioned in the protein between the donor and acceptor will enhance electronic coupling and speed up tunneling was put to test in an unmatched series of measurements done with *Rps. viridis* (Dohse *et al.*, 1995). This careful study on the structure shown in Fig. 7 (see color insert) and results in Table I included direct measurements of ΔG^0 values, time-resolved kinetic analysis of the electron transfer rate, and crystal structure determination of the native Y162 and two mutants Y162F and Y162L. The findings support the conclusions that neither H-bonding nor aromatic π-orbitals of tyrosine contribute significantly to

TABLE I

Rps. viridis *Tyr Mutations*[a] *for Which Crystal Structures and Free Energies are Available*[b]

Mutation	Distance (Å)	ΔG (meV)	Packing	Calculated log rate	Experimental log rate
WT (Tyr)	12.29	−140	0.76	6.50	6.73
Phe	12.28	−190	0.72	6.49	6.76
Thr	11.88	−198	0.62	6.08	6.00

[a] From Dohse *et al.* (1995).

[b] This allows a comparison of packing density sensitive rate with experiment, and the possible special effect of changing an aromatic for nonaromatic residue.

the tunneling reaction beyond that provided by aliphatic replacements. Although crystal coordinates are not available for four other mutants that included methionine and glycine, they confirm this conclusion; indeed methionine proves to be faster than the wild-type protein. These exercises present us with a rudimentary but clear view of protein as a highly robust electronic coupling medium for tunneling between redox cofactor donors and acceptors.

Table I shows that two other parameters change significantly upon mutation of the tyrosine, the ΔG and the packing density, ρ. These changes appear important in accounting for the ~6-fold rate variations that are evident in the family of Y162 mutants. By including these parameters in the four-parameter rate expression of Fig. 6, and assuming λ value of around 0.7 eV, the calculated rate constants match the measured ones rather well (within a factor of 2). In separate studies the same authors specifically altered the ΔG by changing the reduction potential of the heme, first by mutating a charged residue near the heme but outside the tunneling volume, and second by introducing coulombic effects by adjusting the redox states of nearby hemes (Chen *et al.*, 2000). Again, the effect on the rates followed that expected for 0.7 eV reorganization energy.

B. *Robust Natural Design of Charge Separation through Redox Chains*

1. *Heme Chain Archetype in* Rps. viridis

Chains of redox cofactors for long range electron transfer are clearly the way electrons are transferred over the tens of angstroms dimensions of membranes and their proteins. Once again, purple photosynthetic bacterial reaction centers provide an archetype for understanding electron transfer chain design and behavior. The heme chain in *Rps. viridis*

(Fig. 8; see color insert) extends from the $BChl_2$ out into the periplasm some 60–70 Å to a docking site for soluble cytochrome c_2 at the terminus. The chain is characterized by close positioning of the hemes, typically 5 Å, and intrinsically poised to promote very rapid tunneling. The reduction potentials describe a dramatic sequential combination of endergonic and exergonic tunneling steps that is overall from the cytochrome c_2 to the $BChl_2$ modestly exergonic (Alegria and Dutton, 1991; Shopes *et al.,* 1987).

The first measurement of electron transfer in the chain, from heme 3 to heme 1 in 1.8 μs (Shopes *et al.,* 1987), suggested that tunneling could proceed via thermally assisted uphill tunneling to the intervening heme 2 and then exergonically to heme 1, if the intrinsic electron transfer rates between the hemes were in the picosecond time scale, as suggested by our calculations. Using a λ of 0.7 eV, the equilibrium reduction potentials adjusted for charge interactions, and edge-to-edge distances, R, the calculated transit time is $\sim$2 μs. The figure shows several other measurements of the remarkably rapid transit through the five hemes to the $BChl_2$, which again track the calculated rates well. Thus it appears that it is the close proximity of the hemes in the structure that sets the stage for simple sequential electron hopping from heme to heme including substantially unfavorable thermally accessed states. It appears that equilibrium reduction potentials of the hemes (only adjusted for heme–heme coulombic interactions) are operative on the time scale of electron transfer through the heme chain. Additional support for this simple model comes from the activation energy obtained from the temperature dependency of tunneling from heme 3 via heme 2 to heme 1 which matches the free energy obtained from the reduction potentials. This also argues against a model in which the intervening heme 2 acts not as a real redox intermediate in a two-step electron transfer, but as a quantum mechanical virtual superexchange intermediate in a single coherent step tunneling from heme 3 to heme 1. Like the tyrosine between the heme and $BChl_2$, the heme between heme 3 and heme 1 occupies a major part of the intervening space but does not act a dramatic tunneling barrier lowering element.

2. *Primary Chlorin Chains*

The redox chlorins at the core of the various reaction centers form an obvious chain with separations of less than 6 Å that ensure rates of 10 ps or less. The effect of this chain is to apparently make the photoinduced oxidant and reductant capable of residing briefly on any of the chlorins, subject principally to the free energy of that state and the energetic penalty of any uphill reverse electron transfer. When the free energy differences between the various chlorins are small and comparable to the thermal energy, as appears to be the case in the RC of PSII, the light-induced reductant and oxidant may be appropriately described as being shared by all the chlorins. What guides the eventual fate of the oxidant and reductant is simply the availability of the closest nonchlorin centers, which continue charge separation in relatively longer, slower, and larger driving force electron transfers.

The symmetry of the chlorins in the RC means that there are two branches of the chlorin chain extending across the membrane. In the case of PSI and even more so in the RC of green sulfur bacteria (Golbeck, 1994), there may also be a near functional symmetry of electron transfer across these chains to the symmetrically placed quinone acceptors and then to the centrally located iron–sulfur center, F_X (Barber and Archer, 2001; Guergova-Kuras *et al.*, 2001; Jordan *et al.*, 2001). However, in the case of RCs of purple bacteria and PSII the situation is quite different. Only one of the quinones (Q_A) appears to be designed as the functional acceptor. How is the high quantum efficiency of charge separation maintained by guiding electron transfer down the appropriate branch of these RCs? In the case of the purple bacterial RC, if the energy gap between the bacteriopheophytins and the bacteriochlorophylls on both branches were about 0.16 eV, then some electrons could be trapped on the inactive bacteriopheophytin leading to energy-wasting charge recombination to the bacteriochlorophyll dimer ground state. Indeed, under certain experimental conditions (e.g., excess light) the wrong side BPh can be driven reduced (Florin and Tiede, 1987). The active side BPh may be favored kinetically, rather than thermodynamically, by a lower wrong side BChl monomer redox potential that makes electron transfer from the dimer to this side sufficiently uphill; a similar effect can be accomplished by a larger reorganization energy for this electron transfer (Moser *et al.*, 2001; Parson *et al.*, 1990).

Although at first sight the engineering of PSII and that of the purple bacterial RCs are similar, there are distinct differences. In the case of PSII, it appears that the relatively small energy gap of about 0.1 eV between the pheophytins and the chlorophylls allows any electron on the "wrong side" pheophytin to thermally equilibrate with other chlorins on a rapid enough time scale. This similar energetics of the chlorins in PSII means that charge separation need not focus on the central pair of chlorophylls (designated P_A and P_B in Fig. 3 or P_{D1} and P_{D2} in Zouni *et al.*, 2001), as is the case in the purple bacterial reaction center. Indeed, it appears that the accessory chlorophyll (designated Chl_A in Fig. 3 or Chl_{D1} in Zouni *et al.*, 2001) adjacent to the active pheophytin acts as the most favored primary donor (Barber and Archer, 2001; Dekker and

Van Grondelle, 2000; Diner *et al.,* 2001; Prokhorenko and Holzwarth, 2000).

The fate of the resulting "hole" will be governed by the redox differences between the four chlorophylls, which all have high redox potentials in the region of 1V. Since the tyrosine donors Tyr_Z and Tyr_D are oxidized at about the same rate (Faller *et al.,* 2001) and are more or less equidistant between the P_A and P_B chlorophylls respectively (see Fig. 3), then on the first photochemical turnover the hole must be delocalized more or less equally between the two chlorophylls P_A and P_B even at low temperatures. This symmetry in electron donation is broken by the fact that the Mn cluster which catalyzes the water oxidation process, is located only on the D1 side. The consequences of this is that the P_A chlorophyll becomes the active $P680^{\bullet+}$ species, presumably because the long-lived inactive Tyr_D radical cation generated during the first or second photochemical turnover increases the redox potential of P_B relative to P_A. Direct evidence that P_A does indeed generate the $P680^{\bullet+}$ species has come from the recent mutational studies (Diner *et al.,* 2001). The electron transfer from Tyr_D to P_A^+ may, in part be coupled to proton release from the tyrosine which could then facilitate an electron/proton abstraction from the substrate water molecules bound to the Mn cluster (Tommos and Babcock, 2000).

3. Charge Recombination in Chains

One consequence of the common use of redox chains for forward, productive electron transfer is the availability of the same redox chains for reverse, unproductive charge recombination. Indeed, when these recombination reactions are observed in reaction centers, they provide some of the best examples of the manner in which redox chains can mediate uphill electron transfer en route to a final cofactor that confers an overall favorable free energy. Figure 9 presents electron tunneling in the physiologically unproductive direction through the chain from Q_A to $BChl_2$ in *Rba. sphaeroides* and the effect of altering the reduction potential of Q_A on the rate of tunneling and the route taken from Q_A^- to $BChl_2^+$ (Woodbury *et al.,* 1986). Charge recombination from the native ubiquinone-10 is dominated by the slow $\sim 10\ s^{-1}$, temperature independent, direct tunneling over 23 Å from Q_A^- to $BChl_2^+$. The native ubiquinone-10 was replaced with a variety of synthetic analogues of widely differing but mainly lower reduction potentials so as to steadily diminish the free energy gap between Q_A and BPh. The lower reduction potential also served to increase the driving force of direct Q_A^- to $BChl_2^+$ recombination, which causes minor acceleration or slowing depending on whether $-\Delta G$ enters the inverted Marcus region $(-\Delta G > \lambda)$ or

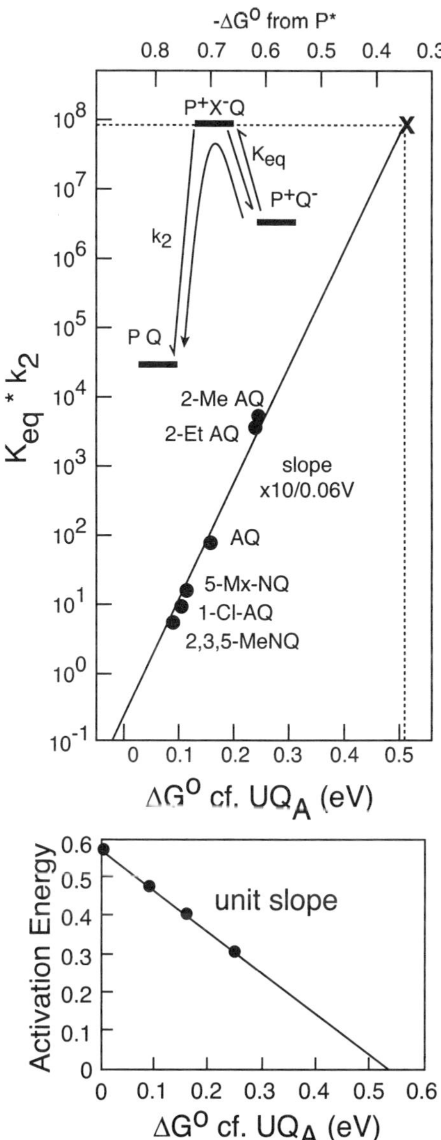

FIG. 9. Substitution of exotic quinones for native ubiquinone at the Q_A site or reaction centers from *Rba. sphaeroides* shows that charge recombination from the $BChl_2^+Q_A^-$ state proceeds uphill via a thermally activated intermediate "X" near the energy level of BPh (Woodbury *et al.,* 1986). Figure after Gunner (1988).

not. However, beyond a certain threshold the route via BPh becomes dramatically dominant. The slope is about a 10-fold change in rate per 0.06 eV change in $-\Delta G$ at room temperature, providing all the signs of a thermally activated electron tunneling between Q_A- and BPh, that transiently forms BPh^- en route to $BChl_2^+$. Further support is provided by the direct correspondence of the activation energy of the recombination reaction and the free energy gap between Q_A and BPh calculated from the redox midpoint potentials for a series of quinones.

Charge recombination does not use the intervening large highly conjugated BPh molecule as a superexchanging element in a single tunneling step, but rather establishes a quasi-equilibrium with the BPh some 0.48 eV uphill. Moreover, at low temperatures when the thermal route is essentially eliminated and only single-step direct recombination is possible (Gunner *et al.*, 1986), there is no evidence that BPh employs superexchange to diminish the effective tunneling barrier and enhance the rate.

Of all the reaction center and photosystem species discussed here, only *Rba. sphaeroides* has an energy gap between the BPh and native Q_A (or their analogues) large enough to eliminate thermally activated charge recombination via the (B)Ph or analogue at room temperature. Yet the submillisecond recombinations in these other species are still sufficiently slower than the forward physiological tunneling to enable stable charge separation.

Figure 10 (see color insert) shows the rate of unproductive charge recombination from flash generated Q_A back to the first and third hemes of *Rps. viridis* through a series of up and down steps covering 100 Å. Again there is solid matching of measurement (Gao *et al.*, 1990) and calculations, even using the unmodified equilibrium redox levels or potentials of the intervening BPh, $BChl_2$ and first second and third hemes. Recombination rates of seconds ensure that productive charge separation will prevail *in vivo.* Chains of relatively closely positioned cofactors appear to be the normal *modus operandi* for very rapid, highly directed electron tunneling over considerable distances. In dramatic contrast to the significant exponential penalty of tunneling long distances through protein media, the tunneling time of transit through the simplest chains in which each step is similar and mildly exergonic is roughly linearly dependent on its length.

C. *Simulating Photosystem Operation with Tunneling Theory*

Enough is known now to begin to estimate all the tunneling rates between all redox centers in the photosystems for both charge separations

TABLE II
Estimated Redox Midpoint Potentials in Photosystems

Cofactor	E_m	Cofactor	E_m [a]	Cofactor	E_m [b]
Rps. viridis RC		PSI		PSII	
$BChl_2/P_{ground}$	0.50 (Gao *et al.*, 1990)	P700	0.43	$P680_{ground}$	1.15 (Klimov *et al.*, 1979)
P*	−0.79	P700*	−1.17	$P680_{anion}$	−0.70
$BChl_A$	−0.71	Chl_{1A}	−1.14	Chl_A*	−0.63
$BChl_B$	−0.85	Chl_{1B}	−1.14	Chl_B	−0.70
BPh_A	−0.63 (Woodbury and Parson, 1984)	Chl_{2A}	−1.1	Ph_A	−0.59 (Klimov *et al.*, 1979)
BPh_B	−0.63	Chl_{2A}	−1.1	Ph_B	−0.59
Q_A	−0.15 (Prince *et al.*, 1976)	Q_A	−0.74	Q_A	−0.08 (Krieger *et al.*, 1995)
Q_B	0.04 (Wraight, 1979)	Q_B	−0.74	Tyr	1.00
Heme 1	0.38 (Alegria and Dutton, 1991)	F_X	−0.67	Mn	0.90
Heme 2	0.02 (Alegria and Dutton, 1991)	F_A	−0.54		
Heme 3	0.32 (Alegria and Dutton, 1991)	F_B	−0.56		
Heme 4	−0.06 (Alegria and Dutton, 1991)				

[a] From Iwaki and Itoh (1994) and Vassiliev *et al.* (2001).
[b] From Rutherford and Krieger-Liszkay (2001).

and charge recombinations and to simulate the electron transfer reactions that take place after a flash of light. We use the tunneling expression in Fig. 6 if the structural resolution permits estimates of the packing density, ρ, or the simpler tunneling expression of Fig. 2 if structural resolution does not permit ρ estimates. Examples follow in Fig. 11 (see color insert and Table II). Even without adjusting the usually unmeasured reorganization energy, these simulations follow kinetic measurements within an order of magnitude of time.

Tunneling distances between chlorins are so close that it is not an essential part of photosystem design that any particular chlorin or chlorin pair be the source of the initial excitation and charge separation. Details of the relative redox levels of each of the excited and ground states will determine which particular chlorin or chlorin pair is first oxidized and which first reduced. Similar concerns will determine which side of the chlorin chains will be reduced. In the *Rps. viridis* simulation, the $BChl_B$ has intentionally been given a midpoint potential slightly more negative than P* so that electron transfer favors the L side. For PSI a simpler case is shown in which both sides have symmetric midpoint potentials; this means both chains participate nearly equally in electron transfer, with no loss in charge separation efficiency. In PSII, asymmetry in the tunneling distances alone is sufficient to favor electron transfer along the chain that eventually leads to Q_A although some temporary reduction of the chlorins on the other side might reasonably be expected. However, the PSII simulation illustrated introduces a slight asymmetry in the midpoint potentials so that A side B* is favored as the principal charge separating center, in recognition of the recent observations (Diner *et al.*, 2001).

Each of the photosystems ejects an electron from the excited chlorin complex to a quinone within a nanosecond, followed by electron transfer along chains leading out of the charge separation center within 100 ns. The high potential reaction of Tyr and Mn in PSII is quite rapid, beginning in the simulations on the same time scale as the quinone reduction reaction. However, it has been suggested that tyrosine oxidation may not be rate limited by tunneling, but by H^+ transfer (Diner *et al.*, 2001).

Charge recombination times will depend on the initial state of the photosystem being simulated. Under physiological conditions we may expect that diffusible species will carry oxidizing and reducing equivalents away so that such recombination is unlikely. However, under certain experimental conditions, these recombination reactions can be observed and can provide information about the engineering tolerances of the electron transfer system. In the *viridis* system, the simulations show a recombination through uphill intermediate redox states in tens of seconds. A similar time scale would be expected for the recombination from Q_A to the Mn cluster, which is here artificially forbidden to oxidize water. For PSI, on the other hand, without any donor to P^+, the simulations anticipate a recombination from the photoreduced FeS chain on the time scale of tens of milliseconds. This is beginning to become an engineering concern and shows the importance of donors to P^+ in stabilizing the charge separation reactions. While close positioning of redox centers in the photosystems allows rapid enough forward electron tunneling for efficient charge separation, the same close positioning offers the possibility of thermally assisted reverse electron transfer, with accompanying loss of efficiency. Donation to P^+ removes the time scale of these recombination reactions to the many seconds, where diffusive reactions can ensure overall efficiency.

VI. Managing Proton Reactions in Photosynthesis

A. *Design Issues for Proton Transfer*

Whether the process is adiabatic or nonadiabatic (tunneling), the range of proton transfer is restricted to distances no more than 1 Å and mechanisms rely exclusively on reactive complex formation. Thus, in contrast to electron transfer, the short range, bond-length nature of proton transfer necessitates considerations of structure.

The charged nature of electron and ion transport functions gives proton transport two distinct roles in bioenergetics—charge compensation and long-distance "chemical" movements. In both cases, proton transfer involves the "normal" acids and bases of Eigen (1964)—oxygen and nitrogen functionalities that can exchange protons at high rates given moderately favorable driving force and no hindrance to reactive complex formation. In solution, the initial encounter complex quickly finds the reactive configuration, and the elementary event of proton transfer is much faster than dissociation. Thus the reaction is diffusion limited:

$$\mathrm{AH} + \mathrm{B} \rightleftharpoons \mathrm{AH} \cdots \mathrm{B} \rightleftharpoons \mathrm{A} \cdots \mathrm{HB} \rightleftharpoons \mathrm{A} + \mathrm{HB}$$

If attainment of the reactive complex is inefficient, so as to be kinetically significant, more elaborate schemes must be considered. Such rate limitation is relatively uncommon in solution reactions of normal acids and bases, but strong internal hydrogen bonding is one significant source of constraint, and the effect can be dramatic (Hibbert, 1986). To the extent that buried polar groups in proteins are largely maintained by internal hydrogen bonding, we might expect this to be an important factor in the design of rapid proton delivery mechanisms in proteins and in the gating of proton conduction paths.

Principles for transmembrane proton transfer have often been sought in the properties of "hydrogen-bonded chains" (HBCs). These can be formed from protonated functional groups with good proton donor *and* acceptor activity (primarily oxygen-containing species: water, hydroxyl, carboxylic acids) and are thought to allow proton transfer by H-bond swapping in a modified Grothuss mechanism. Following Onsager, Nagle has described two distinct processes necessary for net proton transfer through an HBC—hopping of the proton, and turning of the functional groups to reorient themselves back to the original configuration (Nagle, 1987; Nagle and Morowitz, 1978). This has now been well characterized by computational studies of the pore-forming antibiotic gramicidin A (gA), which contains a single file of H-bonded water molecules and exhibits a very high proton conductance (Chiu *et al.*, 1999; Jordan, 1990; Pomes and Roux, 2002).

Molecular dynamics and Brownian dynamics simulations of gA show that proton conduction along the water chain can be an extremely fast one-time event. However, it leaves the hydrogen bond network in the opposite polarity from the starting configuration. Thus, all functional groups (waters) must reorient in order to conduct a second proton. In an effective reduction of dimensionality, the water molecules in the gA pore are restricted to three hydrogen bonds—two to adjacent waters and one to a carbonyl oxygen in the channel wall. This allows the Grothuss mechanism to proceed with no breakage of hydrogen bonds, and proton hopping in gA has very low activation energy. Transfer of a single proton along the whole length of the pore can occur in just a few picoseconds (Pomes and Roux, 2002), with pairwise jump rates faster than 10^{12} s^{-1} (Szczeniak and Scheiner, 1985). Unlike hopping, the turn process involves breaking and making hydrogen bonds for all groups, probably sequentially but in a substantially coordinated way. This normally makes the reorientation part of the cycle the rate limiting process.

Some useful principles of fast proton conduction are evident from studies of gA (Pomes and Roux, 2002). Effective proton conduction relies on pathways or networks with sufficiently strong H-bonds that the barrier in the proton transfer coordinate is not high. Thermal activity can then lower this barrier height sufficiently to allow barrier crossing or facile tunneling on the time scale of solvent fluctuations ($\sim 10^{-12}$ s). At the same time, the H-bonds must not be too strong, so that thermal fluctuations can also cause hydrogen bond breakage to allow reorientation of the chain.

Although the secrets of maximal rates of proton conduction are well illustrated in gA, multifunctional proteins that couple H^+ conduction to other events do not exhibit well-formed, proton-conducting hydrogen bond networks. Indeed, in the bacterial reaction center the putative active path is poorly connected by hydrogen bonds detectable in the best current X-ray structures (2.2 Å resolution; Stowell *et al.*, 1997). Paddock *et al.* (1999) have shown that chemical blockage or a simple mutational lesion of this active path diminishes proton transfer rates by at least 1000-fold. Thus, the several well-connected (but not quite continuous) files of water that are seen in the X-ray structures, reaching toward the Q_B site from the cytoplasmic side, do not conduct protons at significant rates.

The implication is that the proton delivery paths are transient and highly dynamic entities. This is explicitly demonstrated in the elegant X-ray studies of bacteriorhodopsin (bR), where the path for reprotonation of Asp-96 is not in existence in the early part of the photocycle, but forms only after the isomerization-driven, internal translocation has taken place (Sass *et al.*, 2000).

B. *Proton Transfer and Coupled PT/ET*

In biological electron transfer, the role of H^+ depends on the strength of coupling between the electron and proton:

1. Long-range electrostatics—weakly coupled pK shifts, dielectric responses (relaxation); $\Delta G \leq 20$ meV
2. Short-range (or structurally relayed) electrostatics—local charge compensation, chemically specific H-bonding; $\Delta G \leq 150$ meV
3. Strong coupling—direct protonation; H atom transfer; chemical specificity; $\Delta G \sim 100$–1000 meV.

The extremes of these coupling strengths define two distinct functional roles for protons:

1. Dielectric relaxation and distributed charge compensation, based on weak coupling
2. Bond making/breaking in coupled ET/PT to or from a hydrogen carrier, embodying strong coupling

C. *Protein Relaxations*

The light-driven charge separation in the reaction center presents a dramatic impulse to the dark-adapted state of the protein. The *de novo* creation of charge sets in train responses that proceed over many orders of magnitude of time: the fastest of these relaxations (about 0.1 ns) may be crucial in determining the high yield of the primary events, leading to formation of $[P^+Q_A^-]^*$ in about 0.1 ns, whereas relaxation processes at longer times contribute to the thermodynamic and kinetic stabilization of the charge separation and its subsequent utilization (Holzwarth and Muller, 1996; McMahon *et al.*, 1998; Woodbury *et al.*, 1994).

The full extent of relaxation in the $P^+Q_A^-$ state spans about 120 meV over the time range from 100 ps to about 1 ms at room temperature (McMahon *et al.*, 1998). Of this, as much as 80% is achieved prior to 1 μs. Thus, relaxations associated with diffusion-controlled net H^+ uptake do not contribute more than 20–30 meV. However, proton rearrangements (intraprotein H^+ transfer) can certainly be a significant part of the overall response of the protein at much shorter times.

The major fraction of relaxation in the $P^+Q_A^-$ state is frozen out at a temperature of about 100 K, and it is noteworthy that very similar behavior is seen upon dehydration at room temperature (Clayton, 1978). In addition to the known participation of solvent in protein relaxations (Beece *et al.,* 1980) a role for internal water should be expected in the protein dielectric response.

Relaxation phenomena are equally evident at the donor side of the reaction center. The well studied, fast ($\leq 1\ \mu$s) electron transfer reactions from bound *c*-type cytochromes of RCs from such species as *Chromatium* and *Rps. viridis* frequently show a progressive shutoff of electron transfer from the high potential heme, which is closest to P (Gao *et al.,* 1990). It is suggested that this is due to a large (>100 mV) increase in the midpoint potential of the heme, associated with the freezing out of solvent or hydration-related relaxation processes (Kaminskaya *et al.,* 1990). This is consistent with a deuterium solvent isotope effect for this reaction, as reported by Kihara and McCray (1973).

D. *Protonation Archetype: Q_A/Q_B in Purple Bacteria*

In the bacterial RC, export of reducing equivalents is achieved by double reduction of the secondary quinone, Q_B, and release of the hydroquinone into the membrane. Two light-activated turnovers of the RC are required, with the reduction of P^+ by a secondary donor (Okamura *et al.,* 2000; Shinkarev and Wraight, 1993); the change in protons associated with each step is given by the indices m, n, r, and s:

$$Q_AQ_B \underset{m\mathrm{H}^+}{\overset{h\nu}{\longrightarrow}} \underset{(\mathrm{H}^+)_m}{Q_A{}^-Q_B} \underset{(n-m)\mathrm{H}^+}{\rightleftharpoons} \underset{(\mathrm{H}^+)_n}{Q_AQ_B{}^-} \tag{1}$$

$$\underset{(\mathrm{H}^+)_n}{Q_AQ_B{}^-} \underset{(r-n)\mathrm{H}^+}{\overset{h\nu}{\longrightarrow}} \underset{(\mathrm{H}^+)_r}{Q_A{}^-Q_B{}^-} \underset{(1-r+s)\mathrm{H}^+}{\rightleftharpoons} \underset{(\mathrm{H}^+)_s}{Q_AQ_BH^-} \underset{(1-s)\mathrm{H}^+}{\overset{\mathrm{p}K_1}{\rightleftharpoons}} Q_AQ_BH_2 \tag{2}$$

The two one-electron transfer events exhibit the two basic roles for proton uptake and redistribution—weakly coupled charge stabilization for the first electron ($m \sim n \sim r \sim s \ll 1$), and strongly coupled bond formation for the second (net uptake of $2H^+$ per QH_2). However, it is now evident that the two are functionally related, and part of the charge-compensating H^+ uptake, e.g., on the first electron transfer, is destined for delivery to the quinone head group in the second reduction step.

1. *Proton Coupling to the First Electron Transfer*

A description of proton delivery to Q_B has been outlined from a combination of experiment and structure-based calculation of energetics. In *Rba. sphaeroides,* site-specific mutagenesis showed substantial roles for a

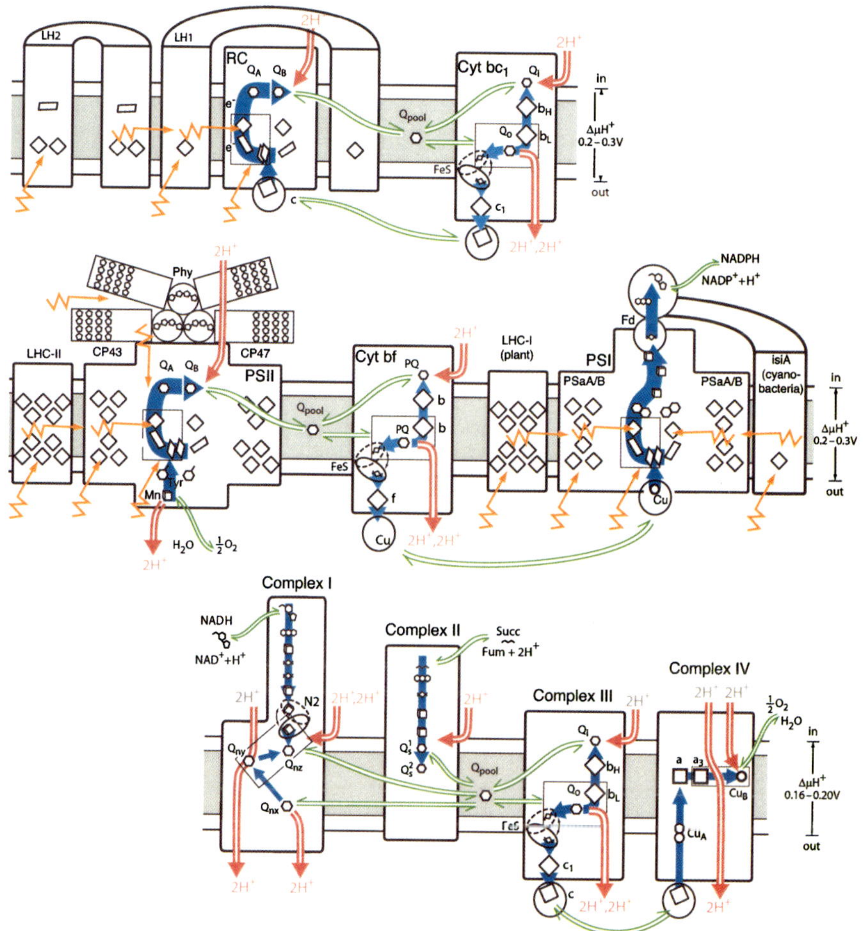

Moser *et al.*, FIG. 1. Schematic representation of cofactors in light harvesting, reaction center, and bc_1 proteins of photosynthetic organisms. First row: purple bacterial membrane; second row: plant membranes with photosystem I (PSI), photosystem II (PSII), and b_6f analogous to bc_1 complex; third row: mitochondrial respiratory membrane with complex III, also known as bc_1. Light energy excitation or energy transfer (orange arrows) between (bacterio)chlorophylls (diamonds) leads eventually to charge separation in reaction centers (green boxes) and electron tunneling (blue arrows) from center to center in a chain across the membrane (gray, low-dielectric region) to create a transmembrane electric field. Proton binding and release (red arrows) occurs at chains terminating in quinones (hexagons), generating a transmembrane proton gradient. In addition, reduced quinone and oxidized cytochrome c or plastocyanin diffuse (green arrows) to connect membrane complexes.

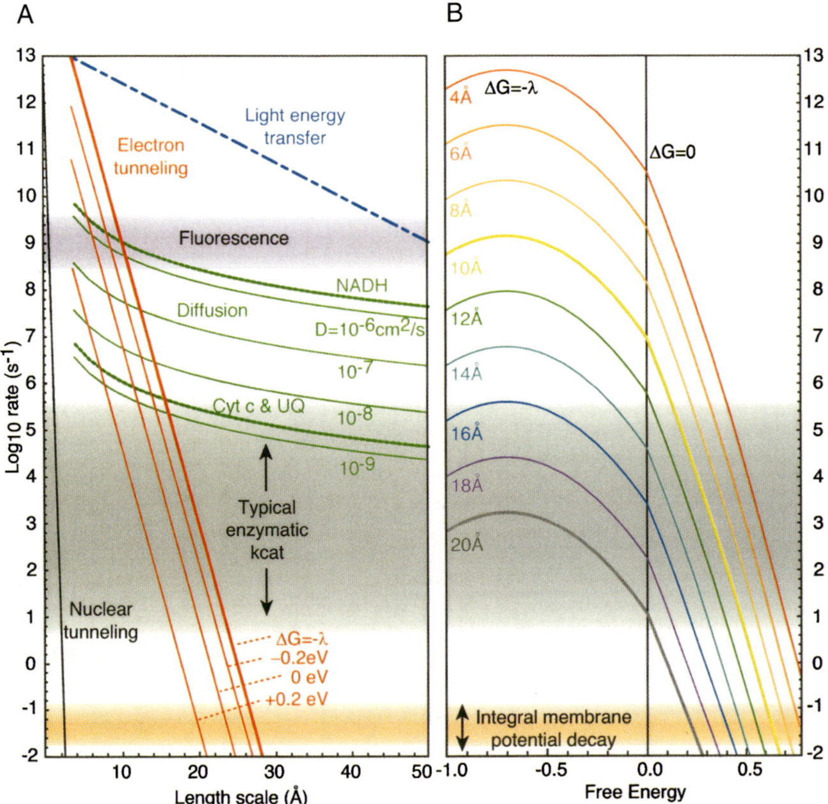

Moser *et al.*, FIG. 2. (A) Fundamental processes in photosynthesis display characteristic rate/distance/free energy dependencies that dictate the basic design of the electron transfer complexes. Light energy transfer (blue), electron tunneling (red), and proton tunneling (black) in protein display an approximately exponential decay with distance but with dramatically different slopes. Rates of diffusion (green) are nonexponential and are the fastest individual process at large distances. Fluorescence decay times (purple), enzymatic turnover times (gray), and membrane potential decay (orange) represent basic engineering time constraints. Fluorescence heralds the end of the energy transfer time domain. (B) The electron tunneling rate dependence on free energy reaches a maximum when the exergonic energy release matches the reorganization energy (λ) for the electron transfer, here illustrated for a typical value of about 0.7 eV. An empirical expression for exergonic electron tunneling in a typical protein medium is $\log k_{et} = 15 - 0.6R - 3.1(\Delta G + \lambda)^2/\lambda$, where k_{et} is in units of s^{-1}, and ΔG and λ are in eV (Moser *et al.*, 1992). Endergonic electron tunneling rates are assumed to be related to the exergonic rates by dividing by $\exp[-\Delta G/kT]$, where k is the Boltzmann constant and T the temperature.

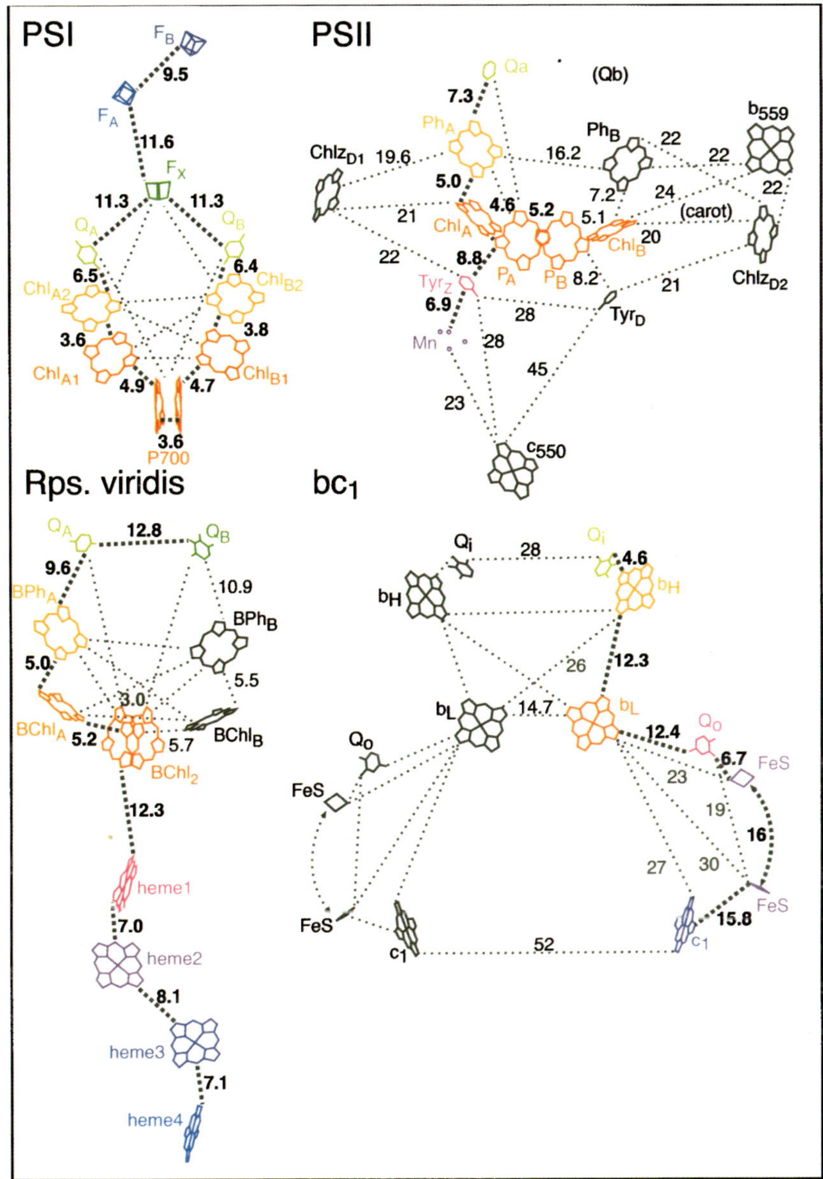

Moser *et al.*, FIG. 3. The arrangement of cofactors in the structure of cyanobacterial reaction centers PSI [1jb0 (Jordan *et al.*, 2001)] and PSII [1Fe1 (Zouni *et al.*, 2001)] from *Syn. elongatus*, purple sulfur bacterial reaction center [1 pcr (Ermler *et al.*, 1994)] from *Rps. viridis* and cytochrome bc_1 complex from chicken heart mitochondria [1bcc and 3bcc (Zhang *et al.*, 1998)]. Bold lines and numbers refer to edge-to-edge distances in likely productive electron transfer reactions. Other distances reflect possible electron-tunneling reactions considered in simulations, and that may be revealed under certain experimental conditions. The positions of Q_B and redox active carotenoid in PSII are uncertain and are reaction centers shown in parentheses. The cytochrome bc_1 reflects that the FeS center takes up different positions in crystal structures with different inhibitors, with relevance for electron transfer distances with the remote redox centers cytochrome c_1 and Q_o. A single quinone is shown at the location of the inhibitor stigmatellin, although multiple quinones may bind in this vicinity and shorten electron transfer distances (Ding *et al.*, 1995).

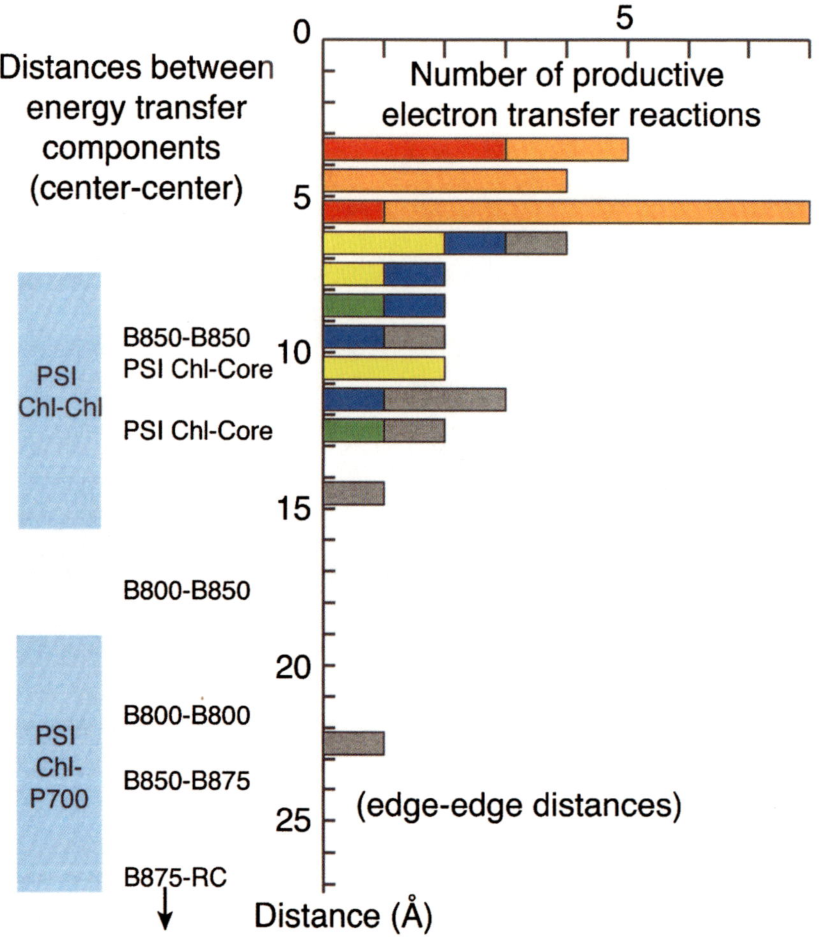

Moser *et al.*, FIG. 4. Distances between cofactors in light harvesting and reaction center proteins. Left: Center-to-center distances between the different antenna BChl types in bacterial LH1 and LH2, and distances between antenna Chl types in light harvesting and reaction center PSI, using the ring structure of *Prochlorococcus*. Right: Edge-to-edge distances for productive electron tunneling between cofactors in the various reaction centers with crystal structures (pdb files 1PRC, 1PCR, 1TEP, 1JB0, 1FE1). Bars are colored according to reaction category: Red, between central chlorin pair; orange, between intermediate chlorins; yellow, between chlorins and nonchlorin acceptor; green, between nonchlorin donor and chlorin pair; blue, between hemes or FeS centers in extended chain; gray, other.

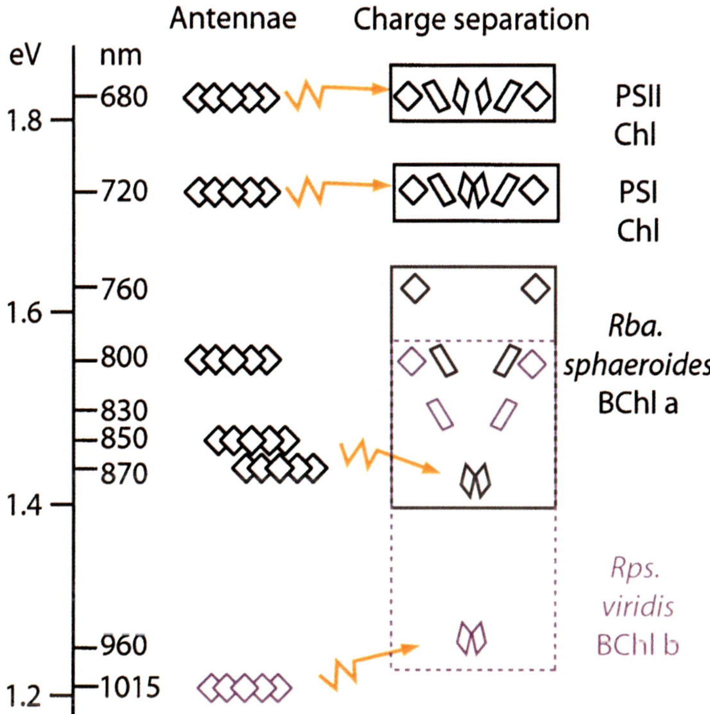

Moser *et al.*, Fig. 5. Long wavelength absorption bands and corresponding energy levels of excited states of chlorins and bacteriochlorins in photosynthetic reaction centers and antenna complexes. The photosystems are nearly isoenergetic, whereas the lower energy purple bacterial systems transfer energy from antenna either downhill (*Rba. sphaeroides*, black) or slightly uphill (*Rps. viridis*, purple).

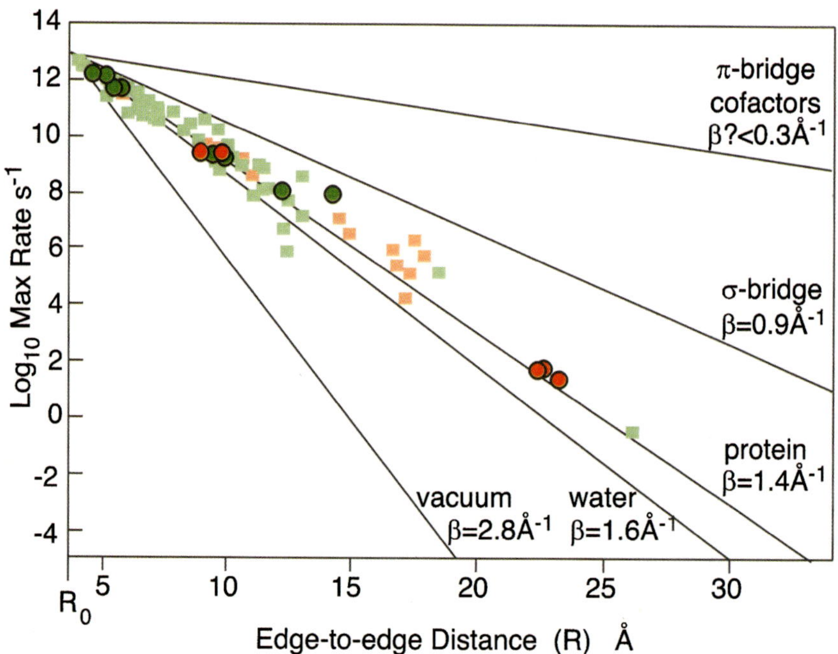

Moser *et al.*, FIG. 6. Free energy-optimized electron tunneling rates expected for different distances in different media with different effective barrier heights for electron tunneling. An empirical expression for the optimized tunneling rate as a function of the distance in Å (R) and the packing density (ρ) of the medium is $\log k_{et} = 13 - (1.2 - 0.8\rho)(R - 3.6)$. ρ varies from 1 when the volume between the redox centers is fully packed with bridging atoms, to 0 when no atoms can be found between the centers. A typical protein has about 75% of the volume within the van der Waals radius of an atom to give a ρ of about 0.75. On a common log scale the distance dependent slope is about 0.6. The green and red circles show the approximate free energy-optimized rate of electron tunneling for productive and counterproductive electron tunneling reactions in reaction centers, while the squares show the expected corresponding rates in natural electron transfer proteins in the pdb database (Page *et al.*, 1999).

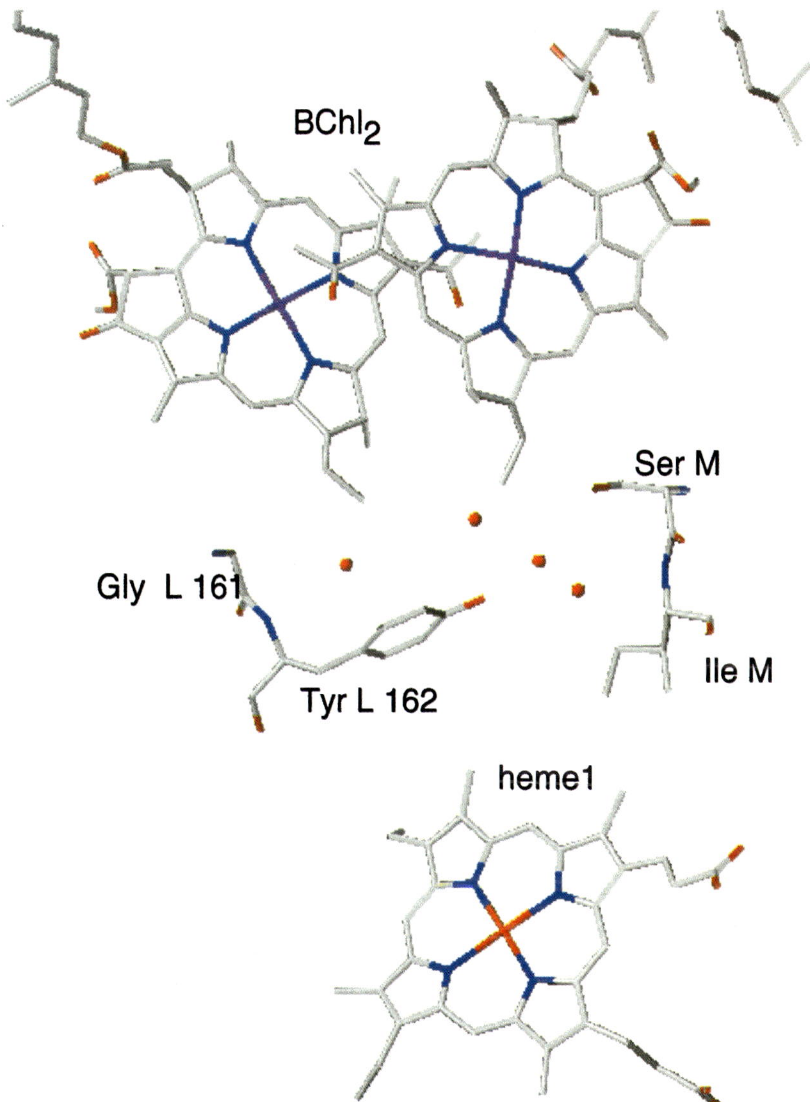

Moser *et al.*, FIG. 7. The region of the *Rps. viridis* reaction center between the $BChl_2$ and heme 1 contains an aromatic tyrosine that has been extensively mutated and characterized (Dohse *et al.*, 1995).

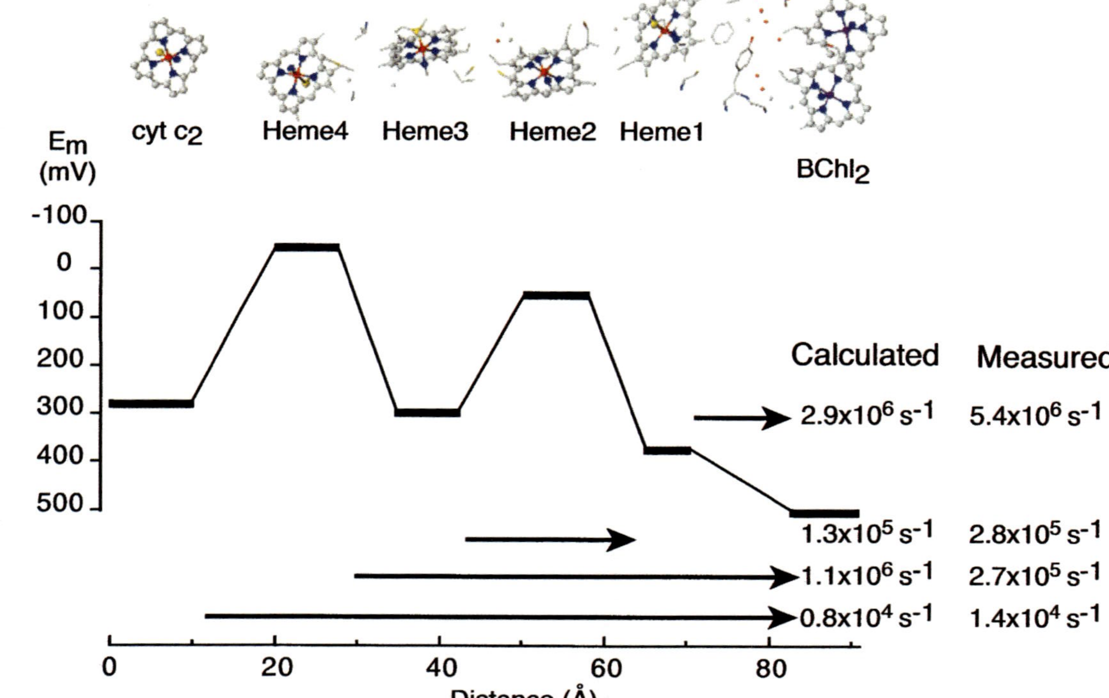

Moser *et al.*, FIG. 8. The cytochrome *c* hemes of the *Rps. viridis* reaction center form an extended chain to which soluble cytochrome c_2 can bind. Light-activated electron transfer under different redox poises and with different mutations allows exploration of endergonic and exergonic electron transfer along the chain and a comparison of measured (Dohse *et al.*, 1995; Ortega *et al.*, 1999; Shopes *et al.*, 1987) with calculated rates.

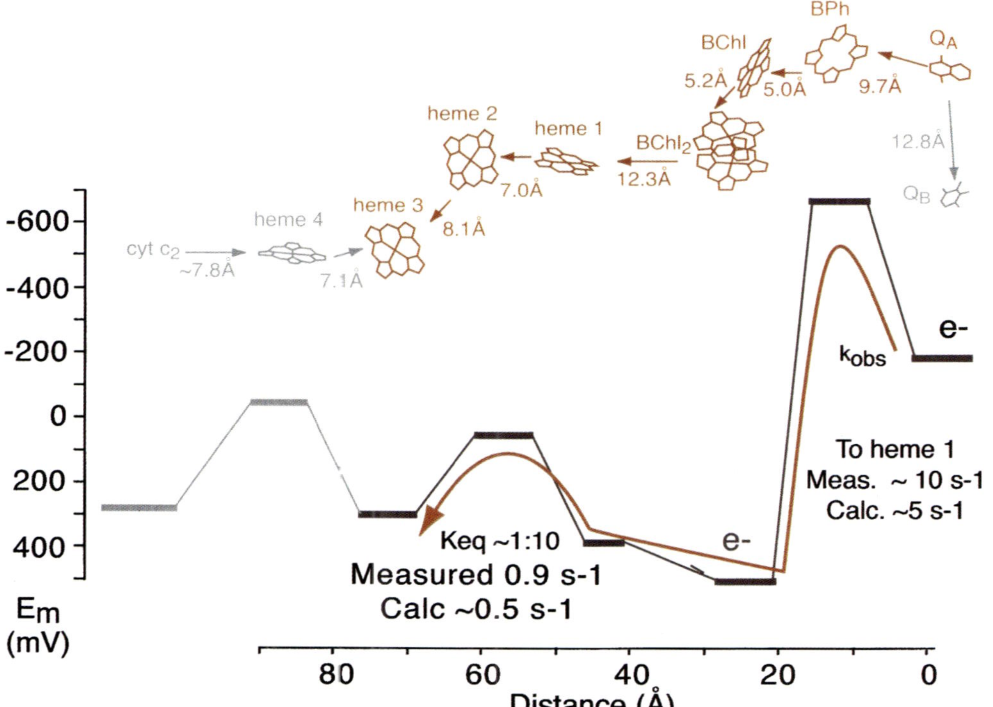

Moser *et al.*, Fig. 10. Charge recombination from Q_A^- in *Rps. viridis* proceeds via several thermally activated steps to reduce photooxidized heme 3 on an approximately second time scale.

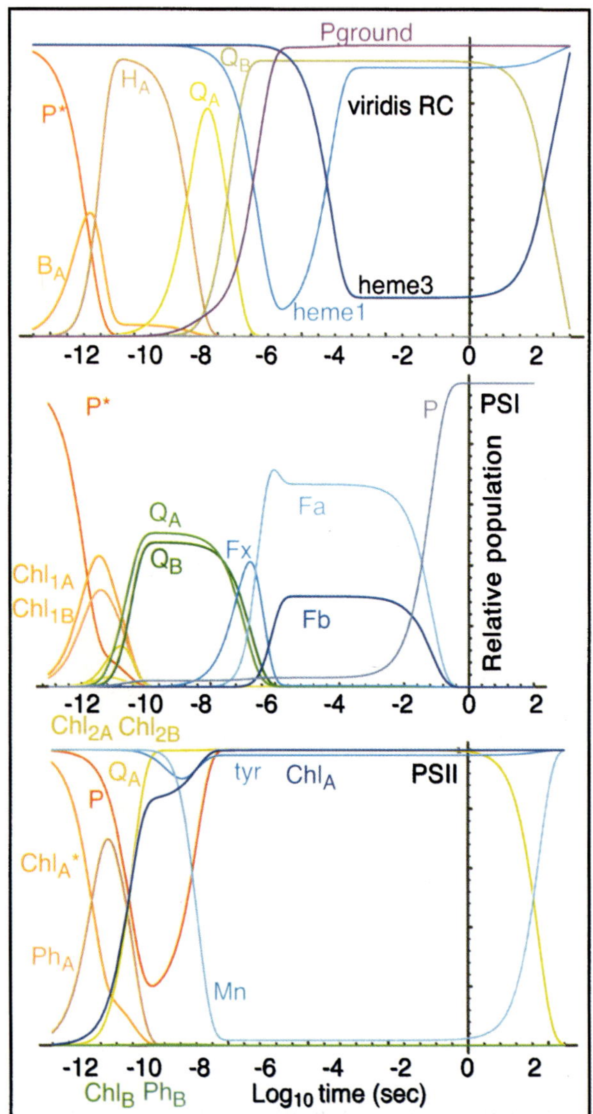

Moser *et al.*, FIG. 11. Simulation of tunneling reactions in three photosystems, using the redox midpoint potential values shown in Table II. Distances are provided by the crystal structures 1PCR, 1FE1, and 1JB0, including distances between non-nearest neighbors (some of which are shown in Fig. 3), such as the distances across the approximate C2 symmetry axis. Reorganization energies assumed a default value of 0.7 eV, except for the initial chlorin charge separation reactions in all photosystems (0.2 or 0.3 eV), recombination from Ph− to P+ (0.5 eV), from Q_A to P+ (1.0 eV), and from Q_B− to P+ (1.3 eV) in *Rps. viridis*; recombination from Fx, Q_A, and Q_B to P700 ground state (1.2 eV). Starting redox state for the systems had the following centers reduced: *viridis* RC: P*, heme 1, and heme 3; PSI: P*; PSII: Chl_A*, Tyr, Mn. Simulations were carried out long enough to allow recombination to restore starting conditions, but with the central light-excitable pigment now in the ground state.

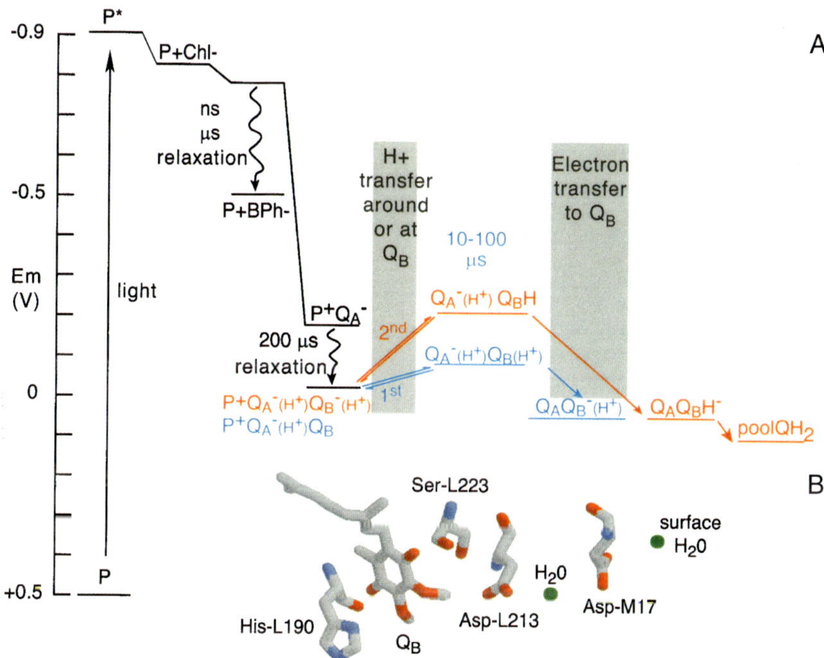

Moser *et al.*, FIG. 12. (A) Uphill electron transfers and relaxations associated with the quinones of purple bacteria. (B) One possible proton transfer chain around Q_B in *Rba. sphaeroides*.

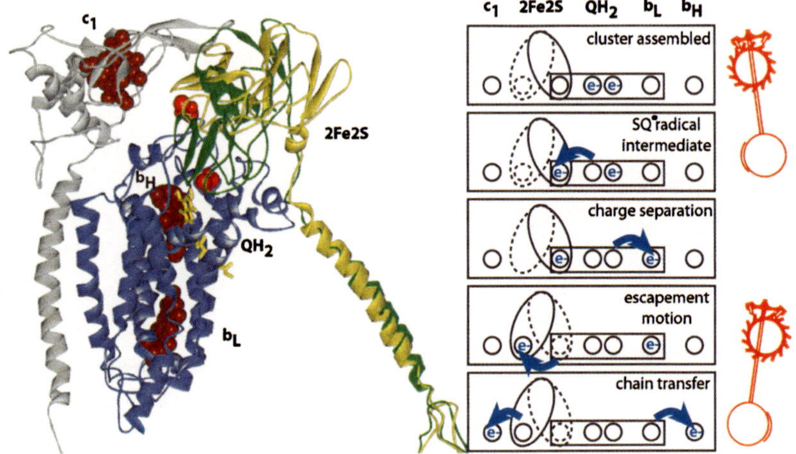

Moser *et al.*, FIG. 13. Motion of the head unit of the FeS subunit of the bc_1 complex may act as an escapement mechanism in the operation of the Q_o site.

small cluster of acidic residues, Asp^{L210}, Glu^{L212}, Asp^{L213}, and Glu^{H173}, but it was not possible to interpret the results uniquely in terms of energetics (electrostatics) or direct functional involvement (as proton carriers) (Paddock *et al.*, 1989, 1994; Takahashi and Wraight, 1992, 1996). By combining conformational sampling with electrostatic energy calculations, Alexov and Gunner (1999) showed that a small number of residues are closely coupled to function in both the energetics and mechanics of proton delivery to Q_B.

At the purely energetic level, the $Q_A^-Q_B \rightleftharpoons Q_AQ_B^-$ electron transfer equilibrium is driven by the differential solvation of Q_B^- versus Q_A^-, which is largely effected by internal proton shifts within the acid cluster. Interactions within the cluster poise the apparent pK values such that, in the ground state, Asp^{L213} is ionized and Asp^{L210} is protonated, even though the latter is more solvent exposed. Arrival of a negative charge on reduction of Q_B—and, to a significant extent, also Q_A—tips the balance in favor of protonation of Asp^{L213} and ionization of Asp^{L210}, which results in a switch of H-bond connections. In the ground state, Ser^{L223} is a H-bond donor to $Asp(-)^{L213}$. Following formation of the semiquinone anion, Ser^{L223} becomes a proton donor to Q_B and $AspH^{L213}$ now donates a H-bond to Ser^{L223} (Fig. 12; see color insert). Although no proton is taken by the semiquinone at this stage, this sets up a connected pathway that could be used for proton delivery in the second reduction step.

More relevant to the first electron process, the proton equilibration capabilities of the acid cluster are largely independent of proton uptake from the medium. Only about 0.3–0.4 H^+ are taken up at neutral pH and, remarkably, this is independent of whether the electron is on Q_A or Q_B, i.e., $m = n \sim 0.3$ (differences appear outside the pH 6–8 range) (Maroti and Wraight, 1988; McPherson *et al.*, 1988; Okamura *et al.*, 2000). This reflects a general lack of ionizable groups near Q_A, leading to a low dielectric response of the immediate environment and consequent long-range influence of the Q_A^- charge on ionizable groups close to the Q_B site. The calculated proton distribution is similar but not identical for both semiquinones, so proton redistribution contributes to the protein relaxation that controls the energetics of the electron transfer. Alexov and Gunner (1999) have described the reduction of Q_A as "preparing" the Q_B site for the subsequent electron transfer, by partially redistributing the protonation states of the acid cluster.

X-Ray structural studies indicate that the ground state of the bacterial RC contains Q_B bound in a distal position that is presumed to be inactive with respect to electron transfer ((Ermler *et al.*, 1994; Graige *et al.*, 1998; Lancaster and Michel, 1997; Stowell *et al.*, 1997). The inactivity would arise from a 2 Å greater distance from Q_A, and a distinctly less polar and less H-bonded environment, which is likely to lower the redox potential, making Q_B unreducible by Q_A^-. However, free energy calculations and molecular dynamics simulations suggest that the equilibrium is finely balanced and the proximal position for Q_B is favored when Q_A^- is present and the acid cluster is suitably protonated (Alexov and Gunner, 1999; Walden and Wheeler, 2001; Zachariae and Lancaster, 2001). Thus, reduction of Q_A appears to prepare the Q_B site by bringing the quinone into the proximal position, as well as by redistributing the protonation states of the protein.

Depending on the degree of preparation of the Q_B site by Q_A^-, the first electron transfer may be downhill or uphill. In the latter case, the electron transfer equilibrium is pulled over by subsequent relaxations, including further internal proton redistributions as well as net H^+ uptake. With artificial quinones as Q_A, however, Gunner and co-workers have reported a very rapid phase of electron transfer to Q_B when Q_A is sufficiently low potential, and have shown this component to speed up with increasing driving force, indicating rate limitation (Li *et al.*, 1998, 2000). This result is controversial (Graige *et al.*, 1998), but a reasonable interpretation is that the initial electron transfer equilibrium, on the 1-μs time scale, is unfavorable and can only be seen when the equilibrium is artificially enhanced by using low potential quinone analogues as Q_A. The slower kinetics, in all cases, would then represent the equilibrium being pulled over by *subsequent* H^+ uptake and redistribution. For the native ubiquinone, the initial equilibrium could be unfavorable by as much as 100 meV (Wraight, 1998).

2. *Proton Coupling to the Second Electron Transfer*

Although the acid cluster plays a distinct and identifiable role in the relaxation response of the protein on the first electron transfer, electrostatic calculations indicate that the full energetic contribution is quite widely distributed, with small ionization state changes of a number of residues. On the other hand, proton delivery to the quinone head group during the second reduction step is a targeted function, with discrete termini at the quinol oxygens.

The pathway for proton uptake associated with the first and second electron seems to be the same (Ädelroth *et al.*, 2000). Both are inhibited by binding of certain divalent transition metal ions, notably Zn^{2+}, Ni^{2+}, Cd^{2+}, and Cu^{2+}, but not Fe^{2+}, or Ca^{2+} or Mg^{2+} (Paddock *et al.*, 1999; Utschig *et al.*, 1998) and by the double mutation of Asp^{L210} and Asp^{M17} (Paddock *et al.*, 2000). The binding site for the inhibitory ions has been identified in X-ray structures as involving one of a surface cluster of histidines, either or both of Asp^{L210} and Asp^{M17}, and solvent waters (Axelrod *et al.*, 2000). The relative insignificance of net H^+ uptake (in contrast to internal redistribution) for the energetics of the first electron transfer is supported by the lack of effect of metal binding on the electron transfer equilibrium.

For the second electron transfer, the pH and free energy dependence of the reaction rate clearly support a mechanism in which protonation of Q_B^- precedes the electron transfer, which is rate limiting (Graige *et al.*, 1996) (Fig. 12). It is noteworthy that in this reaction mechanism, an energetically unfavorable protonation intermediate (Q_BH) is accessed by thermal energies to allow electron transfer to proceed with a large free energy drop. Nevertheless, the electron transfer is still rate limiting, largely because of the substantial distance—at 14.5 Å it is near the limit of the physiological range described by the histogram of Fig. 4.

It seems reasonable to think that the H-bonded connection between Q_B^-, Ser^{L223}, and $AspH^{L213}$, established after the first electron transfer, provides the terminal path for delivery of the first proton to the quinone, ahead of the second electron. From the X-ray structures, a water molecule is positioned between Asp^{L213} and Asp^{M17} and a complete pathway could comprise

$$Q_B^-—Ser^{L223}—AspH^{L213}—H_2O—Asp^{M17}—(aq)$$

With this arrangement, delivery of H^+ from $AspH^{L213}$ to form Q_BH could be in extremely rapid equilibrium—albeit unfavorable because of the low pK of Q_B^- compared to Graige *et al.* (1999)—but equilibration with the bulk phase would require rotation of Asp^{L213}, H_2O, and, probably, Asp^{M17}. None of these presents obvious obstacles and yet mutation of Asp^{M17} (or Asp^{L210}) alone has little or no effect on the second electron transfer and its associated proton deliveries. Only when both Asp^{M17} and Asp^{L210} are substituted (by asparagine) is the proton path inactivated.

It is not clear how this apparent cooperativity between Asp^{M17} and Asp^{L210} works in the proton pathway, nor whether the severe inhibition due to double mutation, or binding of metal ions, arises from an energetic or a kinetic source. It may be that the proton path is actually located between Asp^{L213} and the metal binding site, involving both Asp^{M17} and Asp^{L210}. In the X-ray structures, this region contains significant void volume and appears to be accessible to solvent. This suggests the possibility that a conducting path is dynamically assembled from solvent water, possibly requiring the good hydrogen bonding attributes of the acidic groups. This would put it in a similar category as bacteriorhodopsin, and as proposed for cytochrome oxidase, even though the RC has no evident need to gate its proton transfers.

In fact, transient assembly of H-bonded water files is probably common in enzyme function. In carbonic anhydrase, for example, the rate-limiting step is proton transfer from the active-site Zn^{2+}-OH_2 complex to the surface, via a transient, H-bonded water network that conducts H^+. Analysis of the relationship between rates and free energies (pK differences) by standard Marcus theory shows that the major contribution to the observed activation barrier is in the work term for assembling the water chain (Ren *et al.,* 1995).

Although the well-formed H-bonded file of water in gramicidin may be atypical in the wider realm of proton conduction, the principles of proton conduction gained from gA can probably be generalized to the transient pathways seen in other systems. Even transient networks or files of water can be expected to be quite strongly H-bonded; their transient nature is likely to be more of entropic than enthalpic origin. The symmetry of the acceptor–donor pairs of waters, or otherwise similar functional groups (with similar pK values), will also contribute to keeping a low barrier in the proton transfer coordinate. At the same time, thermal fluctuations are clearly sufficient to disrupt the hydrogen bonds, leading to disassembly of the chain. Studies on gramicidin also illustrate that the presence of an ion, for example a sodium in the channel mouth, can stabilize the file of waters (Jordan, 1990). This suggests the possibility that the proton path in bacterial RCs, or in the channels of cytochrome oxidase, could assemble in response to specific charge states of the active site—Q_B^-, $Q_A^-Q_B^-$, Q_BH^-, etc., in RCs, and various intermediates of the redox cycle in oxidase. Of course, in bacteriorhodopsin the assembly and function of the reprotonation pathway is already being described in structural detail (Lanyi and Luecke, 2001); the energetic drive is provided by the charge and dipole movements associated with the photoisomerization and the initial charge transfer from the Schiff base.

The interesting issue of proton collection at the protein surface is not addressed here, but it does appear that proton antennae are designed surface features of several proton translocating proteins, including the bacterial RC (Ädelroth *et al.,* 2001), cytochrome oxidase (Marantz *et al.,* 1998), and bR (Checover *et al.,* 1997).

E. Protonation in the Oxidation Reactions of Photosystem II

The oxidation of water by Photosystem II in oxygenic photosynthesis is an energetically and mechanistically challenging process, in which proton transfers play a critical role. The oxidizing potential generated by the reaction center chlorophylls (P680) is in excess of 1.1 V and the average potential of the water-oxidizing Mn complex must exceed the operating potential of the O_2/H_2O couple (approximately 870 mV at pH 6). Some of the individual oxidation states of the Mn complex (the S-states) exhibit redox potentials within 150 mV of that of $P680^+$ (Vass and Styring, 1991). To control possible side reactions, $P680^+$ must be rereduced rapidly with minimal free energy loss.

The immediate electron transfer to $P680^+$ occurs from a tyrosine residue, Y_Z, residue 160 in the D1 subunit of PSII, with a lifetime of 20–40 ns under optimal conditions and a free energy drop of only 40 mV (Meyer *et al.,* 1989). The identifiable oxidized form of Y_Z is a neutral radical:

$$Y_ZOH\ P^+ \rightleftharpoons Y_ZO^{\circ}P + H^+$$

The nature of the electron/proton coupling in this reaction is a topic of great current interest. Unfortunately, much of the available experimental data comes from studies on Mn-depleted PSII preparations, which are incapable of water oxidation and which exhibit grossly altered kinetics of $P680^+$ reduction. It is clear for such preparations that abstraction of the proton from Y_Z is rate-limiting for net electron transfer, although, even for this system, it is not known if the proton is removed before or after electron transfer (Hays *et al.,* 1999).

In the intact, oxygen-evolving system, the oxidation of water by sequential electron transfers is accompanied by H^+ release into the bulk phase. The stoichiometry of proton release depends on conditions, including pH. As is well established for the H^+ uptake by bacterial RCs, the released protons are presumed to include weakly coupled "Bohr" protons, arising from pK shifts in amino acid residues under the influence of the charge accumulation in the Mn complex, as well as strongly coupled "chemical" protons released upon oxidation of the Mn complex, including associated ligands. These proton events obscure those associated with the kinetically critical oxidation of Y_ZOH, for which the fate and timing of the proton abstraction is crucial. Thus, our understanding of this process is largely based on thermodynamic and activation energy arguments.

The basic issue is that the oxidation potential for Y_ZOH^+/ Y_ZOH is about 1.5 V in solution, too high to be an effective intermediate between $P680^+$ and the Mn complex. Given that Y_ZOH is the dominant reduced form, is this reactivity problem circumvented by rapid proton trasfer after electron transfer or by a low level of Y_ZO^- produced in a preequilibrium? And can the Y_Z redox potential be brought down to earth by local means, or must the H^+ be released to the bulk phase? Babcock and Tommos (Tommos and Babcock, 1998) consider that H^+ from Y_ZOH^+ must be released to the bulk phase (or at least very far from Y_Z). [This is independent of the proposal that Y_ZO° abstracts a hydrogen atom from the Mn complex (Tommos and Babcock, 1998), rather than an electron (Gilchrist *et al.*, 1995).]

$$Y_ZOH\ P^+ \rightleftharpoons Y_ZOH^+P^+ \rightleftharpoons Y_ZO^\circ P + H^+(\text{bulk})$$

Rappaport and Lavergne (2001), however, have argued that the reactivity of Y_Z is controlled by H-bond formation that results in a low level of Y_ZO^- formation, with rapid local proton transfer preceding electron transfer:

$$Y_ZOH\ P^+ \rightleftharpoons \underset{(H^+)}{Y_ZO^-P^+} \rightleftharpoons \underset{(H^+)}{Y_ZO^\circ P}$$

In particular, they doubt that proton transfer to the bulk phase can occur sufficiently rapidly to account for the electron transfer kinetics (20 ns at pH 5), and hence favor the preequilibrium shown above. Although their energetic analysis is insightful, the kinetic conclusions are unwarranted at the present time. Proton equilibration from many protein surfaces into the bulk phase is known to be slow (Gopta *et al.*, 1999; Haumann and Junge, 1994; Heberle *et al.*, 1994; Maróti and Wraight, 1997) but whether the proton actually makes it to the bulk phase is somewhat moot from the point of view of the Y_Z reaction. There is no doubt that proton transfer over a sufficient distance could be effected in a few picoseconds, given a suitable conduction path, as evidenced by gramicidin. However, whether such a structure might exist must await significant improvements in the PSII X-ray structure. The basic question of timing (proton first or electron first) is fundamental to a mechanistic description, as it was for the second electron transfer to Q_B in bacterial RCs, and it seems unlikely that it will be resolved without similar experimental input.

F. Is Proton Transfer Design Robust?

The charge compensation role of intraprotein proton transfer is an essentially dielectric response and is not structurally specific. Transfer of a proton requires specificity of contact, but a variety of geometries and protein motifs can fulfill the underlying requirement of charge redistribution. Thus, many second site mutations can recover significant function of the first electron transfer in bacterial RCs. On the other hand, efficient delivery of protons to a unique end point requires much greater structural specificity. In spite of the inherently dynamic nature of proteins, their packing is such that even small, neutral molecules (such as water) cannot penetrate deeply at will (Simonson and Perahia, 1995), which makes proton pathways for long-distance transport highly specific. In the bacterial RC, H^+ entry occurs at a single site and intraprotein proton transfer occurs via a pathway that is not yet fully defined but is very sensitive to localized modifications. The Asp^{L213} mutations that inhibit proton delivery to Q_B^- can be reverted by several second site mutations, but the kinetic competency is generally marginal—with the exception of the $Asn^{M44} \rightarrow Asp$ suppressor mutation. The latter is physically very close by and it seems likely that Asp^{M44} could substitute directly for Asp^{L213}, thereby maintaining essentially the same gross structure. Similarly specific design features are evident in the proton channels of oxidase and bR. Thus it appears that the physical contact needed for proton movement toward and away from specific energy coupling sites in proteins can tolerate some degree of mutational change while preserving function, but that it is generally significantly less robust that the long-distance energy and electron transfer functions of these proteins.

The constraints are even more severe at each of the bond making and breaking catalytic sites in photosynthetic systems and their accompanying redox partners. In each catalytic site (the Q_B sites of the purple reaction centers and PSII, the quinone oxidation and reduction sites of bc_1 and b_6f, and in the water splitting site of PSII) multiple electron transfers are coordinated with multiple proton transfers. The electron transfer side of these reactions appears to take place over short enough electron tunneling distances that they are relatively facile and robust. The geometric and kinetic restraints of water and proton management around these catalytic sites, however, could very well be the principal engineering constraint that limits the turnover of these sites and thus turnover of the bioenergetic systems as a whole. Because of this sensitivity, we might expect these sites to be subjected to the most evolutionary selective pressure and to be the least robust with respect to mutational changes.

VII. Managing Diffusion in Photosynthesis

Although diffusion is a slow process compared to energy transfer and electron transfer at the shortest distances, it can be an exceptionally effective way to move electrons and protons over long distances. However, unlike the hard-wired cofactor chains that guide electron transfer in protein complexes, diffusion faces the problem of directing where the diffusing complexes go as well as the problem of avoiding unwanted side reactions with unproductive partners. These engineering problems are addressed in the diffusing cytochrome *c* (or analogous plastocyanin) population by exploiting the protein's surface charge distribution near the electron transfer center to make it more likely that it will dock and come into tunneling distance only with the appropriate reaction partners. The same surface charge distribution also tends to keep the cytochrome associated with the membrane surface. This greatly restricts a large three-dimensional diffusion volume to an essentially two-dimensional diffusion plane, shortening the time between successful encounters. Encounter time is also limited by having a pool of cytochrome in significant excess over its redox partners.

By virtue of its hydrophobicity, quinone is confined to the membrane, which similarly limits the diffusional volume. Steering charges are out of the question given quinone's hydrophobic environment. However, it can control its reactivity with unwanted reaction partners by virtue of the tight coupling of the two-electron oxidation–reduction reaction while in the membrane. Any adventitious redox partner with the potential of transferring electrons must address the high instability of the semiquinone (SQ) product in the free lipid environment. The quinone stability constant, $[SQ]^2/[Q][QH_2]$, has been estimated as 10^{-10} (Dutton *et al.*, 2000); thus even at redox equilibrium with the quinone pool half reduced and half oxidized, the concentration of semiquinone is small, on the order of one part in 100,000. This makes QH_2 a poor reductant and Q a poor oxidant unless the quinone is in a protein site which is designed to either stabilize the half-reduced SQ state, or to provide the bound quinone with two ready and rapid electron transfer partners.

The ultimate way to control diffusion and reactivity with redox partners is to restrict diffusion by anchoring a portion of the redox molecule and allowing essentially only one-dimensional diffusion. This is effectively the case for the FeS center of the bc_1 complex, which has a mobile head group with a surface-exposed FeS center, but also a transmembrane anchor secured to the membrane portion of the bc_1 complex. This severely restricts the range of motion (~16 Å) but perfectly controls the problem of guiding electron transfer. Diffusion over this distance should be on the submicrosecond time scale, much faster than the catalytic turnover of the complex. In a certain sense, this restricted diffusion has properties that lie between unrestricted diffusion and fixed redox cofactor chains: a sort of chain with moving parts.

By exploiting the relative distance dependences of electron tunneling and diffusion, it seems possible that this restricted FeS diffusion may play a role in regulating the action of the $n = 2$ quinone binding site, by allowing the removal of only one electron from the quinone site during the time interval it takes to swing from the quinone site to its other redox partner, fixed cytochrome c_1 (Darrouzet *et al.*, 2000). This allows the other electron to move onto the *b* cytochromes and drive transmembrane field generating electron transfer. Such a regulating swinging motion would be akin to the escapement mechanism of a clock, which regulates the drop of the weights, or the unwinding of the spring according to the extreme movements of the swinging pendulum. Figure 13 (see color insert) shows the essential elements of this action. The FeS subunit is shown in yellow and green as it revealed in analogous crystal structures in the presence or absence of a quinonoid inhibitor myxothiazol (yellow). While the anchoring transmembrane helix remains virtually unchanged, the redox active FeS cluster (red) moves within electron tunneling distance of one or the other redox partner, cytochrome c_1 or the quinone, but it cannot be in tunneling contact with both at the same time.

These various diffusion strategies used by cytochrome *c*, plastocyanin, and quinone are clearly successful, because electron transfer throughput is not limited by pool action, even in tissues that need to support a great deal of electron transfer activity such as bee flight muscle (Suarez *et al.*, 2000). These bioenergetic systems are diffusion-coupled rather than diffusion-limited. It seems that chemistry at catalytic sites is the ultimate throughput restrictor and constrains the power that can be produced by bioenergetic systems.

VIII. Summary

The design of photosynthetic systems reflects the length scales of the fundamental physical processes. Energy transfer is rapid at the few angstrom scale and continues to be rapid even at the 50-Å scale of the membrane thickness. Electron tunneling is nearly as rapid at the shortest distances, but becomes physiologically too slow well before 20 Å. Diffusion, which starts out at a relatively slow nanosecond time scale, has the most modest slowing with distance and is physiologically competent at all biologically relevant distances. Proton transfer always operates on the shortest angstrom scale. The structural consequences of these distance dependencies are that energy transfer networks can extend over large, multisubunit and multicomplex distances and take leaps of 20 Å before entering the domain of charge separating centers. Electron transfer systems are effectively limited to individual distances of 15 Å or less and span the 50 Å dimensions of the bioenergetic membrane by use of redox chains. Diffusion processes are generally used to cover the intercomplex electron transfer distances of 50 Å and greater and tend to compensate for the lack of directionality by restricting the diffusional space to the membrane or the membrane surface, and by multiplying the diffusing species through the use of pools. Proton transfer reactions act over distances larger than a few angstroms through the use of clusters or relays, which sometimes rely on water molecules and which may only be dynamically assembled.

Proteins appear to place a premium on robustness of design, which is relatively easily achieved in the long-distance physical processes of energy transfer and electron tunneling. By placing cofactors close enough, the physical process is relatively rapid compared to decay processes. Thus suboptimal conditions such as cofactor orientation, energy level, or redox potential level can be tolerated and generally do not have to be finely tuned. The most fragile regions of design tend to come in areas of complex formation and catalysis involving proton management, where relatively small changes in distance or mutations can lead to a dramatic decrease in turnover, which may already be limiting the overall speed of energy conversion in these proteins. Light-activated systems also face a challenge to robust function from the ever-present dangers of high redox potential chemistry. This can turn the protein matrix and wandering oxygen molecules into unintentional redox partners, which in the case of PSII requires the frequent, costly replacement of protein subunits.

References

Ädelroth, P., Paddock, M. L., Slagle, L., Feher, G., and Okamura, M. Y. (2000). *Proc. Natl. Acad. Sci. USA* **97,** 13086–13091.

Ädelroth, P., Paddock, M. L., Tehrani, A., Beatty, J. T., Feher, G., and Okamura, M. Y. (2001). *Biochemistry* **40,** 14538–14546.

Alegria, G., and Dutton, P. L. (1991). *Biochim. Biophys. Acta* **1057,** 239–257.

Alexov, E. G., and Gunner, M. R. (1999). *Biochemistry* **38,** 8253–8270.

Axelrod, H. L., Abresch, E. C., Paddock, M. L., Okamura, M. Y., and Feher, G. (2000). *Proc. Natl. Acad. Sci. USA* **97,** 1542–1547.

Barber, J., and Archer, M. D. (2001). *J. Photochem. Photobiol. A* **142,** 97–106.

Beece, D., Eisenstein, L., Frauenfelder, H., Good, D., Marden, M. C., Reinisch, L., Reynolds, A. H., Sorensen, L. B., and Yue, K. Y. (1980). *Biochemistry* **19,** 5147–5157.

Blankenship, R. E., Olson, J. M., and Miller, M. (1995). *Adv. Photosynth.* **2,** 399–435.

Checover, S., Nachliel, E., Dencher, N. A., and Gutman, M. (1997). *Biochemistry* **36,** 13919–13928.

Chen, I. P., Mathis, P., Koepke, J., and Michel, H. (2000). *Biochemistry* **39,** 3592–3602.

Chiu, S.-W., Subramaniam, S., and Jakobsson, E. (1999). *Biophys. J.* **76,** 1939–1950.

Clayton, R. K. (1978). *Biochim. Biophys. Acta* **504,** 255–264.

Cogdell, R. J., Fyfe, P., Fraser, N., Law, C., Howard, T., McLuskey, K., Prince, S., Freer, A., and Isaacs, N. (1998). In: *Microbial Responses to Light and Time* (M. X. Caddick, Ed.), pp. 143–158. Cambridge Univ. Press, Cambridge, UK.

Cogdell, R. J., Isaacs, N. W., Howard, T. D., McLuskey, K., Fraser, N. J., and Prince, S. M. (1999). *J. Bacteriol.* **181,** 3869–3879.

Darrouzet, E., Valkova-Valchanova, M., Moser, C. C., Dutton, P. L., and Daldal, F. (2000). *Proc. Natl. Acad. Sci. USA* **97,** 4567–4572.

Dekker, J. P., and Van Grondelle, R. (2000). *Photosynth. Res.* **63,** 195–208.

Devault, D., and Chance, B. (1966). *Biophys. J.* **6,** 825–847.

Diner, B. A., Schlodder, E., Nixon, P. J., Coleman, W. J., Rappaport, F., Lavergne, J., Vermaas, W. F. J., and Chisholm, D. A. (2001). *Biochemistry* **40,** 9265–9281.

Ding, H., Moser, C. C., Robertson, D. E., Tokito, M. K., Daldal, F., and Dutton, P. L. (1995). *Biochemistry* **34,** 15979–15996.

Dohse, B., Mathis, P., Wachtveitl, J., Laussermair, E., Iwata, S., Michel, H., and Oesterhelt, D. (1995). *Biochemistry* **34,** 11335–11343.

Dutton, P. L., and Moser, C. C. (1994). *Proc. Natl. Acad. Sci. USA* **91,** 10247–10250.

Dutton, P. L., Ohnishi, T., Darrouzet, E., Leonard, M. A., Sharp, R. E., Gibney, B. R., Daldal, F., and Moser, C. C. (2000). In: *Coenzyme Q: From Molecular Mechanisms to Nutrition and Health* (P. Quinn and V. Kagan, Eds.), pp. 65–82. CRC Press, Boca Raton, FL.

Eigen, M. (1964). Proton transfer, acid–base catalysis, and enzymatic hydrolysis. Part I: Elementary processes. *Angew. Chem. Int. Ed.* **3,** 1–19.

Ermler, U., Fritzsch, G., Buchanan, S. K., and Michel, H. (1994). *Structure* **2,** 925–936.

Faller, P., Debus, R. J., Sugiura, M., Rutherford, A. W., and Boussac, A. (2001). *Proc. Natl. Acad. Sci. USA* **98,** 14368–14373.

Florin, S., and Tiede, D. M. (1987). *Prog. Photosynth. Res.* **7,** 205–208.

Frank, H. A., and Cogdell, R. J. (1996). *Photochem. Photobiol.* **63,** 257–264.

Gao, J. L., Shopes, R. J., and Wraight, C. A. (1990). *Biochim. Biophys. Acta* **1015,** 96–108.

Gilchrist, M. L., Ball, J. A., Randall, D. W., and Britt, R. D. (1995). *Proc. Natl. Acad. Sci. USA* **92,** 9545–9549.

Golbeck, J. H. (1994). In: *Advances in Photosynthesis: The Molecular Biology of Cyanobacteria* (D. A. Bryant, Ed.), pp. 319–360. Kluwer Academic, Dordrecht.

Gopta, O. A., Cherepanov, D. A., Junge, W., and Mulkidjanian, A. Y. (1999). *Proc. Natl. Acad. Sci. USA* **96,** 13159–13164.

Graige, M. S., Paddock, M. L., Bruce, J. M., Feher, G., and Okamura, M. Y. (1996). *J. Am. Chem. Soc.* **118,** 9005–9016.

Graige, M. S., Feher, G., and Okamura, M. Y. (1998). *Proc. Natl. Acad. Sci. USA* **95,** 11679–11684.

Graige, M. S., Paddock, M. L., Feher, G., and Okamura, M. Y. (1999). *Biochemistry* **38,** 11465–11473.

Guergova-Kuras, M., Boudreaux, B., Joliot, A., Joliot, P., and Redding, K. (2001). *Proc. Natl. Acad. Sci. USA* **98,** 4437–4442.

Gunner, M. R. (1988). The temperature and $-\Delta G$ dependence of long range electron transfer in reaction center protein from *Rhodobacter sphaeroides*. Univ. of Pennsylvania, Philadelphia.

Gunner, M. R., and Dutton, P. L. (1989). *J. Am. Chem. Soc.* **111,** 3400–3412.

Gunner, M. R., Robertson, D. E., and Dutton, P. L. (1986). *J. Phys. Chem.* **90,** 3783–3795.

Gupte, S., Wu, E.-S., Hoechli, L., Hoechli, M., Jacobson, K., Sowers, A. E., and Hackenbrock, C. R. (1984). *Proc. Natl. Acad. Sci. USA* **81,** 2606–2610.

Haumann, M., and Junge, W. (1994). *Biochemistry* **33,** 864–872.

Hays, A.-M. A., Vassiliev, I. R., Golbeck, J. H., and Debus, R. J. (1999). *Biochemistry* **38,** 11851–11865.

Heberle, J., Riesle, J., Thiedemann, G., Oesterhelt, D., and Dencher, N. A. (1994). *Nature* **370,** 379–382.

Hibbert, F. (1986). *Adv. Phys. Org. Chem.* **22,** 113–212.
Holzwarth, A. R., and Muller, M. G. (1996). *Biochemistry* **35,** 11820–11831.
Iwaki, M., and Itoh, S. (1994). *Plant Cell Physiol.* **35,** 983–993.
Jordan, P. C. (1990). *Biophys. J.* **58,** 1133–1156.
Jordan, P., Fromme, P., Witt, H.-T., Klukas, O., Saenger, W., and Krauss, N. (2001). *Nature* **411,** 909.
Kaminskaya, O., Konstantinov, A. A., and Shuvalov, V. A. (1990). *Biochim. Biophys. Acta* **1016,** 153–164.
Kihara, T., and McCray, J. A. (1973). *Biochim. Biophys. Acta* **292,** 297–309.
Klimov, V. V., Allakhverdiev, S. I., Demeter, S., and Krasnovskii, A. A. (1979). *Dokl. Akad. Nauk Sssr* **249,** 227–230.
Krieger, A., Rutherford, A. W., and Johnson, G. N. (1995). *Biochim. Biophys. Acta* **1229,** 193–201.
Kuehlbrandt, W., Wang, D. N., and Fuijiyoshi, Y. (1994). *Nature* **350,** 614–621.
Lancaster, C. R. D., and Michel, H. (1997). *Structure* **5,** 1339–1359.
Lanyi, J. K., and Luecke, H. (2001). *Curr. Opin. Struct. Biol.* **11,** 415–419.
Law, C. J., and Cogdell, R. J. (1998). *FEBS Lett.* **432,** 27–30.
Li, J. L., Gilroy, D., Tiede, D. M., and Gunner, M. R. (1998). *Biochemistry* **37,** 2818–2829.
Li, J. L., Takahashi, E., and Gunner, M. R. (2000). *Biochemistry* **39,** 7445–7454.
Marantz, Y., Nachliel, E., Aagaard, A., Brzezinski, P., and Gutman, M. (1998). *Proc. Natl. Acad. Sci. USA* **95,** 8590–8595.
Marcus, R. A. (1956). *J. Chem. Phys.* **24,** 966–978.
Maróti, P., and Wraight, C. A. (1988). *Biochim. Biophys. Acta* **934,** 329–347.
Maróti, P., and Wraight, C. A. (1997). *Biophys. J.* **73,** 367–381.
McMahon, B. H., Müller, J. D., Wraight, C. A., and Nienhaus, G. U. (1998). *Biophys. J.* **74,** 2567–2587.
McPherson, P., Okamura, M. Y., and Feher, G. (1988). *Biochim. Biophys. Acta* **934,** 348–368.
Meyer, B., Schlodder, E., Dekker, J. P., and Witt, H. T. (1989). *Biochim. Biophys. Acta* **974,** 36–43.
Michaelis, L. (1951). In: *The Enzymes, Chemistry and Mechanism of Action* (J. B. Sumner and K. Myrback, Eds.), vol. 2, pp. 1–54. New York: Academic Press.
Michel, H., Deisenhofer, J., and Epp, O. (1986). *EMBO J.* **5,** 2445–2451.
Moser, C. C., Keske, J. M., Warncke, K., Farid, R. S., and Dutton, P. L. (1992). *Nature* **355,** 796–802.
Moser, C. C., Page, C. C., and Dutton, P. L. (2001). In: *Electron Transfer in Chemistry* (V. Balzani, Ed.), vol. 3, pp. 24–38. Wiley-VCH, New York.
Nagle, J. (1987). *J. Bioenerg. Biomembr.* **19,** 413–426.
Nagle, J., and Morowitz, H. J. (1978). *Proc. Natl. Acad. Sci. USA* **75,** 298–302.
Nogi, T., Fathir, I., Kobayashi, M., Nozawa, T., and Miki, K. (2000). *Proc. Natl. Acad. Sci. USA* **97,** 13561–13566.
Okamura, M. Y., Paddock, M. L., Graige, M. S., and Feher, G. (2000). *Biochim. Biophys. Acta* **1458,** 148–163.
Ortega, J. M., Drepper, F., and Mathis, P. (1999). *Photosynth. Res.* **59,** 147–157.
Paddock, M. L., Rongey, S. H., Feher, G., and Okamura, M. Y. (1989). *Proc. Natl. Acad. Sci. USA* **86,** 6602–6606.
Paddock, M. L., Rongey, S. H., McPherson, P. H., Juth, A., Feher, G., and Okamura, M. Y. (1994). *Biochemistry* **33,** 734–745.
Paddock, M. L., Graige, M. S., Okamura, M. Y., and Feher, G., (1999). *Proc. Natl. Acad. Sci. USA* **96,** 6183–6188.
Paddock, M. L., Okamura, M. Y., and Feher, G. (2000). *Proc. Natl. Acad. Sci. USA* **97,** 1548–1553.

Page, C. C., Moser, C. C., Chen, X., and Dutton, P. L. (1999). *Nature* **402,** 47–52.
Parson, W. W., Chu, Z. T., and Warshel, A. (1990). *Biochim. Biophys. Acta* **1017,** 251–272.
Pomes, R., and Roux, B. (2002). *Biophys. J.* **82,** 2304–2316.
Prince, R. C., Leigh, J. S., and Dutton, P. L. (1976). *Biochim. Biophys. Acta* **440,** 622–636.
Prokhorenko, V. I., and Holzwarth, A. R. (2000). *J. Phys. Chem. B* **104,** 11563–11578.
Rappaport, F., and Lavergne, J. (2001). *Biochim. Biophys. Acta* **1503,** 246–259.
Ren, X., Tu, C., Laipis, P. J., and Silverman, D. N. (1995). *Biochemistry* **34,** 8492–8498.
Rutherford, A. W., and Krieger-Liszkay, A. (2001). *Trends Biochem. Sci.* **26,** 648–653.
Sass, H. J., Büldt, G., Gessenich, R., Hehn, D., Neff, D., Schlesinger, R., Berendzen, J., and Ormos, P. (2000). *Nature* **406,** 649–652.
Shinkarev, V. P., and Wraight, C. A. (1993). In: *The Photosynthetic Reaction Center* (J. Deisenhofer and J.R. Norris, Eds.), vol. 1, pp. 193–255. Academic Press, New York.
Shinkarev, V. P., Crofts, A. R., and Wraight, C. A. (2001). *Biochemistry* **40,** 12584–12590.
Shopes, R. J., Holten, D., Levine, L. M. A., and Wraight, C. A. (1987). *Photosynth. Res.* **12,** 165–180.
Simonson, T., and Perahia, D. (1995). *Proc. Natl. Acad. Sci. USA* **92,** 1082–1086.
Stowell, M. H. B., McPhillips, T. M., Rees, D. C., Soltis, S. M., Abresch, E., and Feher, G. (1997). *Science* **276,** 812–816.
Suarez, R. K., Staples, J. F., Lighton, J. R. B., and Mathieu-Costello, O. (2000). *J. Exp. Biol.* **203,** 905–911.
Sundstrom, V., Pullerits, T., and van Grondelle, R. (1999). *J. Phys. Chem. B* **103,** 2327–2346.
Szczeniak, M. M., and Scheiner, S. (1985). *J. Phys. Chem.* **89,** 1835–1840.
Takahashi, E., and Wraight, C. A. (1992). *Biochemistry* **31,** 855–866.
Takahashi, E., and Wraight, C. A. (1996). *Proc. Natl. Acad. Sci. USA* **93,** 2640–2645.
Tommos, C., and Babcock, G. T. (1998). *Acc. Chem. Res.* **31,** 18–25.
Tommos, C., and Babcock, G. T. (2000). *Biochim. Biophys. Acta* **1458,** 199–219.
Utschig, L. M., Ohigashi, Y., Thurnauer, M. C., and Tiede, D. M. (1998). *Biochemistry* **37,** 8278–8281.
van Brederode, M.E., and van Grondelle, R. (1999). *FEBS Lett.* **455,** 1–7.
Vass, I., and Styring, S. (1991). *Biochemistry* **30,** 830–839.
Vassiliev, I. R., Antonkine, M. L., and Golbeck, J. H. (2001). *Biochim. Biophys. Acta* **1507,** 139–160.
Walden, S. E., and Wheeler, R. A. (2001). *J. Phys. Chem. B* **106,** 3001–3006.
Woodbury, N. W. T., and Parson, W. W. (1984). *Biochim. Biophys. Acta* **767,** 345–361.
Woodbury, N. W., Parson, W. W., Gunner, M. R., Prince, R. C., and Dutton, P. L. (1986). *Biochim. Biophys. Acta* **851,** 6–22.
Woodbury, N. W., Peloquin, J. M., Alden, R. G., Lin, X., Lin, S., Taguchi, A. K. W., Williams, J. C., and Allen, J. P. (1994). *Biochemistry* **33,** 8101–8112.
Wraight, C. (1979). *Photochem. Photobiol.* **30,** 767–776.
Wraight, C. A. (1998). In: *Photosynthesis: Mechanisms and Effects* (G. Garab, Ed.), vol. 2, pp. 693–698. Kluwer Academic, Dordrecht.
Zachariae, U., and Lancaster, C. R. D. (2001). *Biochim. Biophys. Acta* **1505,** 280–290.
Zhang, Z. L., Huang, L. S., Shulmeister, V. M., Chi, Y. I., Kim, K. K., Hung, L. W., Crofts, A. R., Berry, E. A., and Kim, S. H. (1998). *Nature* **392,** 677–684.
Zouni, A., Witt, H.-T., Kern, J., Fromme, P., Krauss, N., Saenger, W., and Orth, P. (2001). *Nature* **409,** 739.

STRUCTURAL CLUES TO THE MECHANISM OF ION PUMPING IN BACTERIORHODOPSIN

By HARTMUT LUECKE* and JANOS K. LANYI†

*Departments of Biochemistry, Biophysics and Computer Science; †Department of Physiology and Biophysics, University of California, Irvine, California 92697

I. Introduction

Transmembrane ion pumps contain two functional components: the catalytic site for the driving reaction and the machinery for translocating the transported ion from one membrane surface to the other. The coupling of the first to the second during the transport cycle must be, in the various pumps, through displacements of protein subunits, main chains, or side chains and bound water, which will modulate the binding affinity for the transported ion and the continuity of the pathway for its movement across the membrane. Bacteriorhodopsin, a light driven ion pump in halobacteria, is a simpler system than others, because in this small seven-helical protein transport is driven not by a chemical reaction but by the free energy gained upon photoisomerization of the retinal to 13-*cis*,15-*anti*. In this protein therefore, the ion translocation induced by rearrangements in the protein can be studied without complications of the driving reaction.

Over the years, a very large amount of static and time-resolved spectroscopy of various kinds, as well as studies of site-specific mutants of bacteriorhodopsin, has generated kinetic models for the transport cycle ("photocycle") and identified the side chains of importance (Haupts *et al.*, 1999; Lanyi and Varo, 1995; Oesterhelt, 1998). The results had begun to identify the molecular events that underlie the interconversions of the spectroscopically distinct intermediate states termed J, K, L, M, N, and O. Together with low-resolution 3D maps of the protein and some of the intermediate states from cryoelectron microscopy of 2D crystals, these results suggested the beginnings of a mechanistic model

ADVANCES IN PROTEIN CHEMISTRY, Vol. 63

0065-3233/03 $35.00

(Herzfeld and Tounge, 2000; Lanyi, 1998; Luecke, 2000; Subramaniam and Henderson, 2000a). In these models the Schiff base region undergoes local changes in the K to L reaction in response to the retinal isomerization, and the assumption was that these changes cause direct proton transfer from the retinal Schiff base to Asp-85 in the L to M reaction. The latter in turn brings about the release of a proton to the extracellular surface. It was generally held that the directionality of the transport originates from the switch of the connectivity of the unprotonated Schiff base in what was described as the M_1 to M_2 reaction, from the extracellular to the cytoplasmic side, allowing Asp-96 to reprotonate the retinylidene nitrogen in the M_2 to N reaction. Concerted reprotonation of Asp-96 from the cytoplasmic surface and reisomerization of the retinal to all-*trans* occurs in the N to O transition, followed by recovery of the initial state as the O state decays in a strongly unidirectional reaction.

More recently, high-resolution crystal structures of the unilluminated state of the protein (Belrhali *et al.*, 1999; Luecke *et al.*, 1999a) and photointermediates (Edman *et al.*, 1999; Facciotti *et al.*, 2001; Luecke *et al.*, 1998, 1999b, 2000b; Royant *et al.*, 2000) from three-dimensional crystals grown in cubic lipid phase supplied a molecular rationale to this mechanism and identified the way the retinal interacts with the protein and causes its conformational changes. The structures revealed a highly polarized water molecule hydrogen bonded to the positively charged Schiff base and to a pair of negatively charged aspartates, Asp-85 and Asp-212, and the presence of a hydrogen-bonded network of polar residues and bound water in the extracellular region. Coupling of the protonation state of Asp-85 to the protonation state of the as yet unidentified group that releases a proton to the extracellular surface is through the shuttling of the Arg-82 side chain between two alternative positions, inward and outward. Initially, the cytoplasmic region lacks the means to conduct an ion, and the pK_a of the proton donor Asp-96 is high. Lowering this pK_a and building a hydrogen-bonded chain to the Schiff base is accomplished by repacking of side chains between helices F and G and the entry of water molecules, through the relaxation of the polyene retinal chain, which pushes the 13-methyl group against the indole of Trp-182 and distorts the side chain of Lys-216.

The new structural insights have also posed new problems. One of these concerns the first and critical deprotonation/protonation event that involves the Schiff base and Asp-85. What is the direction of the Schiff base N–H bond in L, i.e., after the light-induced isomerization of the retinal but before its deprotonation and the protonation of Asp-85? Does its orientation allow direct proton transfer from the Schiff base to Asp-85, or does it now point toward the cytoplasmic direction as expected from the unstrained configuration of 13-*cis*,15-*anti* retinal? An alternative to direct proton transfer would be the dissociation of the water molecule that bridges the Schiff base and Asp-85. In this case the transiently formed hydroxyl anion would be moving to the cytoplasmic side rather than a proton to the extracellular side (Luecke, 2000). Questions have also arisen about the conformational changes in the cytoplasmic region that first allow proton exchange between the Schiff base and Asp-96 without communication with the bulk, but then allow the reprotonation of Asp-96 from the cytoplasmic surface. Is there a single conformational change in the photocycle (Subramaniam and Henderson, 2000a), or are there several successive (Balashov, 2000; Dioumaev *et al.*, 2001) conformations? Are these transient conformations similar to the altered structures assumed by various mutants even without illumination?

This review will summarize and critically evaluate the crystallographic data for bacteriorhodopsin and its photointermediates. We will attempt to correlate the structural with the nonstructural data in order to explore the various and often contradictory mechanistic conclusions drawn.

Currently, the Protein Data Bank (PDB; Berman *et al.*, 2000) contains 33 atomic coordinate entries of bacteriorhodopsin structures, an increase of over 50% from 2 years earlier, when there were 20 entries (Luecke, 2000). This large number of models attests to the fact that bacteriorhodopsin is one of the most-studied and best-understood integral membrane proteins. Three of these coordinate entries are theoretical models (1BAC, 1BAD, 1I15) and three are NMR structures of fragments (1BCT, 1BHA, 1BHB), which will not be discussed further in this review. Of the remaining 27 entries (Table I), six are electron diffraction structures [1BRD, 2BRD (which is mislabeled as an X-ray diffraction entry), 1AT9, 2AT9, 1FBB, 1FBK], with the first two representing the pioneering work by Henderson and co-workers that produced the first atomic-level models of bacteriorhodopsin at 3.5 Å resolution (Grigorieff *et al.*, 1996; Henderson *et al.*, 1990). The remaining 21 entries are X-ray crystallographic structures of wild type and mutants in various states and to varying resolutions (almost double the 11 X-ray models 2 years ago!). Ten of these describe the ground (light-adapted or resting) state (1AP9, 1BM1, 1BRR, 1BRX, 1C3W, 1C8R, 1F50, 1KGB, 1QHJ, 1QM8). An additional eight models represent various cryotrapped intermediates: two for a low-temperature K intermediate (QKO/1QKP), one for an L intermediate (1E0P), and no fewer than five models describe the structures of various M intermediates (1C8S, late M from D96N mutant; 1F4Z, early M from E204Q mutant; 1CWQ, M from wild type; 1DZE M, M from wild type; 1KG8, early M from wild type). A set of two recent structures describes bacteriorhodopsin mutants that function as light-driven chloride pumps: the D85S single-site mutant (1JV7) and the D85S/F219L

TABLE I
Electron Diffraction and X-Ray Diffraction Models of Bacteriorhodopsin Currently Deposited in the Protein Data Bank[a,b]

PDB code	Description	Resolution (Å)	Method[f]
1BRD	First atomic bacteriorhodopsin structure	3.5	ED
2BRD	Refinement of 1BRD	3.5	ED
1AT9	Wild type	3.0	ED
2AT9	Refinement of 1AT9	3.0	ED
1FBB	Native, based on 2BRD	3.2	ED
1FBK	Triple mutant with cytoplasmically open conformation	3.2	ED
1AP9[c]	First CLP structure, merohedral twinning not recognized	2.35	X-ray, CLP
1BRX	Wild type, ground state, merohedral twinning	2.3	X-ray, CLP
1BM1	Wild type, ground state	3.5	X-ray, via spherical vesicles
1BRR	Wild type, ground state, trimer in the asymmetric unit	2.9	X-ray, detergent
1AP9	Further refinement of 1AP9,[c] taking twinning into account	2.35	X-ray, CLP
1QHJ	Wild type, ground state	1.9	X-ray, CLP
1C3W	Wild type, ground state, twinning	1.55	X-ray, CLP
1C8R	D96N mutant, ground state	1.8	X-ray, CLP
1C8S	*D96N mutant, late M intermediate with 100% occupancy*	*2.0*	*X-ray, CLP*
1QKO/1QKP	*Wild type, low temperature K intermediate with 35% occupancy*	*2.1*	*X-ray, CLP*
1CWQ	*Wild type, mixture of early & late M, and N intermediates with 35% occupancy, twinning*	*2.25*	*X-ray, CLP*
1QM8	Based on 1BM1, wild type, ground state	2.5	X-ray, via spherical vesicles
1DZE[d]	*Wild type, M intermediate with 100% occupancy*	*2.5*	*X-ray, via spherical vesicles*
1E0P[e]	*Wild type, L intermediate with 36% occupancy, twinning*	*2.1*	*X-ray, CLP*
1F50	E204Q mutant, ground state	1.7	X-ray, CLP
1F4Z	*E204Q mutant, early M intermediate with 100% occupancy*	*1.8*	*X-ray, CLP*

(*Continued*)

Luecke and Lanyi, Fig. 1. Comparison of two bacteriorhodopsin structures: the ground state of the wild type (1C3W) is shown in purple, overlaid with the single-site mutant D85S (1JV7), which was shown to pump chloride from the extracellular (bottom) to the cytoplasmic (top) side (Sasaki *et al.*, 1995), colored beige. Note the large relative helical displacements on the extracellular side, which result in a more open half-channel for the chloride-pumping mutant. The green arrow denotes an approximate overall pathway of pumped anions, leading past the retinal Schiff base in the center of the molecule.

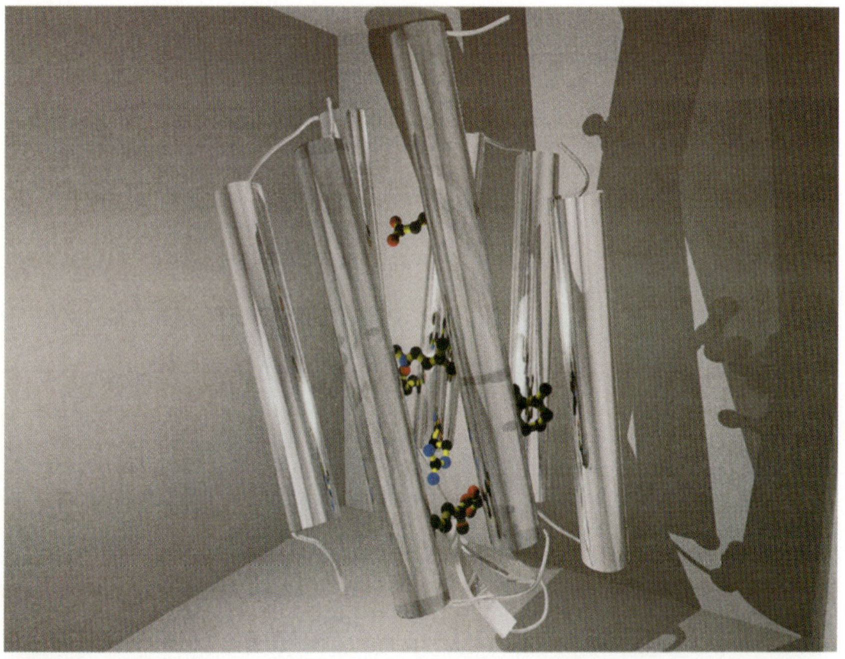

Luecke and Lanyi, FIG. 2. Overall view of bacteriorhodopsin, shown with the retinal and the residues directly implicated in ion transport. Standard orientation with cytoplasmic side on top. The residues rendered explicitly in atom-type colors are (from top to bottom) Asp-96, all-*trans* retinal with Asp-85, Arg-82, and Glu-194/Glu-204. The basic steps after the initial photon-driven all-*trans* to 13-*cis* retinal isomerization are (1) deprotonation of the Schiff base, protonation of nearby Asp-85, yielding the early M intermediate, coupled via downward movement of Arg-82 to (2) proton release from the proton release group near Glu-194/Glu-204 to the extracellular surface, yielding the late M intermediate; (3) deprotonation of Asp-96, reprotonation of the Schiff base, yielding the N intermediate; (4) reprotonation of Asp-96 from the cytoplasmic surface, thermal reisomerization of the retinal, yielding the O intermediate; and (5) deprotonation of Asp-85, reprotonation of the proton release group, regenerating the ground state.

TABLE I (*Continued*)

PDB code	Description	Resolution (Å)	Method[f]
1JV7	D85S mutant without halide, O-like state, head-to-tail dimers	2.25	X-ray, CLP
1JV6	D85S/F219L double mutant without halide, O-like state, head-to-tail dimers	2.0	X-ray, CLP
1KGB	Wild type, ground state	1.65	X-ray, CLP
1KG8	*Wild type, early M intermediate with 100% occupancy*	*2.0*	*X-ray, CLP*
1KME	Wild type, crystallized from bicelles, head-to-tail dimers	2.0	X-ray, from bicelles

[a] In order of publication in the respective category.
[b] Cryo-trapped intermediates are rendered in italics.
[c] As published (Pebay-Peyroula *et al.*, 1997).
[d] No publication since deposition in February 2000, when the title was "Sliding of G-helix in bacteriorhodopsin during proton transport."
[e] For which the original publication and the PDB file specify 70% L occupancy, an estimate that was later reduced to 36% by the same group, along with 12% K and 12% M contamination.
[f] ED, electron diffraction; CLP, cubic lipid phase.

double mutant (1JV6), both of which were determined in the absence of halides and show a significantly opened extracellular half channel with respect to the other BR structures (Fig. 1, see color insert; Facciotti *et al.*, 2001). Because these mutants contain an uncharged residue at position 85 as well as all-*trans* retinal they are thought to reveal a glimpse of the O intermediate of the photocycle. Finally, there is one additional entry being processed at PDB (1KME) with the title "Crystal structure of bacteriorhodopsin crystallized from bicelles" which describes a novel third approach to crystallizing membrane proteins (Faham and Bowie, 2002), distinct from vapor diffusion (Michel, 1991) and the cubic lipid phase method (Landau and Rosenbusch, 1996).

Even though this proliferation of structures has prompted some to call bacteriorhodopsin the "lysozyme of membrane proteins," BR does not yet present a real challenge to X-ray-derived lysozyme models in the database, which currently number 680!

II. THE GROUND, OR RESTING, STATE

The ground state of bacteriorhodopsin, also called the initial, light-adapted, BR, or resting state, refers to the protein with a relatively relaxed, protonated all-*trans* retinal and is characterized by an absorption maximum of 568 nm. Its 3D structure was first reported by the ground-breaking electron diffraction work on 2D crystals of Henderson and co-workers at 7 Å resolution (Henderson and Unwin, 1975). Subsequently, the same group improved the resolution to 3.5 Å, which resulted in the first atomic model of an ion pump (Grigorieff *et al.*, 1996; Henderson *et al.*, 1990). Meanwhile, in an effort to complement the findings of intense spectroscopic research aimed at elucidating the key functional groups and the kinetics of the photocycle, many groups had been working hard to obtain well-diffracting 3D crystals for X-ray diffraction experiments. A breakthrough came with the development of a novel method for the crystallization of membrane proteins, the cubic lipid phase (CLP) method (Landau and Rosenbusch, 1996).

Although the first bacteriorhodopsin structure reported using this new method suffered from severe merohedral twinning that was not accounted for (Pebay-Peyroula *et al.*, 1997), the recent explosion of structural knowledge for bacteriorhodopsin clearly has been driven by this new method (Table I). This method was also employed in the determination of the high-resolution structures of two related integral membrane proteins: halorhodopsin, a light-driven chloride pump (Kolbe *et al.*, 2000), and sensory rhodopsin II, a photoreceptor (Luecke *et al.*, 2001; Royant *et al.*, 2001a).

With merohedral twinning taken into account, crystals from CLP allowed the first localization of key waters, including highly polarized water 402 between the positively charged Schiff base and two negatively charged aspartic acids, Asp-85 and Asp-212 (Luecke *et al.*, 1998). More recently, crystals from CLP have yielded very high resolution structures that describe accurately the conformations of the retinal, the main chain and side chains, as well as the locations of all ordered internal waters (Belrhali *et al.*, 1999; Luecke *et al.*, 1999a).

In the extracellular half-channel an extensive three-dimensional hydrogen-bonded network of protein residues and seven water molecules leads from the buried, positively charged retinal Schiff base via water 402, a pair of negatively charged aspartic acids (Asp-85 and Asp-212), positively charged Arg-82, and finally a pair of glutamic acids (Glu-194 and Glu-204) to the extracellular surface. In the ground state the long, flexible guanidinium side chain of Arg-82 is in a position roughly midway between the two aspartates above it (Asp-85 and Asp-212) and the two glutamates below it (Glu-194 and Glu-204) (Fig. 2, see color insert; Luecke *et al.*, 1999a). The proton that is released to the extracellular surface after the Schiff base has deprotonated and Asp-85 has protonated in the L to M transition is thought to initially reside either on the pair of glutamic acids (Essen *et al.*, 1998) or on a network of waters in their vicinity, collectively called the proton release group (Spassov *et al.*, 2001).

Near Lys-216, to which the retinal is covalently attached, transmembrane helix G contains a π-bulge that causes a nonproline kink in the helix. This bulge causes the peptide plane between Ala-215 and Lys-216 to tilt away from the helix axis, and locally disrupts the α-helical hydrogen-bonding pattern. In the ground state this π-bulge is stabilized by hydrogen bonding of the main-chain carbonyl groups of Ala-215 and Lys-216 with two buried water molecules located in the otherwise very hydrophobic region between the Schiff base and Asp-96 in the cytoplasmic region. The water that hydrogen bonds to the C=O of Ala-215 in turn accepts a hydrogen bond from the indole nitrogen of Trp-182, a residue in van der Waals contact with the polyene chain of the retinal, in particular with the C-13 methyl group (Luecke *et al.*, 1999a).

Asp-96, a key residue in the middle of the relatively hydrophobic cytoplasmic half-channel about 11 Å from the protonated Schiff base, is protonated in the ground state. Its only polar interaction stems from a hydrogen bond to the hydroxyl of Thr-46. Otherwise it is surrounded by a hydrophobic barrel whose walls are formed by residues Ile-45, Leu-223, and Leu-224, and the lids by Phe-42, Leu-99, and Leu-100 on the cytoplasmic side, and by Val-49, Leu-93, and Phe-219 Schiff base side. In the M to N transition of the photocycle, reprotonation of the Schiff base from Asp-96 can occur only after this hydrophobic region between the Schiff base and Asp-96 has been populated by a chain of water molecules (Luecke *et al.*, 2000b).

Between the molecules 18 tails of ordered lipids were identified, constituting near-complete bilayers in the 3D crystals. During analysis a surprising amount of shape complementarity between the hydrophobic lipid tails and the membrane-embedded hydrophobic surface of bacteriorhodopsin became evident (Fig. 3, see color insert). This specificity of protein–lipid interactions contradicts older views that bilayer lipids simply provide a hydrophobic fluid compartment for integral membrane proteins. Each layer of the hexagonal 3D crystals has the same arrangement as the naturally occurring 2D crystals of bacteriorhodopsin, called "purple membrane," and the photocycle has been shown to proceed with similar kinetics in the 3D crystals (Heberle *et al.*, 1998). Additional evidence for the importance of specific lipids for protein–lipid contacts comes from the observation that highly purified bacteriorhodopsin preparations fail to crystallize in the cubic lipid phase (unpublished results). Our interpretation is that native lipids are stripped away during excessive purification, lipids that are essential in forming the highly ordered 2D layers in the CLP 3D crystals. Furthermore, a squalene molecule was identified in the middle of the bilayer near the π-bulge of helix G (Luecke *et al.*, 1999a), providing a structural angle to reports that squalene is present in purple membrane and that it affects the kinetics of the photocycle (Joshi *et al.*, 1998). A review by Hendler and Dracheva (2000) favors a location of the entirely hydrophobic squalene at the cytoplasmic surface near Asp-36 and Asp-38. Although the photocycle in the crystals is only slightly perturbed, the review mentions the possibility that detergent extraction and CLP crystallization in the presence of nonnative lipids (monoolein) might give rise to perturbations of the native protein–lipid arrangement.

Recent improvements in the details of the CLP crystallization procedure and crystal treatment have led to even better diffracting crystals. At 1.4 Å resolution, refinement, including anisotropic B factors, illustrates, the thermal motion anisotropy of the retinal and key neighbors (Fig. 4, see color insert; manuscript in preparation).

III. Early Photocycle Intermediates (K and L)

Within a few picoseconds after absorption of a photon, which deposits about 50 kcal/mol of energy into the retinal, and with a quantum yield of about 67% (Govindjee *et al.*, 1990), the all-*trans* retinal of the ground state is isomerized to a strained 13-*cis*,15-*anti* configuration, with the strain giving rise to large-amplitude hydrogen-out-of-plane vibrations (Braiman and Mathies, 1982; Rothschild *et al.*, 1986; Siebert and Mantele, 1983). The resulting K intermediate displays a slightly red-shifted absorption maximum ($\lambda_{max} = 590$ nm). The K intermediate has a reported ΔH of 11.6 kcal/mol (Birge *et al.*, 1991), thus about 20% of the photon energy is converted to enthalpy. At room temperature the pump cycle proceeds from the K intermediate without further energy input, i.e., thermally. In contrast, at or below 100 K motions of the protein and internal waters are severely restricted, arresting the protein in the K intermediate (cryotrapping) even after illumination has ceased. However, because of the large spectral overlap between the ground state and the K intermediate (λ_{max} difference 22 nm), it is not possible to achieve K intermediate occupancies higher than about 50% (Balashov *et al.*, 1991; Balashov and Ebrey, 2001). With cryotrapping there is also always the possibility of trapping unnatural intermediates, in other words, conformations that do not normally occur during the room-temperature photocycle (Balashov and Ebrey, 2001).

In 1999 Landau and co-workers published a set of 2.1 Å structures for a 35%-occupied, low-temperature K intermediate (Edman *et al.*,

1999), which, for unknown reasons, required continuous green light illumination during X-ray diffraction data collection. The main finding was that "a key water molecule [water 402] is dislocated, allowing the primary proton acceptor, Asp 85, to move." Other groups, including ours, have also attempted to obtain high-resolution structures of the K intermediate, but up to now decided that meaningful interpretation was too difficult because of the combination of limited resolution, twinning, and less than 50% occupancy, compounded by the nearly complete spatial overlap of all-*trans* and 13-*cis*,15-*anti* retinal (Fig. 5, see color insert). Furthermore, despite Raman and FTIR evidence of significant strain in the K intermediate (Braiman and Mathies, 1982; Maeda *et al.*, 1991; Rothschild *et al.*, 1986; Siebert and Mantele, 1983), the retinals in the published K models were forced into planar (i.e., nonstrained) all-*trans* and 13-*cis*,15-*anti* configurations as described in the Methods section of Edman *et al.* (1999):

> At 2.1 Å resolution, the two closely overlapping retinal configurations (all-*trans*,15-*anti* and 13-*cis*,15-*anti*) could not be refined free from bias resulting from stereochemical constraints. The retinal backbone and Schiff base linkage up to Cε of Lys 216 were constrained to be planar for both the all-*trans*,15-*anti* and the 13-*cis*,15-*anti* configurations.

The structure of the L intermediate is arguably the most interesting one. In the L state, the strain of the retinal has relaxed considerably, and the active site is now primed for the decisive event in the photocycle, the protonation of Asp-85, coupled to the deprotonation of the Schiff base (L to M reaction). The key question in this point is the orientation of the N–H bond (charge dipole) of the protonated Schiff base just before it gives up its proton. **Does the N–H point toward the hydrophobic cytoplasmic side, as a relaxed 13-*cis*,15-*anti* configuration would suggest?** And if this were the case, which entity would be the acceptor of the Schiff base proton? **Or is there still considerable strain in the retinal (particularly around the C_{13}=C_{14}, C_{14}–C_{15}; and C_{15}=NZ bonds), enough to keep the orientation of the N–H bond toward the extracellular side, in the same direction that it pointed to in the ground state?** More details regarding this issue can be found in Luecke (2000).

Only a high-resolution structural method that overcomes the difficulties of deconvoluting multiple, spatially highly overlapping structures (Fig. 4) is likely to be able to accurately answer this question. The spectral overlap problem is a principal one since even under ideal conditions it is not possible to achieve occupancies of the K or L intermediate higher than about 50% and 70%, respectively, because of the extensive spectral overlap between the ground state and both the K and the L intermediate (λ_{max} differences of +22 nm and −28 nm, respectively) (Balashov and Ebrey, 2001).

Landau and co-workers reported a merohedrally twinned 2.1 Å structure, claimed to be of a 70%-occupied, low-temperature L intermediate (Royant *et al.*, 2000). The claimed L-intermediate occupancy was later corrected to 36%, in a mixture with 40% ground state, 12% K and 12% M intermediate (Royant *et al.*, 2001b). The main interpretation of this structure was that

> a bend of this helix [helix C] enables the negatively charged primary proton acceptor, Asp 85, to approach closer to the positively charged primary proton donor, the Schiff base. The primary proton transfer event would then neutralize these two groups, canceling their electrostatic attraction and facilitating a relaxation of helix C to a less strained geometry.

In other words, an approach of Asp-85 toward the Schiff base was reported, presumably resulting in direct proton transfer of the Schiff base proton to Asp-85. Even with a decreased donor–acceptor distance, for this proton transfer to occur, the Schiff base N–H bond would have to point toward Asp-85, i.e., mostly in the extracellular direction. This in turn would require severe rotational distortions about one or more of the three bonds between C_{13} and NZ (as discussed above) since the retinal is in a 13-*cis*,15-*anti* configuration at this point. However, the retinal of this L structure (PDB entry 1E0P), like the retinals of the K structures, appears to be in a planar, relaxed 13-*cis*,15-*anti* configuration, with the Schiff base N–H bond pointing toward the hydrophobic cytoplasmic side, and thus maximally away from the acceptor, Asp-85. Unlike in the K paper, the L paper contains no information regarding retinal planarity restraints during refinement.

Furthermore, the distances measured for PDB entry 1E0P do not appear to support the claim that a bending of helix C causes an approach (i.e., decrease in distance) of Asp-85 toward the Schiff base. The distance from the NZ of the all-*trans* (i.e., ground state) retinal to the C_α of Asp-85 of the ground state is 6.37 Å, whereas the same pair in the L intermediate structure is 6.52 Å apart. This relative increase does not take into account the extra increase one would obtain if one were to measure the distance from Asp-85 to the Schiff base N–H hydrogen instead of the NZ nitrogen. Because of the reorientation of the N–H bond this would further increase the distance by up to 2 Å.

There remain concerns about the actual amount of trapped L in the studies, based in part on the published visible spectra of the crystals (Balashov and Ebrey, 2001). In particular, the contamination by the M intermediate (deprotonated Schiff base, λ_{max} near 410 nm, at least 12% of sample) would tend to dominate the changes in diffraction intensities due to the relatively larger structural changes of the M and later photocycle intermediates. In this case difference electron density maps

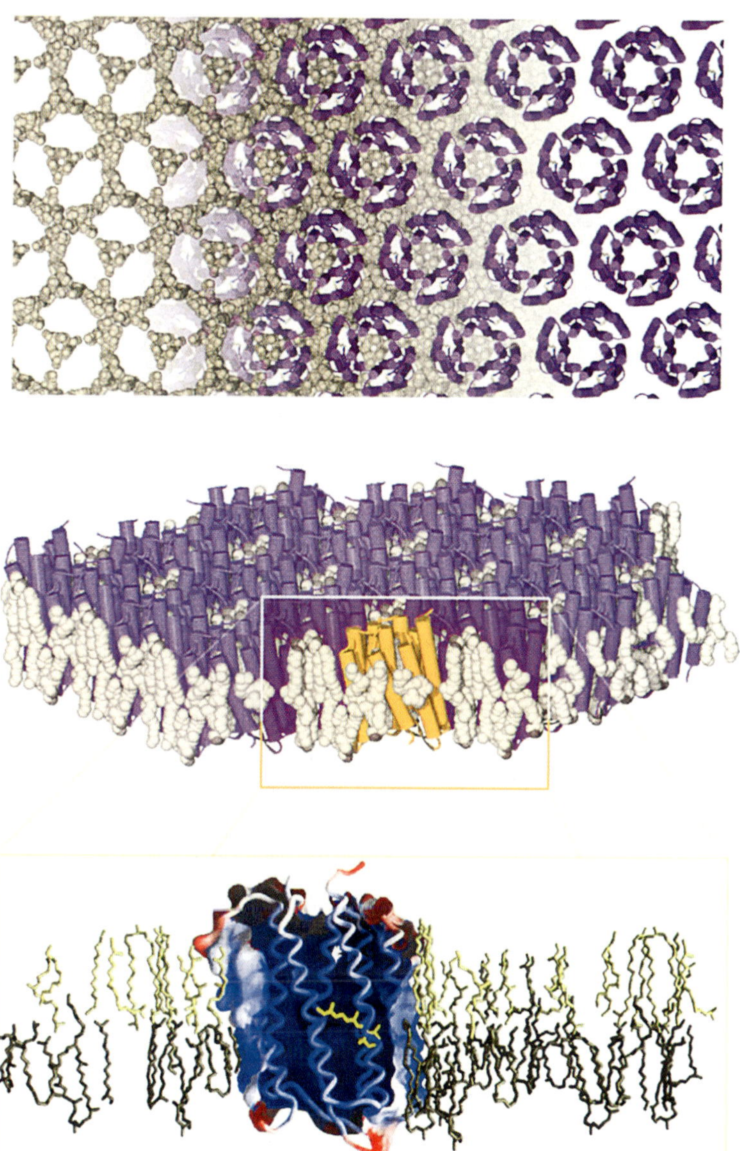

Luecke and Lanyi, FIG. 3. Arrangement of protein and lipids in each layer of the 3D crystals grown in the cubic lipid phase. The 3D crystals consist of stacked layers of the naturally occurring "purple membrane" 2D crystals, with each layer rotated 60° with respect to the neighboring ones. Complementarity between the protein trimers (purple) and the lipids (beige) is illustrated in the top panel. The lipids (in space-filling representation, on the left) and the bacteriorhodopsin trimers (on the right) are essential for formation of the complete crystal lattice (center). The center panel is tilted to show the edge of the bilayer with one bacteriorhodopsin molecule highlighted in yellow. The bottom panel provides a magnified view of one bacteriorhodopsin molecule embedded in the surrounding lipid bilayer. The surface colors of the protein molecule correspond to the isotropic temperature factors (blue denoting low thermal motion, red denoting high thermal motion).

A

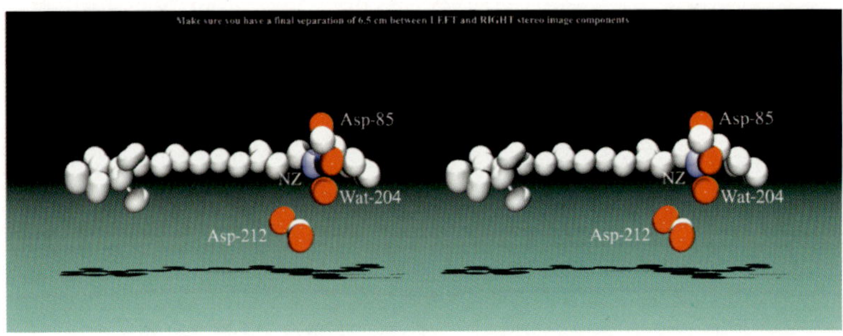

B

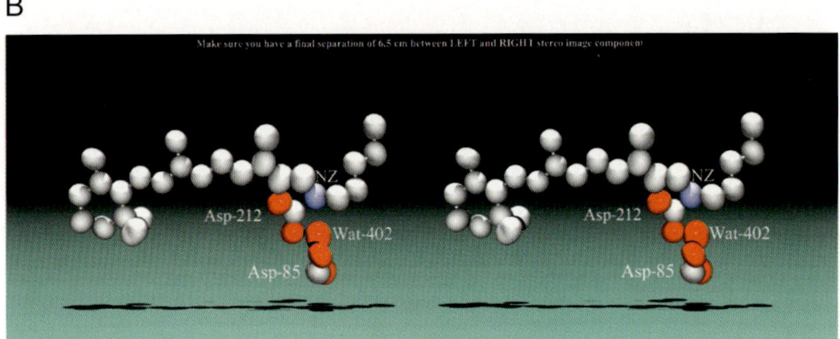

Luecke and Lanyi, FIG. 4. Thermal motion anisotropy of the retinal and key neighbors. Shown are thermal motion probability ellipsoids at the 20% level for the retinal, two aspartic acid side chains (Asp-85 and Asp-212), and water 402 in stereo. (A) Edge-on view, showing that the largest anisotropic motions are located in the β-ionone ring and its methyl groups. The largest displacements occur approximately parallel to the plane of the membrane bilayer. (B) The side view shows some anisotropy at retinal carbons C-14 and C-15 as well as water 402. This stereo figure was produced from anisotropic temperature factors after SHELX-L refinement (Sheldrick and Schneider, 1997) of a 1.38 Å data set (manuscript in preparation) using the programs ORTEP-III (Burnett and Johnson, 1996) and POV-Ray version 3.1g (http://www.povray.org/). The refinement employed 46,275 structure factors between 1.38 Å and 20 Å with a crystallographic R factor of 11.1% and an R_{free} of 16.5%. Using only isotropic B factors these values were 13.3% and 17.9%, respectively.

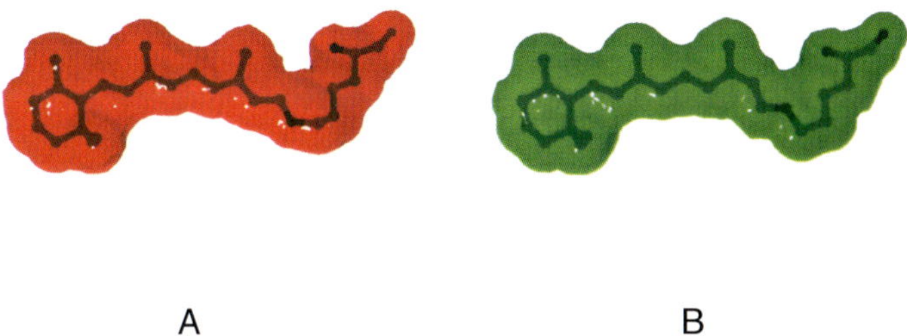

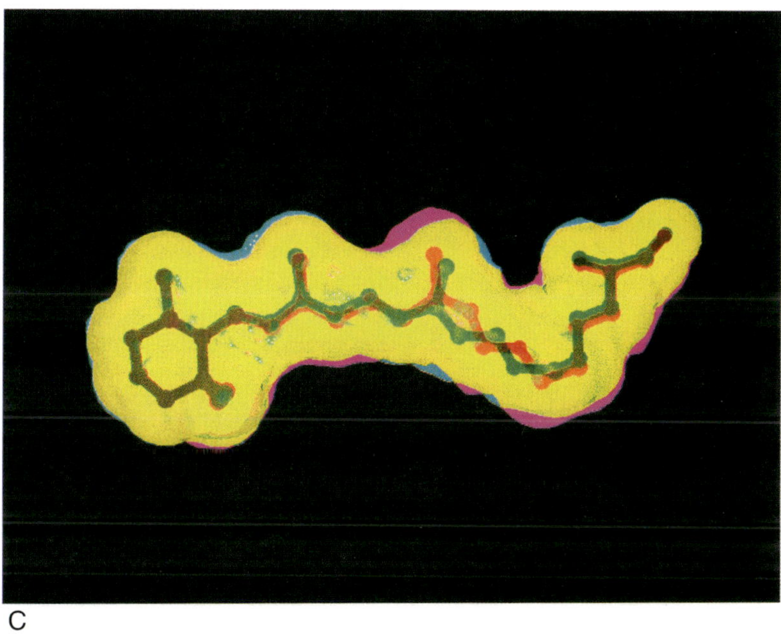

Luecke and Lanyi, FIG. 5. Illustration of the similarity in shape (and electron density) of the all-*trans* and 13-*cis*,15-*anti* configurations of retinal. Both retinal configurations were taken from the PDB file 1QKP (Edman *et al.*, 1999), which describes the 2.1 Å resolution structure of a low-temperature K intermediate of bacteriorhodopsin. The intermediate is reported at 35% occupancy, spatially overlapping with the remaining 65% constituting the ground or resting state. Panel (A) shows the molecular envelope of the light-isomerized 13-*cis*,15-*anti* K state and panel (B) shows the all-*trans* or ground state, both with Lys-216 on the right. Panel (C) shows the molecular envelopes of the two isomers overlaid. Yellow depicts regions where both envelopes coincide, blue depicts regions occupied only by the all-*trans* isomer, and magenta depicts regions occupied by only the 13-*cis*,15-*anti* isomer. This figure was produced with the programs Swiss-PdbViewer (Guex and Peitsch, 1997) and POV-Ray version 3.1g (http://www.povray.org/).

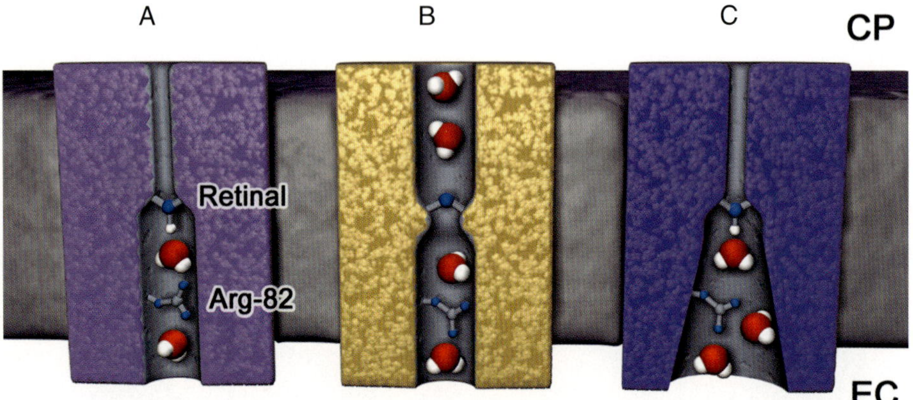

Luecke and Lanyi, Fig. 6. Schematic representation of the major conformations during the bacteriorhodopsin transport mechanism. The extracellular medium is at the bottom. Three states are shown to illustrate the most important changes in the extracellular and the cytoplasmic regions: (A) the ground or resting or initial state with an all-*trans* retinal, (B) an M photointermediate with a 13-*cis* retinal and a deprotonated Schiff base, and (C) an O-like state with an all-*trans* retinal. In M, on the extracellular side, downward movement of the side-chain of Arg-82 in response to rearrangements at the Schiff base causes the release of a proton at the EC surface. Reprotonation of the retinal Schiff base by Asp-96 is facilitated by a conformational change and the formation of a hydrogen-bonded chain of water molecules in the cytoplasmic region. To close the cycle, in the proton pumping scheme a proton has to be translocated from Asp-85 to the proton release group, or, in the anion-pumping scheme (Luecke, 2000), a chloride or hydroxide ion (equivalent to a water molecule coupled to a proton moving in the opposite direction) must move from the release group region toward the Schiff base. The latter mechanism would be expected to require much larger conformational changes (and thus be significantly slower than proton transport through a hydrogen-bonded network) to allow the anion/water to travel 14Å past Arg-82.

would reveal a convolution of features of the K, L, and M states. Furthermore, it is now clear that the determination of the partial occupancy of the L state by crystallographic refinement alone, which yielded the initial 70% value (Royant *et al.*, 2000), against a background of structurally highly overlapping states (BR, K, M) at medium resolution and in the presence of merohedral twinning led to an overestimation by a factor of at least 2, as the authors themselves have acknowledged (Royant *et al.*, 2001b).

IV. M Intermediates

The first photocycle intermediate to be determined exploited the fact that decay of the M intermediate is dramatically slowed down ($T_{1/2}$ of several seconds versus tens of microseconds) when the proton donor Asp-96 in the cytoplasmic half-channel is replaced with asparagine, a nonionizable analogue (D96N mutant; Luecke *et al.*, 1999b; Otto *et al.*, 1989; Tittor *et al.*, 1989). In this mutant the Schiff base eventually reprotonates from the distant cytoplasmic surface for lack of a nearby proton donor. Illumination of the D96N mutant crystals with red or yellow light at ambient temperature caused the initially deep purple (ground state) crystals to turn light yellow in less than 1 s. Once full conversion was obtained, the crystals were flash-cooled in the 100 K cryostream for data collection, yielding a fully occupied (i.e., full depletion of ground state absorption at 570 nm) cryotrapped M intermediate that persists even after the illumination is turned off (Luecke *et al.*, 1999b). In contrast, turning off the illumination at ambient temperature will cause the crystals to become deep purple again in a few seconds, indicating thermal completion of the photocycle to the ground state.

When working with any mutant, even in the case of a conservative single-residue replacement, there is always a possibility of changes due to the mutation, in particular changes to the local water structure. For that reason, a second single-site mutant with slow M decay was also crystallized and investigated, but this time the mutation was located on the extracellular side. The mutation of Glu-204, which is part of the proton release group, to a glutamine (E204Q mutant; Brown *et al.*, 1995; Luecke *et al.*, 2000b) results in markedly slower M decay in the 3D crystals, but, for reasons not understood, displays slow O decay in suspension. The M structure obtained from this mutant represents an M earlier in the photocycle than that obtained from the D96N mutant, presumably because the E204Q mutation uncouples the protonation of Asp-85 from proton release to the extracellular surface, a step that in the native photocycle occurs prior to the reprotonation of the Schiff base from Asp-96.

To exclude possible interference from mutations, various other groups used wild-type protein, yielding M intermediate occupancies ranging from 35% to 100%, depending on the illumination and cryotrapping protocol. Illumination at room temperature followed by rapid cooling resulted in a mixture of ground state (35%), early M (30%), late M (35%), and some N (not specified) (Sass *et al.,* 2000). A different protocol first cooled the wild-type crystals to 100 K, then increased the temperature to 230 K, followed by 3 min of illumination and subsequent cooling back to 100 K for data collection (Facciotti *et al.,* 2001). A third PDB entry deposited 2 years ago is as yet without an accompanying paper and cannot be evaluated for lack of information (PDB entry 1DZE, Kouyama *et al.*).

All five structures of M intermediates (Table I) emphasize the crucial role of discrete internal water molecules in the pump cycle. In all of these structures, water 402, which in the ground state is bridging the protonated Schiff base and the two nearby anionic aspartates, is no longer detectable. In several structures, additional waters not observed in the ground state were detected in the cytoplasmic half-channel, extending from the proton donor during the M to N transition, Asp-96, toward the deprotonated Schiff base. In addition, one of these studies describes a

> crystallographically unusual water molecule, $W740_M$, which is surrounded by mostly hydrophobic residues (Thr 170, Phe 42, Leu 100, Leu 223, Ile 229) without hydrogen-bonding partners in its vicinity

(Sass *et al.,* 2000). This water is located between Asp-96 and the cytoplasmic surface and the authors suggest

> that the contact to Asp 96 might be achieved in the millisecond time domain by a fluctuating water molecule and not by a continuous water channel.

Near the Schiff base, the protein begins to react to the kink in the retinal caused by photoisomerization about the C_{13}=C_{14} double bond. This allows progressive upward movement of the C_{13}-methyl group in the direction of the nearby indole moiety of Trp-182. Early M structures (1F4Z, 1KG8) show less displacement and a limited effect on the Trp-182 side chain, whereas late M structures (1C8S, 1CWQ) show a 1.5 Å movement of the C-13 methyl group with a concomitant displacement of the indole by over 1 Å, leading to the disruption of the water bridge (water 502) between the Trp-182 indole nitrogen and the C=O of Ala-215 of helix G (Luecke *et al.,* 1999b, 2000b).

The early part of the M phase is dominated by reactions in the extracellular region in response to the electrostatic changes at the Schiff base. The titration behavior of Asp-85 revealed that there is coupling between the protonation state of this aspartate and another protonatable group (Govindjee *et al.,* 1996). All five M structures show that the means of this coupling is the shuttling of the positive charge of Arg-82 from a ground state position roughly midway between the two pairs Asp-85/Asp-212 and Glu194/Glu204 toward Glu194/Glu204 in M, when Asp-85 is protonated. Protonation of Asp-85 greatly diminishes its electrostatic interaction with the positively charged guanidinium of Arg-82, which is now free to approach the region of Glu-194 and Glu-204, the two acidic residues implicated, directly or indirectly, in proton release. The approach of a positive charge is thought to stabilize the deprotonated from of the proton release group, leading to proton release at the extracellular surface. This type of long-range electrostatic coupling (in this case over a distance of 14 Å) could be a general mechanism in systems in which conformational or p*K* changes need to be propagated rapidly.

V. Large-Scale Conformational Changes in the M, N, and O Intermediates

As discussed above, crystallographic structures have been reported for the M state of the wild-type protein at ambient temperature and at 230 K, as well as in the D96N and E204Q mutants. Because the overlap between the spectra of the M state and the unilluminated state is small, and in all cases listed but the first the decay of M is slow, the occupancy in photostationary states is nearly 100%. This is a unique advantage and makes the structures of M the best defined of all the intermediates. The results confirm the earlier hints from projection maps (Dencher *et al.*, 1989; Han *et al.*, 1994; Nakasako *et al.*, 1991) that the largest changes involve helices F and G. The various M states studied are not equivalent. The M produced at 230 K might correspond approximately to the earliest M, the M_1 state that is in equilibrium with L. The E204Q mutation inhibits proton release to the extracellular surface during the rise of the M state, but the D96N mutation inhibits the reprotonation of the Schiff base and thus M decay. For this reason, the M intermediates in these mutants were assumed to correspond to an "early" (but later than the earliest M) and a "late" M state, respectively, in the wild-type photocycle. Indeed, there are progressive changes in the three structures. In the earliest M there is little change in the positions of any of the helices (Facciotti *et al.*, 2001). In the M presumed to follow it, the cytoplasmic end of helix F, from approximately residue 182, tilts outward by about 0.7 Å. In the late M the cytoplasmic end of helix F, from residue 177, is disordered (Luecke *et al.*, 1999b), presumably because the magnitude of the tilt is greater and locally disrupts crystal contacts. Helix G also undergoes progressive changes, locally at the π-bulge. The kink in helix G associated with the π-bulge at Lys-216 becomes less extreme as its connection with helix

F through the hydrogen bond of the C=O of Ala-215 to water 501 and from there to the indole N of Trp-182 is lost.

The causes of these changes are (1) the relaxation of the polyene chain of the photoisomerized 13-*cis*,15-*anti* retinal, constrained into an extended form up to this time in its binding site, that moves the 13-methyl group against the indole ring of Trp-182, and (2) the displacements of the Lys-216 side chain. These changes are more pronounced in the late M than in the earlier M. The coordinated movement of the chain of covalent and hydrogen bonds linking Trp-182 to the main chain of helix G and through water 502 to helix C and thus Asp-96, and the repacking of the side chains between helices F and G that ensue are related to the entry of two additional water molecules into the cytoplasmic region. In the M state of the E204Q mutant (where this region is not changed by the mutation), water 504 is now interposed between Asp-96 and Thr-46. Water 503 forms hydrogen bonds with water 502 and 504, as well with the C=O of Lys-216.

There is very little direct information about the crystallographic structures of the N and O states. Projection maps of M and N, in the wild type (for M), D96N (for M), or F171C (for N), exhibit the same kind of electron density changes at helices C, G, and F (Subramaniam and Henderson, 2000b). Low-resolution electron microscopy of tilted two-dimensional crystals of the F219L mutant with long-living N confirmed that there is a considerable outward tilt of the cytoplasmic end of helix F (Vonck, 2000). With occupancies greater than can be achieved for the N intermediate (about 30%) whose spectrum strongly overlaps the spectrum of the unilluminated state, the tilt should create steric conflicts between trimers. This is likely to be the case for the M state of the D96N mutant, where it may be the cause of the local disorder in the three-dimensional crystals at helix F.

A novel approach that attempts to overcome the problems of both low and high occupancy is to search for mutations that, for one reason or another, assume stable structures similar to the transient photocycle intermediates. Many residue replacements in the cytoplasmic region cause drastic slowing of the reisomerization of the retinal and/or protonation reactions in the second half of the photocycle. When three of these mutations were introduced together into the protein (D96G/F171C/F219L), helix F was seen to tilt outward as in the M and N states, but without illumination (Subramaniam and Henderson, 2000a,b). There was no additional structural change in the photocycle of this mutant. It appears that when some degree of stabilization conferred on it by the mutations, this M-like (or N-like) conformation represents a second energy minimum to the normal bacteriorhodopsin conformation.

Changing the ionization state of key residues in the protein to resemble their states in the photocycle is another approach to simulating structural changes in intermediates. Mutants in which Asp-85 is replaced by a nonionizable residue, such as asparagine or serine, exist as three species in pH dependent equilibrium (Turner *et al.*, 1993). Their absorption maxima resemble those of the M, N, and O states. The yellow, M-like state, at pH > 9, contains deprotonated retinal Schiff base, and the density changes at helices F and G characteristic of M are indeed evident in projection X-ray maps of unilluminated samples (Brown *et al.*, 1997; Vonck, 2000). The purple, N-like state, at pH between 7 and 9, contains protonated Schiff base but deprotonated Asp-96, and the retinal is 13-*cis*,15-*anti* even without illumination (Dioumaev *et al.*, 1998). It resembles the N state in its changed infrared amide bands. The blue, O-like state, at pH < 7, contains all-*trans* retinal and protonated Asp-96. The structure of this state, from the D85S mutant, has been solved to 2.25 Å resolution (Rouhani *et al.*, 2001). Its global conformation, induced evidently by missing the negative charge from the Schiff base region from replacement of Asp-85 with a neutral residue, is considerably different from both the unilluminated state and the M and N intermediates.

In the O-like state the extracellular ends of helices A, B, C, and D are tilted outward, but their cytoplasmic ends are not displaced. Helix E is tilted also, but around a pivot point near its middle, so its extracellular and cytoplasmic ends are displaced outward and inward, respectively. If this structure is indeed like that of the O state, the implication is that the protein undergoes a scissoring motion in the second half of the photocycle. It begins with a splaying of the cytoplasmic side of the seven helical bundle in M, which continues in N but reverses in O and opens the extracellular cavity instead. These suggested large-scale global motions are in sharp contrast with the relatively small (1–2 Å) and more local atomic displacements in the first half of the photocycle. The rationale must be that the structure of the protein in the unilluminated state predisposes it to the early reactions in the cycle, but the later reactions require drastically different conformations.

VI. Protonation Pathways in the M to N and the N to O Reactions

In the second half of the photocycle the Schiff base is reprotonated in a protonation equilibrium with Asp-96, and this is followed by reprotonation of Asp-96 from the cytoplasmic surface. Unlike in the extracellular region, where a three-dimensional hydrogen-bonded network already exists to facilitate the earlier events, involving transfer of protons (or hydroxyl ions), in the photocycle, the cytoplasmic region contains no such features. Lacking polar side chains or coordinated water molecules to conduct protons, this region will have to rely on the entry and specific binding of water to provide a pathway for ion transfers, as well as to lower the initially very high pK_a of Asp-96. How would large-scale conformational changes in N and O accomplish this?

At pH > 8 the kinetic relationship of M, N, and O is simpler than at lower pH and can be described with the scheme $M \leftrightarrow N \rightarrow O$. The protonation equilibrium of the Schiff base with Asp-96, the $M \leftrightarrow N$ reaction, is pH independent consistent with internal proton transfer, but the reprotonation of Asp-96, the $N \rightarrow O$ reaction, is pH dependent consistent with proton uptake from the bulk. Thus, one should expect a two-stage conformational shift, to enable ion transfer first between the Schiff base and Asp-96, and then between Asp-96 and the cytoplasmic surface.

The network of hydrogen-bonded water in M extends from Asp-96 toward the Schiff base but does not reach it (Luecke *et al.*, 2000b). It is likely that the rise of the N state depends on completing the proton transfer pathway by several more water additional molecules. The large influence of osmotic agents (Cao *et al.*, 1991) on the $M \leftrightarrow N$ equilibrium had suggested that this reaction, uniquely in the photocycle, involves increase in bound water. However, to date there is no crystallographic structure available for the N state that would reveal the positions of the putative additional water molecules and the groups that coordinate them.

The M to N and the N to O reactions both occur on the millisecond time scale, yet Asp-96 cannot be in communication with the cytoplasmic surface at the time its protonation equilibrium with the Schiff base is established. It is likely that this is ensured by the hydrophobic shield between Asp-96 and the aqueous interface that contains the side chains of Phe-42, Leu-100, Phe-171, and Leu-223. At the time Asp-96 is reprotonated from the bulk in the N to O reaction, this shield must be made permeable, while the pathway between Asp-96 and the Schiff base is abolished. The crystal structure of the O-like state suggests how these two goals are accomplished (Fig. 6, see color insert; Rouhani *et al.*, 2001).

Acknowledgments

We thank Jean-Philippe Cartailler for assistance in preparing Figs. 1, 2, and 3. We are also grateful to the beamline staff at the Advanced Light Source in Berkeley (ALS) and at beamline ID13 at the European Synchrotron Radiation Facility in Grenoble (ESRF). This work was supported, in part, by grants from the NIH to H.L. (R01-GM59970) and J.K.L. (R01-GM29498).

References

Balashov, S. P. (2000). Protonation reactions and their coupling in bacteriorhodopsin. *Biochim. Biophys. Acta* **1460,** 75–94.

Balashov, S. P., and Ebrey, T. G. (2001). Trapping and spectroscopic identification of the photointermediates of bacteriorhodopsin at low temperatures. *Photochem. Photobiol.* **73,** 453–462.

Balashov, S. P., Imasheva, E. S., Govindjee, R., and Ebrey, T. G. (1991). Quantum yield ratio of the forward and back light reactions of bacteriorhodopsin at low temperature and photosteady-state concentration of the bathoproduct K. *Photochem. Photobiol.* **54,** 955–961.

Belrhali, H., Nollert, P., Royant, A., Menzel, C., Rosenbusch, J. P., Landau, E. M., and Pebay-Peyroula, E. (1999). Protein, lipid and water organization in bacteriorhodopsin crystals: a molecular view of the purple membrane at 1.9 Å resolution. *Structure Fold. Des.* **7,** 909–917.

Berman, H. M., Westbrook, J., Feng, Z., Gilliland, G., Bhat, T. N., Weissig, H., Shindyalov, I. N., and Bourne, P. E. (2000). The Protein Data Bank. *Nucleic Acids Res.* **28,** 235–242.

Birge, R. R., Cooper, T. M., Lawrence, A. F., Masthay, M. B., Zhang, C. F., and Zidovetzki, R. (1991). Revised assignment of energy storage in the primary photochemical event in bacteriorhodopsin. *J. Am. Chem. Soc.* **113,** 4327–4328.

Braiman, M. S., and Mathies, R. A. (1982). Resonance Raman spectra of bacteriorhodopsin's primary photoproduct: evidence for a distorted 13-*cis* retinal chromophore. *Proc. Natl. Acad. Sci. USA* **79,** 403–407.

Brown, L., Sasaki, J., Kandori, H., Maeda, A., Needleman, R., and Lanyi, J. K. (1995). Glutamic acid 204 is the terminal proton release group at the extracellular surface of bacteriorhodopsin. *J. Biol. Chem.* **270,** 27122–27126.

Brown, L., Kamikubo, H., Zimanyi, L., Kataoka, M., Tokunaga, F., Verdegem, P., Lugtenburg, J., and Lanyi, J. K. (1997). A local electrostatic change is the cause of the large-scale protein conformation shift in bacteriorhodopsin. *Proc. Natl. Acad. Sci. USA* **94,** 5040–5044.

Burnett, M. N., and Johnson, C. K. (1996). *ORTEP-III: Oak Ridge Thermal Ellipsoid Plot Program for Crystal Structure Illustrations* (Rep. ORNL-6895). Oak Ridge National Laboratory, Oak Ridge, TN.

Cao, Y., Varo, G., Chang, M., Ni, B., Needleman, R., and Lanyi, J. K. (1991). Water is required for proton transfer from aspartate-96 to the bacteriorhodopsin Schiff base. *Biochemistry* **30,** 10972–10979.

Dencher, N. A., Dresselhaus, D., Zaccai, G., and Bueldt, G. (1989). Structural changes in bacteriorhodopsin during proton translocation revealed by neutron diffraction. *Proc. Natl. Acad. Sci. USA* **86,** 7876–7879.

Dioumaev, A. K., Brown, L. S., Needleman, R., and Lanyi, J. K. (1998). Partitioning of free energy gain between the photoisomerized retinal and the protein in bacteriorhodopsin. *Biochemistry* **37,** 9889–9893.

Dioumaev, A. K., Brown, L. S., Needleman, R., and Lanyi, J. K. (2001). Coupling of the reisomerization of the retinal, proton uptake, and reprotonation of Asp-96 in the N photointermediate of bacteriorhodopsin. *Biochemistry* **40,** 11308–11317.

Edman, K., Nollert, P., Royant, A., Beirhali, H., Pebay-Peyroula, E., Hajdu, J., Neutze, R., and Landau, E. M. (1999). High-resolution x-ray structure of an early intermediate in the bacteriorhodopsin photocycle. *Nature* **401,** 822–826.

Essen, L., Siegert, R., Lehmann, W. D., and Oesterhelt, D. (1998). Lipid patches in membrane protein oligomers: crystal structure of the bacteriorhodopsin–lipid complex. *Proc. Natl. Acad. Sci. USA* **95,** 11673–11678.

Facciotti, M. T., Rouhani, S., Burkard, F. T., Betancourt, F. M., Downing, K. H., Rose, R. B., McDermott, G., and Glaeser, R. M. (2001). Structure of an early intermediate in the M-state phase of the bacteriorhodopsin photocycle. *Biophys. J.* **81,** 3442–3455.

Faham, S., and Bowie, J. U. (2002). Bicelle crystallization: a new method for crystallizing membrane proteins yields a monomeric bacteriorhodopsin structure. *J. Mol. Biol.* **316,** 1–6.

Govindjee, R., Balashov, S. P., and Ebrey, T. G. (1990). Quantum efficiency of the photochemical cycle of bacteriorhodopsin. *Biophys. J.* **58,** 597–608.

Govindjee, R., Misra, S., Balashov, S. P., Ebrey, T. G., Crouch, R. K., and Menick, D. R. (1996). Arginine-82 regulates the pK_a of the group responsible for the light-driven proton release in bacteriorhodopsin. *Biophys. J.* **71,** 1011–1023.

Grigorieff, N., Ceska, T. A., Downing, K. H., Baldwin, J. M., and Henderson, R. (1996). Electron-crystallographic refinement of the structure of bacteriorhodopsin. *J. Mol. Biol.* **259,** 393–421.

Guex, N., and Peitsch, M. C. (1997). SWISS-MODEL and the Swiss-PdbViewer: an environment for comparative protein modeling. *Electrophoresis* **18,** 2714–2723.

Han, B. G., Vonck, J., and Glaeser, R. M. (1994). The bacteriorhodopsin photocycle: direct structural study of two substates of the M-intermediate. *Biophys. J.* **67,** 1179–1186.

Haupts, U., Tittor, J., and Oesterhelt, D. (1999). Closing in on bacteriorhodopsin: progress in understanding the molecule. *Annu. Rev. Biophys. Biomol. Struct.* **28,** 367–399.

Heberle, J., Buldt, G., Koglin, E., Rosenbusch, J. P., and Landau, E. M. (1998). Assessing the functionality of a membrane protein in a three-dimensional crystal. *J. Mol. Biol.* **281,** 587–592.

Henderson, R., and Unwin, P. N. (1975). Three-dimensional model of purple membrane obtained by electron microscopy. *Nature* **257,** 28–32.

Henderson, R., Baldwin, J. M., Ceska, T. A., Zemlin, F., Beckmann, E., and Downing, K. H. (1990). Model for the structure of bacteriorhodopsin based on high-resolution electron cryo-microscopy. *J. Mol. Biol.* **213,** 899–929.

Hendler, R. W., and Dracheva, S. (2001). Importance of lipids for bacteriorhodopsin structure, photocycle, and function. *Biochemistry* (Mosc.) **66,** 1311–1314.

Herzfeld, J., and Tounge, B. (2000). NMR probes of vectoriality in the proton-motive photocycle of bacteriorhodopsin: evidence for an "electrostatic steering" mechanism. *Biochim. Biophys. Acta* **1460,** 95–105.

Joshi, M. K., Dracheva, S., Mukhopadhyay, A. K., Bose, S., and Hendler, R. W. (1998). Importance of specific native lipids in controlling the photocycle of bacteriorhodopsin. *Biochemistry* **37,** 14463–14470.

Kolbe, M., Besir, H., Essen, L. O., and Oesterhelt, D. (2000). Structure of the light-driven chloride pump halorhodopsin at 1.8 Å resolution. *Science* **288,** 1390–1396.

Landau, E. M., and Rosenbusch, J. P. (1996). Lipidic cubic phases: a novel concept for the crystallization of membrane proteins. *Proc. Natl. Acad. Sci. USA* **93,** 14532–14535.

Lanyi, J. K. (1998). Understanding structure and function in the light-driven proton pump bacteriorhodopsin. *J. Struct. Biol.* **124,** 164–178.

Lanyi, J. K., and Varo, G. (1995). The photocycles of bacteriorhodopsin. *Isr. J. Chem.* **35,** 365–385.

Luecke, H. (2000). Atomic resolution structures of bacteriorhodopsin photocycle intermediates: the role of discrete water molecules in the function of this light-driven ion pump. *Biochim. Biophys. Acta* **1460,** 133–156.

Luecke, H., Richter, H. T., and Lanyi, J. K. (1998). Proton transfer pathways in bacteriorhodopsin at 2.3 angstrom resolution. *Science* **280,** 1934–1937.

Luecke, H., Schobert, B., Richter, H. T., Cartailler, J.-P., and Lanyi, J. K. (1999a). Structural changes in bacteriorhodopsin during ion transport at 2 angstrom resolution. *Science* **286,** 255–261.

Luecke, H., Schobert, B., Richter, H. T., Cartailler, J.-P., and Lanyi, J. K. (1999b). Structure of bacteriorhodopsin at 1.55 Å resolution. *J. Mol. Biol.* **291,** 899–911.

Luecke, H., Schobert, B., Richter, H. T., Cartailler, J.-P., Rosengarth, A., Needleman, R., and Lanyi, J. K. (2000). Coupling photoisomerization of the retinal in bacteriorhodopsin to directional transport. *J. Mol. Biol.* **300,** 1237–1255.

Luecke, H., Schobert, B., Lanyi, J. K., Spudich, E. N., and Spudich, J. L. (2001). Crystal structure of sensory rhodopsin II at 2.4 angstroms: insights into color tuning and transducer interaction. *Science* **293,** 1499–1503.

Maeda, A., Sasaki, J., Pfefferle, J. M., Shichida, Y., and Yoshizawa, T. (1991). Fourier transform infrared spectral studies on the Schiff base mode of all-*trans* bacteriorhodopsin and its photointermediates, K and L. *Photochem. Photobiol.* **54,** 911–921.

Michel, H. (1991). *Crystallization of Membrane Proteins.* CRC Press, Boca Raton, FL.

Nakasako, M., Kataoka, M., Amemiya, Y., and Tokunaga, F. (1991). Crystallographic characterization by x-ray diffraction of the M-intermediate from the photo-cycle of bacteriorhodopsin at room temperature. *FEBS Lett.* **292,** 73–75.

Oesterhelt, D. (1998). The structure and mechanism of the family of retinal proteins from halophilic archaea. *Curr. Opin. Struct. Biol.* **8,** 489–500.

Otto, H., Marti, T., Holz, M., Mogi, T., Lindau, M., Khorana, H. G., and Heyn, M. P. (1989). Aspartic acid–96 is the internal proton donor in the reprotonation of the Schiff base of bacteriorhodopsin. *Proc. Natl. Acad. Sci. USA* **86,** 9228–9232.

Pebay-Peyroula, E., Rummel, G., Rosenbusch, J. P., and Landau, E. M. (1997). X-ray structure of bacteriorhodopsin at 2.5 angstroms from microcrystals grown in lipidic cubic phases. *Science* **277,** 1676–1681.

Rothschild, K. J., Roepe, P., Ahl, P. L., Earnest, T. N., Bogomolni, R. A., Das Gupta, S. K., Mulliken, C. M., and Herzfeld, J. (1986). Evidence for a tyrosine protonation change during the primary phototransition of bacteriorhodopsin at low temperature. *Proc. Natl. Acad. Sci. USA* **83,** 347–351.

Rouhani, S., Cartailler, J.-P., Facciotti, M., Walian, P., Needleman, R., Lanyi, J., Glaeser, R. M., and Luecke, H. (2001). Crystal structure of the D85S mutant of bacteriorhodopsin: a model of an O-like photocycle intermediate. *J. Mol. Biol.* **313,** 615–628.

Royant, A., Edman, K., Ursby, T., Pebay-Peyroula, E., Landau, E. M., and Neutze, R. (2000). Helix deformation is coupled to vectorial proton transport in the photocycle of bacteriorhodopsin. *Nature* **406,** 645–648.

Royant, A., Edman, K., Ursby, T., Pebay-Peyroula, E., Landau, E. M., and Neutze, R. (2001a). Spectroscopic characterization of bacteriorhodopsin's L-intermediate in 3D crystals cooled to 170 K. *Photochem. Photobiol.* **74,** 794–804.

Royant, A., Nollert, P., Edman, K., Neutze, R., Landau, E. M., Pebay-Peyroula, E., and Navarro, J. (2001b). X-ray structure of sensory rhodopsin II at 2.1-Å resolution. *Proc. Natl. Acad. Sci. USA* **98,** 10131–10136.

Sasaki, J., Brown, L. S., Chon, Y. S., Kandori, H., Maeda, A., Needleman, R., and Lanyi, J. K. (1995). Conversion of bacteriorhodopsin into a chloride ion pump. *Science* **269,** 73–75.

Sass, H. J., Buldt, G., Gessenich, R., Hehn, D., Neff, D., Schlesinger, R., Berendzen, J., and Ormos, P. (2000). Structural alterations for proton translocation in the M state of wild-type bacteriorhodopsin. *Nature* **406,** 649–653.

Sheldrick, G. M., and Schneider, T. (1997). *Methods Enzymol.* **277,** 319.

Siebert, F., and Mantele, W. (1983). Investigation of the primary photochemistry of bacteriorhodopsin by low-temperature Fourier-transform infrared spectroscopy. *Eur. J. Biochem.* **130,** 565–573.

Spassov, V. Z., Luecke, H., Gerwert, K., and Bashford, D. (2001). pK(a) Calculations suggest storage of an excess proton in a hydrogen-bonded water network in bacteriorhodopsin. *J. Mol. Biol.* **312,** 203–219.

Subramaniam, S., and Henderson, R. (2000a). Crystallographic analysis of protein conformational changes in the bacteriorhodopsin photocycle. *Biochim. Biophys. Acta* **1460,** 157–165.

Subramaniam, S., and Henderson, R. (2000b). Molecular mechanism of vectorial proton translocation by bacteriorhodopsin. *Nature* **406,** 653–657.

Takeda, K., Matsui, Y., Sato, H., Hino, T., Kanamori, E., Okumura, T., Yamane, T., Iizuka, T., Kamiya, N., Adachi, S., and Kouyama, T. Sliding of G-helix in bacteriorhodopsin during proton transport. To be published.

Tittor, J., Soell, C., Oesterhelt, D., Butt, H. J., and Bamberg, E. (1989). A defective proton pump, point-mutated bacteriorhodopsin Asp96 → Asn is fully reactivated by azide. *EMBO J.* **8,** 3477–3482.

Turner, G. J., Miercke, L. J. W., Thorgeirsson, T. E., Kliger, D. S., Betlach, M. C., and Stroud, R. M. (1993). Bacteriorhodopsin D85N: three spectroscopic species in equilibrium. *Biochemistry* **32,** 1332–1337.

Vonck, J. (2000). Structure of the bacteriorhodopsin mutant F219L N intermediate revealed by electron crystallography. *EMBO J.* **19,** 2152–2160.

THE STRUCTURE OF *Wolinella succinogenes* QUINOL:FUMARATE REDUCTASE AND ITS RELEVANCE TO THE SUPERFAMILY OF SUCCINATE:QUINONE OXIDOREDUCTASES

By C. ROY D. LANCASTER

Department of Molecular Membrane Biology, Max Planck Institute of Biophysics, D-60528 Frankfurt am Main, Germany

I. INTRODUCTION

Succinate:quinone oxidoreductases (EC 1.3.5.1; Hägerhäll, 1997; Lancaster, 2002a,b) are enzymes that couple the two-electron oxidation of succinate to fumarate (reaction 1) to the two-electron reduction of quinone to quinol (reaction 2).

$$\text{Succinate} \rightleftharpoons \text{fumarate} + 2\text{H}^+ + 2\text{e}^- \qquad (1)$$

$$\text{Quinone} + 2\text{H}^+ + 2\text{e}^- \rightleftharpoons \text{quinol} \qquad (2)$$

They can also catalyze the opposite reaction, the coupling of quinol oxidation to quinone to the reduction of fumarate to succinate (Lemma *et al.*, 1991). The *cis*-configuration isomer of fumarate, maleinate, is neither produced in the oxidation reaction nor consumed as a substrate in the reduction reaction, i.e, the reaction is stereospecific in both directions. Depending on the direction of the reaction catalyzed *in vivo*, the members of the superfamily of succinate:quinone oxidoreductases

ADVANCES IN PROTEIN CHEMISTRY, Vol. 63

0065-3233/03 $35.00

can be classified as either succinate:quinone reductases (SQR) or quinol:fumarate reductases (QFR). SQR and QFR can be degraded to form succinate dehydrogenase and fumarate reductase (both EC 1.3.99.1), which no longer react with quinone and quinol, respectively.

SQR and QFR complexes are anchored in the cytoplasmic membranes of archaebacteria and eubacteria and in the inner mitochondrial membrane of eukaryotes with the hydrophilic domain extending into the cytoplasm and the mitochondrial matrix, respectively.

SQR (respiratory complex II) is involved in aerobic metabolism as part of the citric acid cycle and of the aerobic respiratory chain (Saraste, 1999). QFR participates in anaerobic respiration with fumarate as the terminal electron acceptor (Kröger, 1978; Kröger *et al.*, 2002) and is part of the electron transport chain catalyzing the oxidation of various donor substrates (e.g., H_2 or formate) by fumarate. These reactions are coupled via an electrochemical proton potential (Δp) to ADP phosphorylation with inorganic phosphate by ATP synthase (Mitchell, 1979).

Succinate:quinone oxidoreductases generally contain four protein subunits referred to as A, B, C, and D. Subunits A and B are hydrophilic, whereas the subunits C and D are integral membrane proteins. Among species, subunits A and B have high sequence homology, while that for the hydrophobic subunits is much lower. Most of the SQR enzymes of gram-positive bacteria and the QFR enzymes from ε-proteobacteria contain only one larger hydrophobic polypeptide (C), which is thought to have evolved from a fusion of the genes for the two smaller subunits C and D (Hägerhäll and Hederstedt, 1996; Hederstedt, 1999; Lancaster *et al.*, 1999). While subunit A harbors the site of fumarate reduction and succinate oxidation, the hydophobic subunit(s) contain the site of quinol oxidation and quinone reduction.

Based on their hydrophobic domain and heme content (Hägerhäll and Hederstedt, 1996; Hederstedt, 1999), succinate:quinone oxidoreductases can be classified in five types (cf. Fig. 1; Lancaster and Kröger, 2000; Lancaster, 2001a). Type A enzymes contain two hydrophobic subunits and two heme groups, e.g., SQR from the archaea *Archaeoglobus fulgidus, Natronomonas pharaonis,* and *Thermoplasma acidophilum.* Type B enzymes contain one hydrophobic subunit and two heme groups, as is the case for SQR from the gram-positive bacteria *Bacillus subtilis* and *Paenibacillus macerans* and QFR from the ε-proteobacteria *Campylobacter jejuni, Helicobacter pylori,* and *Wolinella succinogenes.* Examples for type C enzymes, which posses two hydrophobic subunits and one heme group, are SQR from mammalian mitochondria and from the proteobacteria *Paracoccus denitrificans* and *Escherichia coli* and QFR from the nematode *Ascaris suum.* The QFR of *E. coli* is an example of a type D enzyme,

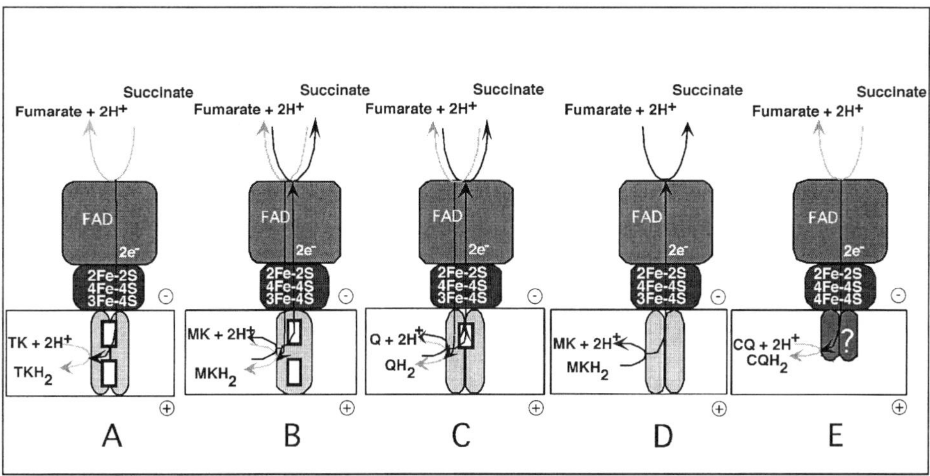

FIG. 1. Classification (A to E) of succinate:quinone oxidoreductases (Lancaster and Kröger, 2000; Lancaster 2001a) based on their hydrophobic domain and heme content (Hägerhäll and Hederstedt, 1996; Hederstedt, 1999). The hydrophilic subunits A and B are drawn schematically in gray and dark gray, respectively, the hydrophobic subunits C and D in light gray or gray. Heme groups are symbolized by small rectangles. The directions of the reactions catalyzed by SQR and QFR are indicated by gray and black arrows, respectively. White rectangles symbolize the respective cytoplasmic or inner mitochondrial membrane bilayer. The positive (+) and negative (−) sides of the membrane are indicated. In bacteria, the negative side is the cytoplasm ("inside"), the positive side the periplasm ("outside"). For mitochondrial systems, these are the mitochondrial matrix and the intermembrane space, respectively. The type of quinone transformed *in vivo* is not necessarily unique for each type of enzyme. The examples given are thermoplasma-quinone (TK), menaquinone (MK), ubiquinone (Q), and caldariella quinone (CQ; Collins and Jones, 1981; Hägerhäll, 1997; Lancaster, 2001b). See text for further details. Modified from Lancaster (2001). *FEBS Lett.* **504,** 133–141, with permission from Elsevier Science.

which contains two hydrophobic subunits and no heme group. Finally, type E enzymes, such as SQRs from the archaea *Acidianus ambivalens* and *Sulfolobus acidocaldarius,* but also from the proteobacterium *C. jejuni* and the cyanobacterium *Synechocystis,* also contain no heme, but have two hydrophobic subunits very different from the other four types and more similar to those of heterodisulfide reductase from methanogenic archaea (Schäfer *et al.,* 1999). The phylogenetic analyses presented recently (Lemos *et al.,* 2002; Schäfer *et al.,* 2002) corroborate the above classification scheme.

Generally, succinate:quinone oxidoreductases contain three iron–sulfur clusters, which are exclusively bound by the B subunit. Enzyme types A-D contain one $[2Fe\text{-}2S]^{2+,1+}$, one $[4Fe\text{-}4S]^{2+,1+}$, and one $[3Fe\text{-}4S]^{1+,0}$ cluster, whereas an additional [4Fe-4S] cluster apparently replaces the [3Fe-4S] in the type E enzyme (Gomes *et al.*, 1999). The A subunit of all described membrane-bound succinate:quinone oxidoreductase complexes contains a covalently bound FAD prosthetic group (Singer and McIntire, 1984). The chemical structure of the linkage as 8α-[Nε-histidyl]-FAD was first established for mammalian SQR (Walker and Singer, 1970) and subsequently for the QFR enzymes of *W. succinogenes* (Kenny and Kröger, 1977) and *E. coli* (Weiner and Dickie, 1979).

II. Overall Description of the Structure

The currently available crystal structures of succinate:quinone oxidoreductases are those of two prokaryotic quinol:fumarate reductases, both since 1999. The *E. coli* QFR, determined at 3.3 Å (Iverson *et al.*, 1999), belongs to the type D enzymes, and the QFR of *W. succinogenes*, refined at 2.2 Å resolution (Lancaster *et al.*, 1999), is of type B. Three structures of the latter enzyme, based on three different crystal forms, are available. The first two, PDB entries 1QLA and 1QLB (Lancaster *et al.*, 1999), are considerably better defined and more accurate than the structure of the third crystal form, PDB entry 1E7P (Lancaster *et al.*, 2000, 2001). Therefore, the first two crystal forms of *W. succinogenes* QFR will be used for the description of structural features, and that of the third crystal form will be referred to for comparison.

In all three *W. succinogenes* QFR crystal forms, two heterotrimeric complexes of A, B, and C subunits are associated in an identical fashion, thus forming a dimer (Fig. 2, see color insert). *Wolinella succinogenes* QFR has an overall length of 120 Å in the direction perpendicular to the membrane. Parallel to the membrane, the maximum width is 130 Å for the dimer, and 70 Å for the monomer. Approximately 3665 Å^2 (8%) of the *W. succinogenes* QFR monomer surface is buried upon dimer formation. As derived from analytical gel filtration experiments, this dimer is apparently also present in the detergent-solubilized state of the enzyme (Lancaster and Kröger, 2000; Unden *et al.*, 1980), implying that it is unlikely to be an artifact of crystallization.

III. The Hydrophilic Subunits

A. *Subunit A, the Flavoprotein, and Interdomain Movement at the Site of Fumarate Reduction*

Wolinella succinogenes QFR subunit A, of 73 kDa (Lauterbach *et al.*, 1990), is composed of four domains (Fig. 3a, see color insert), the bipartite FAD binding domain (blue, residues A1–260 and A366–436, with "A"

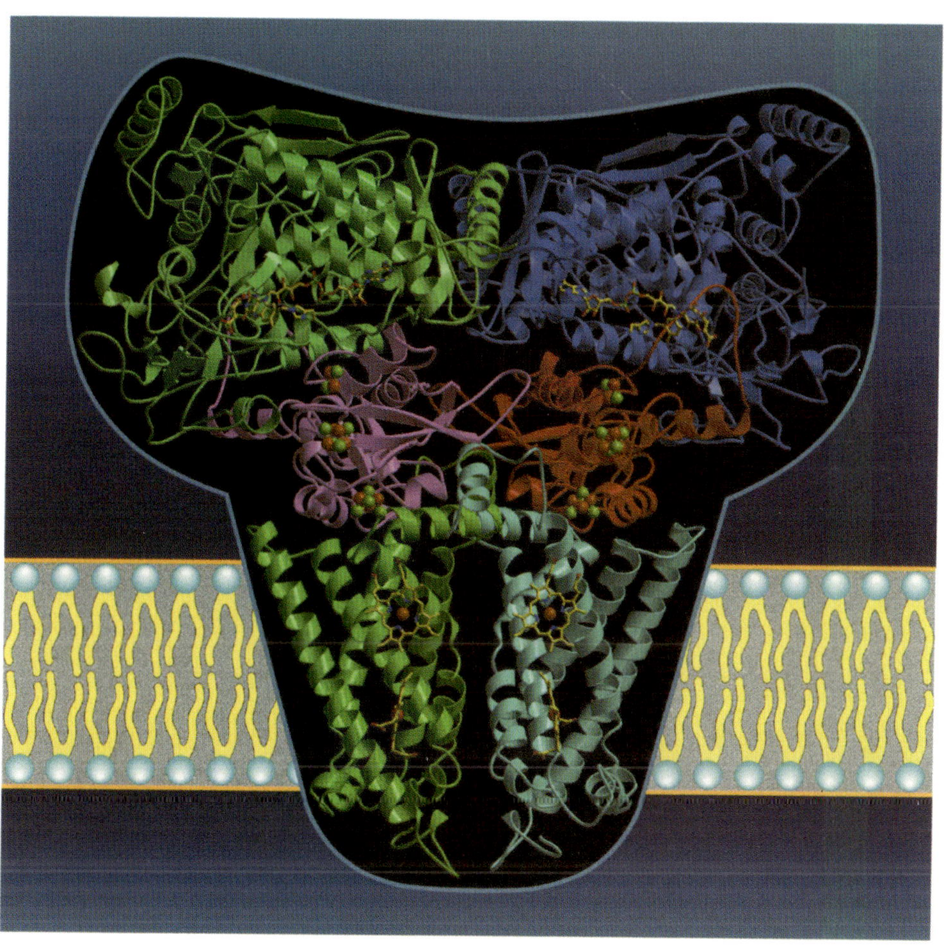

Lancaster, FIG. 2. Three-dimensional structure of the *W. succinogenes* QFR dimer of heterotrimeric complexes of A, B, and C subunits. The Cα traces of the two A subunits are shown in blue and blue-green, those of the two B subunits in red and purple, and those of the two C subunits in green and light blue. The atomic structures of the six prosthetic groups per heterotrimer are superimposed for better visibility. From top to bottom, these are the covalently bound FAD, the [2Fe-2S], the [4Fe-4S], and the [3Fe-4S] iron-sulfur clusters, and the proximal and the distal heme *b* groups. Atomic color coding is as follows: C, N, O, P, S, and Fe are displayed in yellow, blue, red, light green, green, and orange, respectively. The figure is drawn from the PDB coordinate set 1QLA (Lancaster *et al.*, 1999). The position of bound fumarate close to the isoalloxazine ring of FAD is taken from the coordinate set 1QLB (Lancaster *et al.*, 1999). Figures with atomic models were prepared with a version of Molscript (Kraulis, 1991) modified for color ramping (Esnouf, 1997) and rendered with the program Raster3D (Merrit and Bacon, 1997). Modified from Lancaster (2001), *FEBS Lett.* **504**, 133–141, with permission from Elsevier Science.

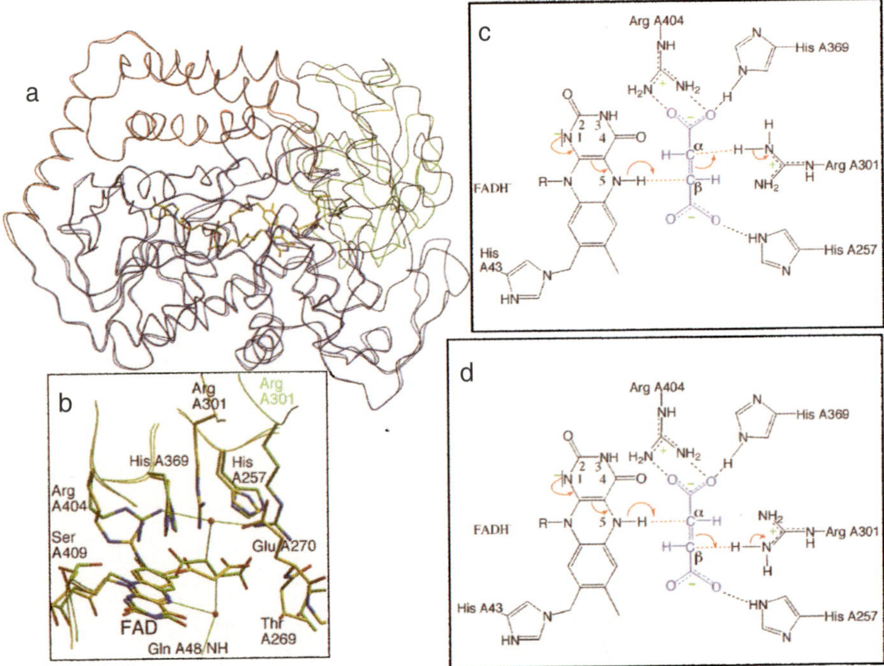

Lancaster, FIG. 3. *Wolinella succinogenes* QFR subunit A and the possible mechanism of fumarate reduction. Figures modified from Lancaster *et al.* (2001) and Lancaster (2001a). (A) The first three domains of subunit A and the different relative orientations of the capping domain in crystal forms B [PDB entry 1QLB (Lancaster *et al.*, 1999); black Cα trace] and C [PDB entry 1E7P (Lancaster *et al.*, 2001); colored Cα trace]. In the latter case, the FAD-binding domain, the capping domain, and the helical domain are drawn in blue, green, and red, respectively. In the center, the histidyl-FAD, the bound fumarate of crystal form "B," and the location of residue Arg A301 in both structures are indicated as stick models. Carbon, nitrogen, oxygen, and phosphorus atoms are shown in yellow, blue, red, and light green, respectively. The QFR crystal form "B" Arg A301 carbon atoms are drawn in green. For a better view of these structural differences, short animations with the structures of the two crystal forms as starting and ending structures are available at http://www.mpibp-frankfurt.mpg.de/lancaster/febs2001/fig2a.swf and fig2b.swf, respectively. (B) Comparison of crystal forms "C" (PDB entry 1E7P, carbon atoms in yellow, complex with malonate) and "B" (PDB entry 1QLB, carbon atoms in green, complex with fumarate) at the site of fumarate reduction in subunit A. The isolated red spheres correspond to the oxygen atoms of two water molecules in PDB entry 1QLB. The dicarboxylate binding site in the form C crystal for which a diffraction data set could be obtained contained the smaller competitive inhibitor malonate rather than fumarate, but this structural difference is negligible compared to the other large structural differences shown here. (C) and (D) Alternate possible mechanisms of fumarate reduction in *W. succinogenes* QFR involving the residues shown in panel (B) for the crystal form C. Since the precise location of the bound fumarate molecule in this crystal form is not yet known, it could be either the β-methenyl group (C) or the α-methenyl group (D), which is reduced by hydride transfer from the N-5 position of $FADH^-$. This is coupled to proton transfer to the respective other methenyl group from the side chain of Arg A301. Modified from Lancaster (2001), *FEBS Lett.* 504, 133–141, with permission from Elsevier Science.

indicating the A subunit), into which the capping domain (green, A260–366) is inserted, the helical domain (red, A436–554), and the C-terminal domain (A554–656, not shown in Fig. 3a.). The FAD is covalently bound as 8α-[Nε-histidyl]-FAD (Kenny and Kröger, 1977) to the residue His A43. The capping domain contributes to burying the otherwise solvent-exposed FAD isoalloxazine ring from the protein surface.

A *W. succinogenes* QFR crystal grown in the presence of fumarate was found to be of crystal form "B." The structure was refined at 2.33 Å resolution (PDB entry 1QLB; Lancaster *et al.,* 1999). This allowed the localization of the fumarate binding site between the FAD binding domain and the capping domain next to the plane of the FAD isoalloxazine ring (Fig. 3b). The structure of the enzyme in the third crystal form, "C" (Lancaster *et al.,* 2000), was refined at 3.1 Å resolution (PDB entry 1E7P; Lancaster *et al.,* 2001). Compared with the previous crystal forms, the altered crystal packing (Lancaster, 2002c,d) results in the capping domain being in a different arrangement relative to the FAD-binding domain (Fig. 3a). This leads to interdomain closure at the fumarate reducing site, suggesting that the structure encountered in this crystal form represents a closer approximation to the catalytically competent state of the enzyme (Fig. 3b). The *trans* hydrogenation of fumarate to succinate could occur by the combination of the transfer of a hydride ion and of a proton from opposite sides of the fumarate molecule. One of the fumarate methenyl carbon atoms could be reduced by direct hydride transfer from the N-5 position of the reduced FADH$^-$, while the other fumarate methenyl carbon is protonated by the side chain of Arg A301 (Figs. 3c and 3d). The latter residue replaces the water molecule previously suggested to be the proton donor (Lancaster *et al.,* 1999) based on the structure in crystal form B (Fig. 3b). The assignment as to which of the fumarate methenyl carbon atoms accepts the hydride and which the proton is currently ambiguous (Fig. 3c versus 3d), because data of sufficient completeness and quality for this crystal form have so far only been obtained for the complex with malonate and not yet in the presence of fumarate. Release of the product could be facilitated by movement of the capping domain away from the dicarboxylate site (Lancaster *et al.,* 1999, 2001). All residues implicated in substrate binding and catalysis are conserved throughout the superfamily of succinate:quinone oxidoreductases, so that this reversible mechanism is considered generally relevant for all succinate:quinone oxidoreductases.

This mechanistic interpretation of the structure is supported by the results from site-directed mutagenesis, where Arg A301 was replaced relatively conservatively by a Lys (Lancaster *et al.,* 2001). Strain FrdA-R301K contained a variant enzyme, very similar to the wild-type enzyme in terms of cofactor and subunit composition, in particular a fluorescence typical for FAD covalently bound to the A subunit, but which lacked succinate dehydrogenase and fumarate reductase activity (Lancaster *et al.,* 2001). The loss in enzymatic activity is tentatively attributed to the fact that Lys ($pK_{sol} = 10.8$) cannot substitute for Arg ($pK_{sol} = 12.5$) in protonating the fumarate methenyl carbon, possibly because the protonating group is no longer close enough to protonate the fumarate methenyl group.

B. Subunit B, the Iron–Sulfur Protein

The Cα trace of *W. succinogenes* subunit B is shown in Fig. 4. This subunit of 27 kDa (Lauterbach *et al.,* 1990) consists of two domains, an N-terminal "plant ferredoxin" domain (B1-106) binding the [2Fe-2S] iron–sulfur cluster and a C-terminal "bacterial ferredoxin" domain (B106-239) binding the [4Fe-4S] and the [3Fe-4S] iron–sulfur clusters. The [2Fe-2S] iron–sulfur cluster is coordinated by the Cys residues B57, B62, B65, and B77 as proposed on the basis of sequence alignments (Lauterbach *et al.,* 1990). All four Cys residues are within segments that are in contact with the A subunit. The [4Fe-4S] iron–sulfur cluster is ligated to the protein through Cys residues B151, B154, B157, and B218,

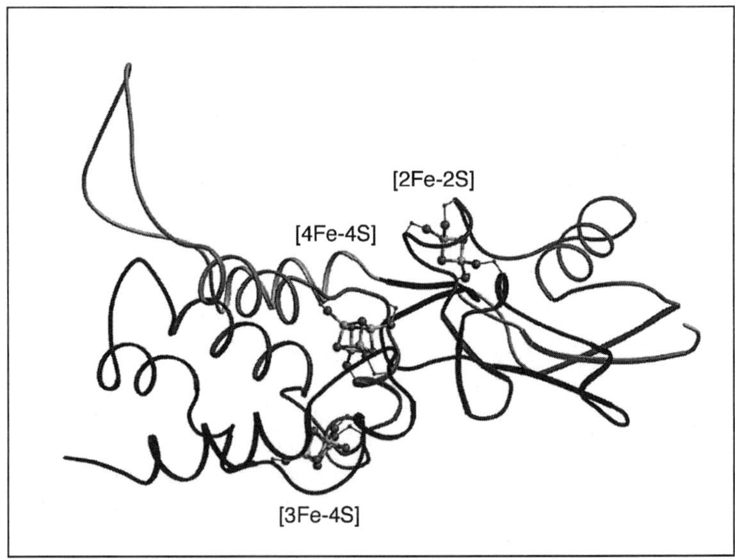

FIG. 4. Subunit B, the iron–sulfur protein of *W. succinogenes* QFR. The Cα trace is color-coded from light gray to dark gray for the amino-terminal [2Fe-2S] domain on the right (residues B1–106) and again from light gray to dark gray for the carboxy-terminal [7Fe-8S] domain on the left (B106–B239). Modified from Lancaster *et al.* (1999). *Nature* **402,** 377–385, with permission from Nature Publishing Group.

and the [3Fe-4S] cluster is coordinated by Cys residues B161, B208, and B214. The latter three residues are within segments that are in contact with the C subunit. At the position corresponding to the fourth Cys of the [4Fe-4S] cluster, the *W. succinogenes* QFR [3Fe-4S] cluster contains a Leu. Whereas the introduction of a Cys into *E. coli* QFR (Mandori *et al.*, 1992) could replace the native [3Fe-4S] by a [4Fe-4S] cluster, this was not the case for *B. subtilis* SQR (Hägerhäll *et al.*, 1995).

IV. Subunit C, the Integral Membrane Diheme Cytochrome *b*

The Cα trace of *W. succinogenes* subunit C is shown in Fig. 5a. This subunit of 30 kDa (Körtner *et al.*, 1990) contains five membrane-spanning segments with preferentially helical secondary structure. For systematic reasons within the superfamily of succinate:quinone oxidoreductases, these segments are labeled [according to Hägerhäll and Hederstedt (1996)] I (C22–52), II (C77–100), IV (C121–149), V (C169–194), and VI (C202–237). To a varying degree, all five transmembrane segments are tilted with respect to the membrane normal, and helix IV is strongly kinked at position C137 (Lancaster *et al.*, 1999). This kink is stabilized by the side-chain γ-hydroxyl of Ser C141, which, instead of its backbone NH, donates a hydrogen bond to the carbonyl oxygen of Phe C137 (Fig. 5b). As pointed out earlier (Lancaster and Kröger, 2000), this feature is very similar to that found for helix F of bacteriorhodopsin [bR, PDB entry 1C3W (Luecke *et al.*, 1999), Figs. 5c and 5d].

The planes of both heme molecules bound by the *W. succinogenes* enzyme are approximately perpendicular to the membrane surface and their interplanar angle is 95° (Lancaster *et al.*, 1999). The axial ligands to the "proximal" heme b_P are His C93 of transmembrane segment II and His C182 of transmembrane segment V (Fig. 6a). This causes heme b_P to be located toward the cytoplasmic surface of the membrane, and thus toward the [3Fe-4S] iron–sulfur cluster. Hydrogen bonds and salt bridges with the propionate groups of heme b_P are formed with the side chains of residues Gln C30, Ser C31, Trp C126, and Lys C193 (Fig. 6a). Thus, side chains from the residues of the first four transmembrane segments are involved in the binding of heme b_P (Lancaster *et al.*, 1999), which underscores the structural importance of the bound heme (Simon *et al.*, 1998). The axial ligands to the "distal" heme b_D are His C44 of transmembrane segment I and His C143 of transmembrane segment IV (Fig. 6b), demonstrating that all four heme axial ligands had been correctly predicted by sequence alignment (Körtner *et al.*, 1990) and site-directed mutagenesis (Simon *et al.*, 1998). Residues of *W. succinogenes* QFR subunit C conserved among the succinate:quinone oxidoreductases from

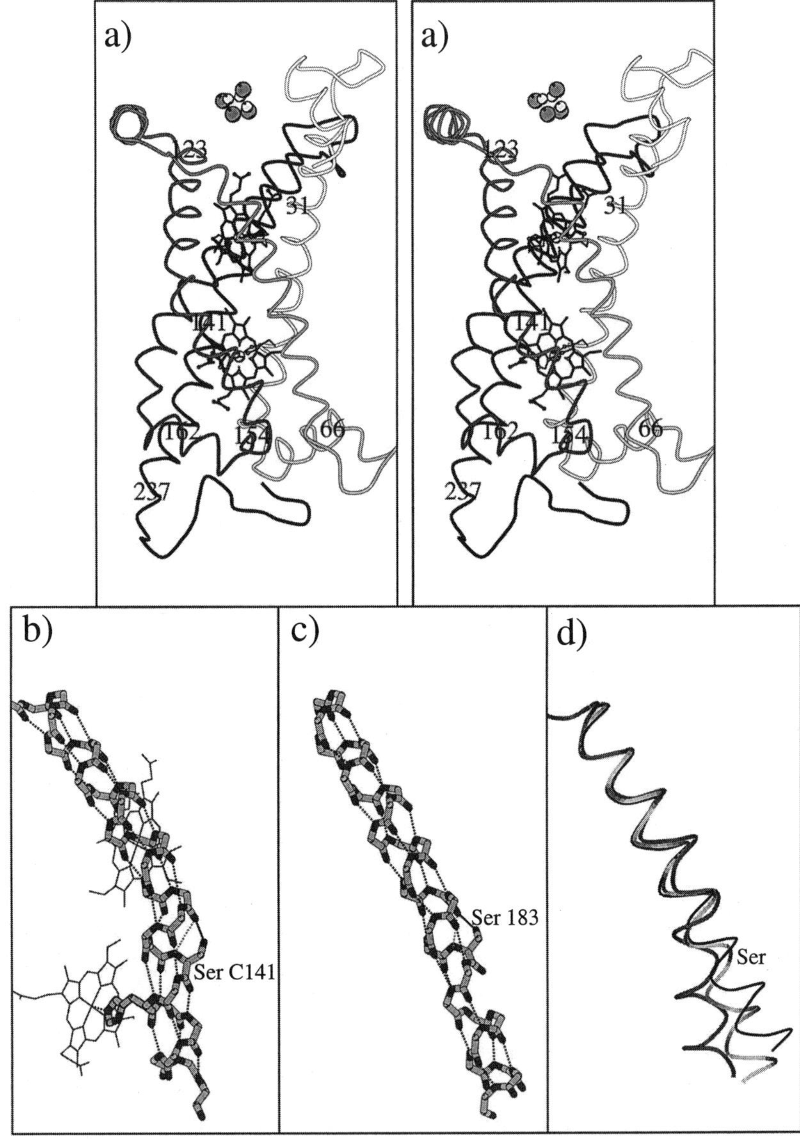

FIG. 5. *Wolinella succinogenes* QFR subunit C. (a) Subunit C, the diheme cytochrome *b* of *W. succinogenes* QFR (stereo views). Also shown are the proximal (upper) and distal (lower) heme groups and the position of the [3Fe-4S] cluster (top), which is bound by the B subunit (Fig. 4). Selected C subunit residues are labeled by their residue number. The Cα trace is drawn from light gray (amino terminus) to dark gray (carboxy terminus). (Modified from Lancaster and Simon, 2002, *Biochim. Biophys. Acta Bioenerget.* **1553,** 84–101, with

ε-proteobacteria (Lancaster and Simon, 2002) are concentrated around the heme groups and the contact surface with subunit B (Fig. 6a). However, a distal "rim" of conserved residues is also apparent involving residues Glu C66, Ile C154, Ser C159, and Arg C162 (Fig. 6b). While the latter two residues interact with a propionate of heme b_D, Glu C66 and Ile C154 are likely to line the oxidation site of the menaquinol substrate, as discussed below.

V. General Comparison of Membrane-Integral Diheme Cytochrome *b* Proteins

As noted earlier (Lancaster *et al.*, 1999), the binding of the two heme *b* molecules by an integral membrane protein four-helix bundle described here is very different from that described for the four-helix bundle of the cytochrome bc_1 complex (Xia *et al.*, 1997). In the latter complex, only two transmembrane segments provide two axial heme *b* ligands each. This we refer to as a "two-helix motif" (Lancaster, 2002e; Fig. 7a). Examples for a "three-helix motif," where one transmembrane helix provides two heme *b* ligands, and two others provide one heme *b* ligand each (Fig. 7b), may be found in the cases of membrane-bound hydrogenases (Berks *et al.*, 1995; Gross *et al.*, 1998), and formate dehydrogenases (Berks *et al.*, 1995; Jormakka *et al.*, 2002). As described above, the axial ligands for heme binding in *W. succinogenes* QFR are located on four different transmembrane segments ("four-helix motif"; Fig. 7c). One consequence of this difference is that the distance between the two heme iron centers is distinctly shorter in *W. succinogenes* QFR (15.6 Å) than it is in the mitochondrial cytochrome bc_1 complex (21 Å; Xia *et al.*, 1997) and in *E. coli* formate dehydrogenase-N (20.5 Å; Jormakka *et al.*, 2002).

---

permission from Elsevier Science.) (b)–(d) Comparison of (b) transmembrane helix IV from *W. succinogenes* QFR (PDB entry 1QLA; Lancaster *et. al.*, 1999) and (c) transmembrane helix F from *Halobacterium salinarum* bacteriorhodopsin (bR) (PDB entry 1C3W; Luecke *et al.*, 1999). In both panels hydrogen bonding interactions are indicated by black dashed lines. Highlighted hydrogen bonds donated from the Ser side chain are indicated by black solid lines. The corresponding interaction donated by the backbone NH for a standard α-helix is indicated as a gray dashed line. (b) The distance between Ser C141 N and Phe C137 O (indicated in gray dashes) is 4.0 Å; the distance between Ser C141 Oγ and Phe C137 O (indicated in black) is 2.8 Å. (c) The distance between Ser 183 N and Val 179 O (gray) is 3.0 Å; the distance between Ser 183 Oγ and Val 179 O (black) is 2.6 Å. (d) Superposition of QFR transmembrane helix IV (dark gray, from panel b), bR transmembrane helix F (light gray, from panel c), and an idealized α-helix (black). Modified from Lancaster and Kröger (2000). *Biochim. Biophys. Acta Bioenerget.* **1459,** 422–431, with permission from Elsevier Science.

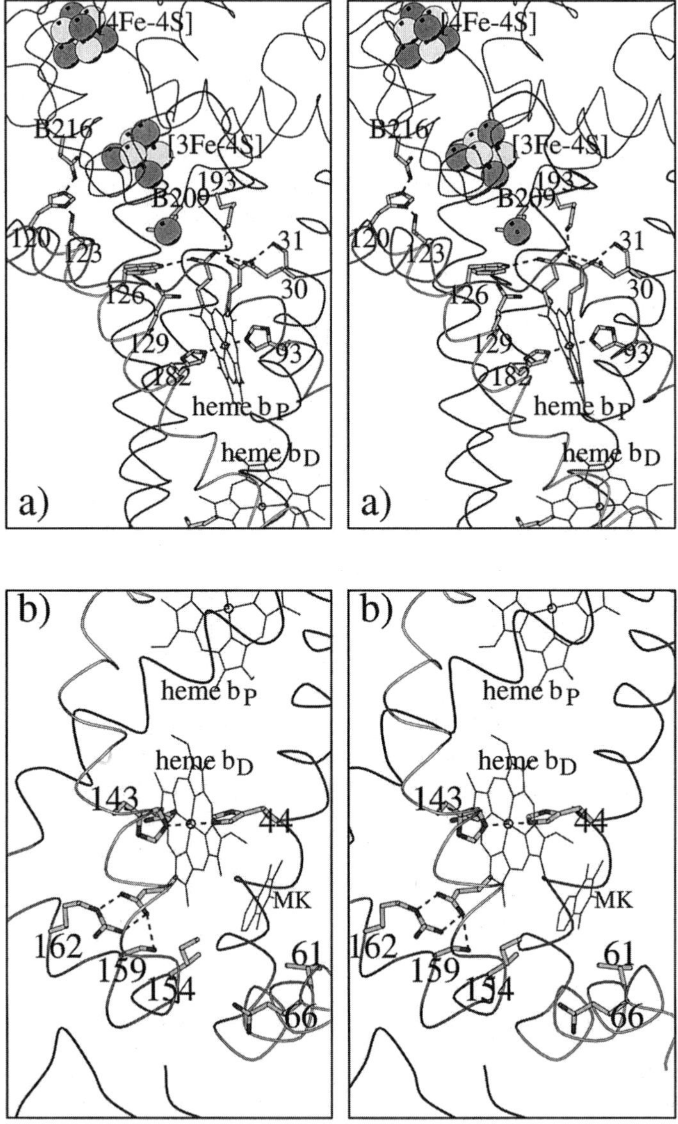

FIG. 6. Proximal (a) and distal (b) residues in the structure of *W. succinogenes* QFR that are conserved in the QFR enzymes from other ε-proteobacteria (stereo views). The Cα trace of subunit C is also shown. (a) In addition of the Cα trace of subunit C, that of subunit B is shown at the top. Also indicated are two conserved B subunit residues (B209 and B216). All other labeling refers to selected proximal C subunit residues. The following prosthetic groups are included from the top left to the bottom right: the [4Fe-4S] cluster, the [3Fe-4S] cluster, the proximal heme, and the distal heme. (b) Selected distal C subunit residues are labeled. The proximal (upper) and distal (lower) heme are included as is a tentative menaquinol binding position. Transmembrane helices II and VI have been omitted for clarity. Modified from Lancaster and Simon (2002). *Biochim. Biophys. Acta Bioenerget.* **1553,** 84–101, with permission from Elsevier Science.

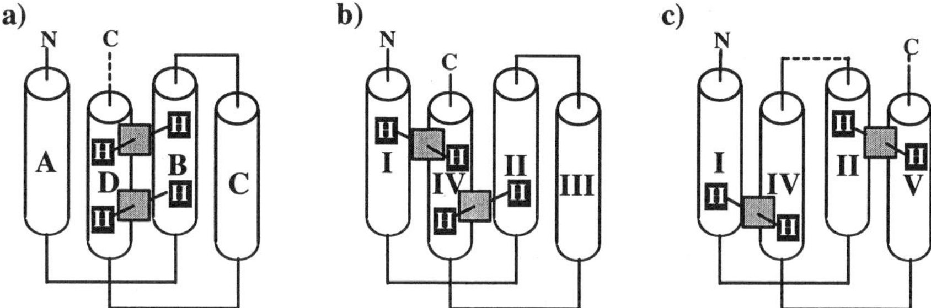

FIG. 7. Diheme binding by integral membrane protein four helix bundles (Lancaster, 2002e). (a) "Two-helix motif": His ligands from two transmembrane helices (mitochondrial cytochrome bc_1 complex). (b) "Three-helix motif": His ligands from three transmembrane helices (e.g., hydrogenase, formate dehydrogenase). (c) "Four-helix motif": His ligands from four transmembrane helices (dihemeic succinate:quinone oxidoreductases). Modified from Lancaster (2002). *Biochim. Biophys. Acta Biomembr.* **1565,** 215–231, with permission from Elsevier Science.

VI. RELATIVE ORIENTATION OF SOLUBLE AND MEMBRANE-EMBEDDED QFR SUBUNITS

The structure of *E. coli* QFR can be superimposed on the structure of *W. succinogenes* QFR based on the hydrophilic subunits A and B (Figs. 3c and 6b in Lancaster *et al.,* 1999). This similarity in structure was expected based on sequence comparisons. However, in this superimposition, the membrane embedded subunits cannot be aligned. In an alternate superimposition, the transmembrane subunits C and D of the *E. coli* enzyme can be overlayed on to the *W. succinogenes* C subunit (Figs. 3d and 6c in Lancaster *et al.,* 1999). Compared to the former superimposition, the latter involves a rotation around the membrane normal of approximately 180° and an orthogonal 25° rotation. This immediately leads to two important conclusions (Lancaster *et al.,* 1999). First, the structures of the transmembrane subunits carrying no hemes and two hemes, respectively, can be aligned to a significant degree, although only 11 of the aligned residues are identical. Second, the relative orientation of the soluble subunits and the transmembrane subunits is different in the QFR complexes from the two species.

VII. THE SITE OF MENAQUINOL OXIDATION/MENAQUINONE REDUCTION

The site of menaquinol oxidation on the diheme cytochrome *b* subunit of *W. succinogenes* QFR is not known. No convincing density for a quinol or quinone could be found in any of the three crystal forms of the oxidized enzyme. No specific inhibitor of menaquinol oxidation by *W. succinogenes* QFR has been identified. The *E. coli* QFR coordinate set 1FUM [Iverson *et al.*, 1999; now replaced by coordinate set 1L0V (Iverson *et al.*, 2002)] contains models for two menaquinone molecules per ABCD monomer. Although some of the atomic temperature factors of the quinone ring atoms were larger than 100 Å^2, indicating that these quinone models may not be well defined, these models were included in Figs. 3d and 6c in Lancaster *et al.* (1999) for comparison. This structural alignment showed that the *E. coli* QFR menaquinone models are at positions occupied by heme propionates in *W. succinogenes* QFR (Lancaster *et al.*, 1999). Consequently, no conclusions regarding quinone binding in *W. succinogenes* QFR could be drawn from the menaquinone positions in *E. coli* QFR.

In the *W. succingenes* QFR crystal structure, a cavity which extends from the hydrophobic phase of the membrane, close to the distal heme b_D, to the periplasmic aqueous phase could accommodate a menaquinol molecule, after minor structural alterations (Lancaster *et al.*, 2000), which are consistent with experimentally observed structural differences for the presence and absence of a quinone substrate (Lancaster and Michel, 1997). A glutamate residue (Glu C66) lines the cavity and could be involved in the acceptance of the protons liberated upon oxidation of the menaquinol (Fig. 6b). Replacement of Glu C66 by a glutamine residue resulted in a mutant which did not catalyze quinol oxidation by fumarate, whereas the activity of fumarate reduction was not affected by the mutation (Lancaster *et al.*, 2000). X-Ray crystal structure analysis of the Glu C66 → Gln variant enzyme ruled out significant structural alterations. The midpoint potentials of the two heme groups of subunit C were not significantly affected. These results indicate that the inhibition of quinol oxidation activity in the mutant enzyme is due to absence of the carboxyl group of Glu C66. Thus it was concluded that Glu C66, which is conserved in the QFR enzymes from the ε-proteobacteria *C. jejuni* and *H. pylori*, is an essential constituent of the menaquinol oxidation site (Lancaster *et al.*, 2000) close to heme b_D (Fig. 6b).

VIII. Electron and Proton Transfer and the *Wolinella succinogenes* Paradox

For the function of QFR, electrons have to be transferred from the quinol-oxidizing site in the membrane to the fumarate-reducing site, protruding into the cytoplasm. The arrangement of the prosthetic groups in the QFR dimer is displayed in Fig. 8a together with the

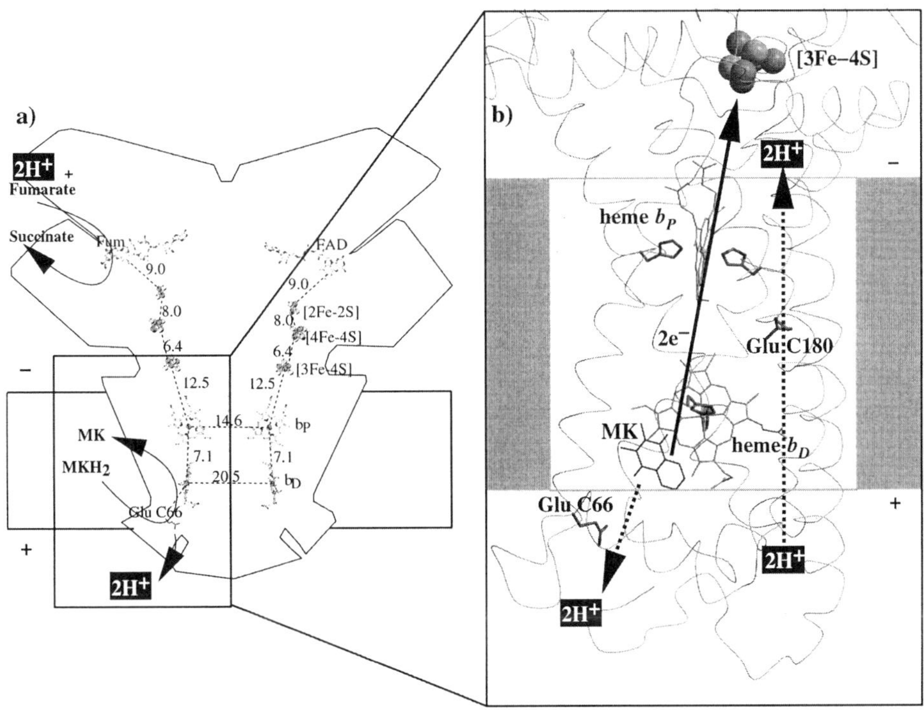

FIG. 8. The coupling of electron and proton flow in *W. succinogenes* QFR. Positive and negative sides of the membrane are as described for Fig. 1. (a) Hypothetical transmembrane electrochemical potential as suggested by the essential role of Glu C66 for menaquinol oxidation by *W. succinogenes* QFR (Lancaster *et al.*, 2000). The prosthetic groups of the *W. succinogenes* QFR dimer are displayed (coordinate set 1QLA; Lancaster *et al.*, 1999). Distances between prosthetic groups are edge-to-edge distances in angstroms as defined by Page *et al.* (1999). Also indicated are the side chain of Glu C66 and a tentative model of menaquinol (MKH_2) binding. The position of bound fumarate (Fum) is taken from PDB entry 1QLB (Lancaster *et al.*, 1999). (b) Hypothetical cotransfer of one H^+ per electron across the membrane ("E-pathway hypothesis"). The two protons liberated on oxidation of menaquinol (MKH_2) are released to the periplasm (bottom) via the residue Glu C66. In compensation, coupled to electron transfer via the two heme groups, protons are transferred from the periplasm via the ring C propionate of the distal heme b_D and the residue Glu C180 to the cytoplasm (top), where they replace those protons which are bound during fumarate reduction. In the oxidized state of the enzyme, the "E-pathway" is blocked. Modified from Lancaster (2002). *Biochim. Biophys. Acta Biomembr.* **1565,** 215–231, with permission from Elsevier Science.

edge-to-edge distances relevant for electron transfer as defined by Page *et al.* (1999). It has been shown for other electron transfer proteins that physiological electron transfer occurs if such distances are shorter than 14 Å, but not if they are longer than 14 Å (Page *et al.*, 1999). In the case of *W. succinogenes* QFR, this indicates that physiological electron transfer can occur between the six prosthetic groups of one QFR heterotrimeric complex, but not between the two QFR complexes in the dimer (Lancaster *et al.*, 2000).

The fumarate molecule is in van der Waals contact with the isoalloxazine ring of FAD. The linear arrangement of the prosthetic groups in one QFR complex therefore provides one straightforward pathway by which electrons could be transferred efficiently from the menaquinol oxidizing site via the two heme groups, the three iron–sulfur clusters, and the FAD to the site of fumarate reduction.

The two heme groups have different oxidation–reduction potentials (Kröger and Innerhofer, 1976): one is the "high-potential" heme b_H, the other the low-potential "low-potential" heme b_L. For the membrane-bound enzyme, these potentials are $E_M = -20$ mV and $E_M = -200$ mV, respectively (Kröger and Innerhofer, 1976). For the detergent-solubilized QFR enzyme, the respective values are -15 mV and -150 mV (Lancaster *et al.*, 2000). It has not yet been established which of the hemes b_P and b_D corresponds to b_L and b_H in *W. succinogenes* QFR.

Because of its very low midpoint potential ($E_m < -250$ mV; Unden *et al.*, 1984), the [4Fe-4S] iron–sulfur cluster has a very low potential and has been suggested not to participate in electron transfer [see Hägerhäll (1997) for a discussion]. However, the determined low potential may be an artifact due to anti-cooperative electrostatic interactions between the redox centers (Salerno, 1991). The position of the [4Fe-4S] cluster as revealed in the structures of *W. succinogenes* QFR and *E. coli* QFR is highly suggestive of its direct role in electron transfer from the [3Fe-4S] cluster to the [2Fe-2S] cluster. Despite this major thermodynamically unfavorable step, the calculated rate of electron transfer is on a microsecond scale, demonstrating that this barrier can easily be overcome by thermal activation as long as the electron transfer chain components are sufficiently close to promote intrinsically rapid electron tunneling (Dutton *et al.*, 1998).

In addition to the transfer of electrons, two protons are bound on fumarate reduction (see reaction 1) and two protons are liberated on menaquinol oxidation (see reaction 2). The protons consumed on fumarate reduction are undoubtably bound from the cytoplasm (see Fig. 8a). The experimental results on intact bacteria, with inverted vesicles or liposomes containing *W. succinogenes* QFR (Kröger *et al.*, 2002; Biel *et al.*, 2002), suggest that the oxidation of menaquinol by fumarate as catalyzed by *W. succinogenes* QFR is an electroneutral process. The protons formed by menaquinol oxidation have therefore been assumed to be released to the cytoplasmic side of the membrane where they balance the protons consumed by fumarate reduction.

However, the essential role of Glu C66 for menaquinol oxidation demonstrated by Lancaster *et al.* (2000) contrasts this interpretation. There is no conceivable proton transfer pathway from the inferred menaquinol oxidation site to the cytoplasmic phase. This strongly suggests that the protons liberated during menaquinol oxidation are released on the periplasmic side of the membrane. In summary, the location of the catalytic sites of fumarate reduction and menaquinol oxidation in the structure suggests that quinol oxidation by fumarate should be an electrogenic process in *W. succinogenes,* in contrast to the results of experimental measurements for *W. succinogenes* QFR.

Succinate oxidation by menaquinone, an endergonic reaction under standard conditions, is catalyzed by a B-type succinate:quinone oxidoreductase in gram-positive bacteria, e.g., *B. subtilis.* There is experimental evidence indicating that succinate oxidation by menaquinone in *B. subtilis* is driven by the electrochemical proton potential (Schirawski and Unden, 1998). This is the analogous reaction to that suggested for *W. succinogenes* QFR (Fig. 8a), but in the opposite direction (Ohnishi *et al.,* 2000; Lancaster and Kröger, 2000). In *B. subtilis* SQR, the site of menaquinone reduction, has been proposed to be located close to heme b_D (Matsson *et al.,* 2000). Recent experimental results indeed indicate that *B. subtilis* SQR generates a proton potential when functioning as a QFR (Schnorpfeil *et al.,* 2001). Consequently, there are discrepancies, first between the experimental results for *B. subtilis* SQR (Schirawski and Unden, 1998; Schnorpfeil *et al.,* 2001) and *W. succinogenes* QFR (Geisler *et al.,* 1994; Kröger *et al.,* 2002; Biel *et al.,* 2002) and second also between the latter and the implications of the essential role of Glu C66 in *W. succinogenes* QFR. We refer to the latter discrepancy as the "*W. succinogenes* paradox."

IX. THE "E-PATHWAY HYPOTHESIS" OF COUPLED TRANSMEMBRANE ELECTRON AND PROTON TRANSFER

Previously, there was no satisfactory explanation for the above apparent discrepancies. A working hypothesis has been proposed to explain them satisfactorily and to resolve the "*W. succinogenes* paradox" (Lancaster, 2002e). According to this hypothesis, the quinol oxidation process in *W. succinogenes* QFR is coupled to the compensating, parallel transfer of protons from the periplasm to the cytoplasm. The pathway of this proton transfer is transiently established during reduction of the heme groups of the enzyme and is not obvious from the available crystal structure of the oxidized enzyme. Possible constituents of such a proton transfer pathway are the ring C propionate of the distal heme and the transmembrane helix V residue Glu C180 (the "E-pathway," Fig. 8b). The latter residue is conserved in the QFR enzymes from *H. pylori* and *C. jejuni,* but not in *B. subtilis* SQR. This "E-pathway hypothesis" explains why the net reaction catalyzed by *W. succinogenes* QFR does not contribute directly to the generation of a transmembrane electrochemical potential. Such a process is consistent with both a distal quinol oxidation site and the observed apparent electroneutrality of the net reaction. An electroneutral reaction is more consistent with the small oxidation–reduction potential difference between the fumarate/succinate couple ($E_M = +25$ mV; Ohnishi *et al.,* 2000) and the menaquinone/menaquinol couple ($E_M = -74$ mV; Thauer *et al.,* 1977). Initial experiments providing support for the E-pathway hypothesis have been performed (Lancaster, C. R. D., Sauer, U. S., Gross, R., Haas. A., Mäntele, W., and Simon, J., manuscript in preparation; Haas, A., Sauer, U. S., Gross, R., Simon, J., Mäntele, W., and Lancaster, C. R. D., manuscript in preparation).

X. Concluding Remarks

An overview of the current status of discussion of electron and proton transfer in succinate:quinone oxidoreductases is shown in Figs. 9a–9d. In mitochondrial complex II and other C-type enzymes, such as SQR from *P. denitrificans* and *E. coli,* electron transfer from succinate to ubiquinone does not lead to the generation of a transmembrane electrochemical potential Δp, since the protons released by succinate oxidation are on the same side of the membrane as those consumed by quinone reduction [Fig. 9a; see Ohnishi *et al.* (2000) for a review]. As noted earlier (Lancaster *et al.,* 1999), it is unlikely that transmembrane electron transfer occurs in the *E. coli* QFR, because of the large edge-to-edge distance of ~25 Å between the two quinone models. Therefore, it is most likely that quinol oxidation occurs at a proximal site (Fig. 9b). This is supported by crystallographic studies of inhibitor binding to *E. coli* QFR (Iverson *et al.,* 2002). The indications that the endergonic oxidation of succinate by menaquinone as catalyzed by B-type SQRs in *B. subtilis* and other gram-positive bacteria is driven by Δp (Fig. 9c) have been summarized in Section VIII. In the opposite direction, the reduction of fumarate by menaquinol does not seem to provide enough energy to allow for the generation of Δp. In this case, enzymes with a compensatory "E-pathway" are apparently preferred (Fig. 9d).

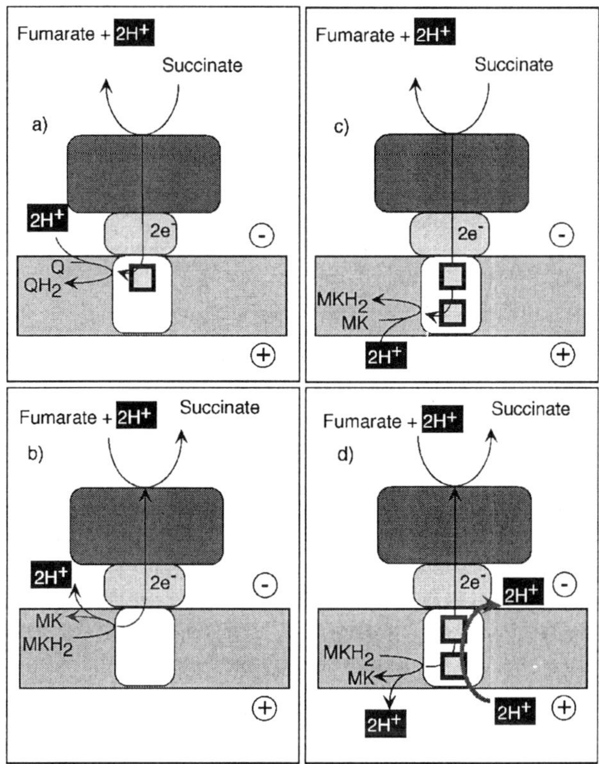

FIG. 9. The coupling of electron and proton flow in succinate:quinone oxidoreductases in aerobic (a,c) and anaerobic respiration (b,d), respectively. Positive and negative sides of the membrane are as described for Fig. 1. (a) and (b) Electroneutral reactions as catalyzed by C-type SQR enzymes (a) and D-type *E. coli* QFR (b). (c) Utilization of a transmembrane electrochemical potential Δp as possibly catalyzed by A-type and B-type enzymes. (d) Electroneutral fumarate reduction by B-type QFR enzymes with a proposed compensatory "E-pathway."

ACKNOWLEDGMENTS

The author thanks H. Michel and the late A. Kröger for support and the coauthors of his cited publications for their contributions. Support by DFG Sonderforschungsbereich 472 ("Molecular Bioenergetics," P19) is gratefully acknowledged.

REFERENCES

Berks, B. C., Page, M. D., Richardson, D. J., Reilly, A., Cavill, A., Outen, F., and Ferguson, S. J. (1995). *Mol. Microbiol.* **15,** 319–331.

Biel, S., Simon, J., Gross, R., Ruiz, T., Ruitenberg, M., and Kröger, A. (2002). *Eur. J. Biochem.* **269,** 1974–1983.

Collins, M. D., and Jones, D. (1981). *Microbiol. Rev.* **45,** 316–354.
Dutton, P. L., Chen, X., Page, C. C., Huang, S., Ohnishi, T., and Moser, C. C. (1998). In: Biological Electron Transfer Chains: Genetics, Composition and Mode of Operation (G. W. Canters and E. Vijgenboom, Eds.), pp. 3–8. Kluwer Academic, Dordrecht.
Esnouf, R. M. (1997). *J. Mol. Graphics Mod.* **15,** 132–134.
Geisler, V., Ullmann, R., and Kröger, A. (1994). *Biochim. Biophys. Acta Bioenerget.* **1184,** 219–226.
Gomes, C. M., Lemos, R. S., Teixeira, M., Kletzin, A., Huber, H., Stetter, K. O., Schäfer, G., and Anemüller, S. (1999). *Biochim. Biophys. Acta Bioenerget.* **1411,** 134–141.
Gross, R., Simon, J., Lancaster, C. R. D., and Kröger, A. (1998). *Mol. Microbiol.* **30,** 639–646.
Hägerhäll, C. (1997). *Biochim. Biophys. Acta Bioenerget.* **1320,** 107–141.
Hägerhäll, C., and Hederstedt, L. (1996). *FEBS Lett.* **389,** 25–31.
Hägerhäll, C., Sled, V., Hederstedt, L., and Ohnishi, T. (1995). *Biochim. Biophys. Acta* **1229,** 356–362.
Hederstedt, L. (1999). *Science* **284,** 1941–1942.
Iverson, T. M., Luna-Chavez, C., Cecchini, G., and Rees, D. C. (1999). *Science* **284,** 1961–1966.
Iverson, T. M., Luna-Chavez, C., Croal, L. R., Cecchini, G., and Rees, D. C. (2002). *J. Biol. Chem.* **277,** 16124–16130.
Jormakka, M., Tornroth, S., Byrne, B., and Iwata, S. (2002). *Science* **295,** 1863–1868.
Kenny, W. C., and Kröger, A. (1977). *FEBS Lett.* **73,** 239–243.
Körtner, C., Lauterbach, F., Tripier, D., Unden, G., and Kröger, A. (1990). *Mol. Microbiol.* **4,** 855–860.
Kraulis, P. J. (1991). *J. Appl. Crystallogr.* **24,** 946–950.
Kröger, A. (1978). *Biochim. Biophys. Acta* **505,** 129–145.
Kröger, A., and Innerhofer, A. (1976). *Eur. J. Biochem.* **69,** 497–506.
Kröger, A., Biel, S., Simon, J., Gross, R., Unden, G., and Lancaster, C. R. D. (2002). *Biochim. Biophys. Acta Bioenerget.* **1553,** 23–38.
Lancaster, C. R. D. (2001a). *FEBS Lett.* **504,** 133–141.
Lancaster, C. R. D. (2001b). In: *Handbook of Metalloproteins* (A. Messerschmidt, R. Huber, T. Poulos, and K. Wieghardt, Eds.), **vol. 1,** pp. 379–401. Wiley, Chichester, UK.
Lancaster, C. R. D., Ed. (2002a). *Biochim. Biophys. Acta Bioenerget.* **1553,** 1–176. (Special issue on fumarate reductases and succinate dehydrogenases)
Lancaster, C. R. D. (2002b). *Biochim. Biophys. Acta Bioenerget.* **1553,** 1–6.
Lancaster, C. R. D. (2002c). In: *Membrane Protein Purification and Crystallization: A Practical Guide* (C. Hunte, H. Schägger, and G. von Jagow, Eds.), 2nd ed., pp. 219–228. Academic Press, San Diego.
Lancaster, C. R. D. (2002d). In: *Methods and Results in Membrane Protein Crystallization* (S. Iwata, Ed.). International Univ. Line, La Jolla, CA. (in press).
Lancaster, C. R. D. (2002e). *Biochim. Biophys. Acta Biomembr.* **1565,** 215–231.
Lancaster, C. R. D., and Kröger, A. (2000). *Biochim. Biophys. Acta Bioenerget.* **1459,** 422–431.
Lancaster, C. R. D., and Michel, H. (1997). *Structure* **5,** 1339–1359.
Lancaster, C. R. D., and Simon, J. (2002). *Biochim. Biophys. Acta Bioenerget.* **1553,** 84–101.
Lancaster, C. R. D., Kröger, A., Auer, M., and Michel, H. (1999). *Nature* **402,** 377–385.
Lancaster, C. R. D., Gross, R., Haas, A., Ritter, M., Mäntele, W., Simon, J., and Kröger, A. (2000). *Proc. Natl. Acad. Sci. USA* **97,** 13051–13056.
Lancaster, C. R. D., Gross, R., and Simon, J. (2001). *Eur. J. Biochem.* **268,** 1820–1827.
Lauterbach, F., Körtner, C., Albracht, S. P., Unden, G., and Kröger, A. (1990). *Arch. Microbiol.* **154,** 386–393.

Lemma, E., Hägerhäll, C., Geisler, V., Brandt, U., von Jagow, G., and Kröger, A. (1991). *Biochim. Biophys. Acta* **1059,** 281–285.

Lemos, R. S., Fernandes, A. S., Pereira, M. M., Gomes, C. M., and Teixeira, M. (2002). *Biochim. Biophys. Acta* **1553,** 158–170.

Luecke, H., Schobert, B., Richter, H. T., Cartailler, J. P., and Lanyi, J. K. (1999). *J. Mol. Biol.* **291,** 899–911.

Mandori, A., Cecchini, G., Schröder, I., Gunsalus, R. P., Werth, M. T., and Johnson, M. K. (1992). *Biochemistry* **31,** 2703–2712.

Matsson, M., Tolstoy, D., Aasa, R., and Hederstedt, L. (2000). *Biochemistry* **39,** 8617–8624.

Merrit, E. A., and Bacon, D. J. (1997). *Methods Enzymol.* **277,** 505–524.

Mitchell, P. (1979). *Science* **206,** 1148–1159.

Ohnishi, T., Moser, C. C., Page, C. C., Dutton, P. L., and Yano, T. (2000). *Structure* **8,** R23–R32.

Page, C. C., Moser, C. C., Chen, X., and Dutton, P. L. (1999). *Nature* **402,** 47–52.

Salerno, J. C. (1991). *Biochem. Soc. Trans.* **19,** 599–605.

Saraste, M. (1999). *Science* **283,** 1488–1493.

Schäfer, G., Engelhard, M., and Müller, V. (1999). *Microbiol. Mol. Biol. Rev.* **63,** 570–620.

Schäfer, G., Anemüller, S., and Moll, R. (2002). *Biochim. Biophys. Acta* **1553,** 57–73.

Schirawski, J., and Unden, G. (1998). *Eur. J. Biochem.* **257,** 210–215.

Schnorpfeil, M., Janausch, I. G., Biel, S., Kröger, A., and Unden, G. (2001). *Eur. J. Biochem.* **268,** 3069–3074.

Simon, J., Gross, R., Ringel, M., Schmidt, E., and Kröger, A. (1998). *Eur. J. Biochem.* **251,** 418–426.

Singer, T. P., and McIntire, W. S. (1984). *Methods Enzymol.* **106,** 369–378.

Thauer, R. K., Jungermann, K., and Decker, K. (1977). *Bacteriol. Rev.* **41,** 100–180.

Unden, G., Hackenberg, H., and Kröger, A. (1980). *Biochim. Biophys. Acta* **591,** 275–288.

Unden, G., Albracht, S. P. J., and Kröger, A. (1984). *Biochim. Biophys. Acta* **767,** 460–469.

Walker, W. H., and Singer, T. P. (1970). *J. Biol. Chem.* **245,** 4224–4225.

Weiner, J. H., and Dickie, P. (1979). *J. Biol. Chem.* **254,** 8590–8593.

Xia, D., Yu, C. A., Kim, H., Xian, J. Z., Kachurin, A. M., Zhang, L., Yu, L., and Deisenhofer, J. (1997). *Science* **277,** 60–66.

STRUCTURE AND FUNCTION OF QUINONE BINDING MEMBRANE PROTEINS

By MOMI IWATA,* JEFF ABRAMSON,† BERNADETTE BYRNE,† and SO IWATA*,†

*Division of Biomedical Sciences, Imperial College of Science, Technology and Medicine London SW7 2AZ, UK, and †Department of Biological Sciences, Imperial College of Science, Technology, and Medicine, London SW7 2AZ, UK

I. Introduction

The mitochondria of eukaryotic cells are the organelles that convert energy to forms which can be used in the cell. In the inner membrane of the mitochondria, four respiratory enzymes perform a series of redox reactions by transferring electrons through electron carriers to O_2, the terminal electron acceptor (Fig. 1A). NADH dehydrogenase (Complex I) and succinate dehydrogenase (Complex II) reduce membrane-bound ubiquinone using NADH and succinate, respectively, which are provided by the TCA cycle in the matrix space. The reduced ubiquinone (ubiquinol) can freely diffuse within the membrane to other complexes. Using this ubiquinol, the cytochrome bc_1 complex (ubiquinol–cytochrome *c* oxidoreductase or Complex III) reduces cytochrome *c*, the sole soluble protein component in the respiratory chain. The reduced cytochrome *c* is then utilized by cytochrome *c* oxidase (Complex IV) to reduce oxygen molecules to water molecules. The free energy released by electron transfer is captured as chemical energy in the form

ADVANCES IN PROTEIN CHEMISTRY, Vol. 63

0065-3233/03 $35.00

A

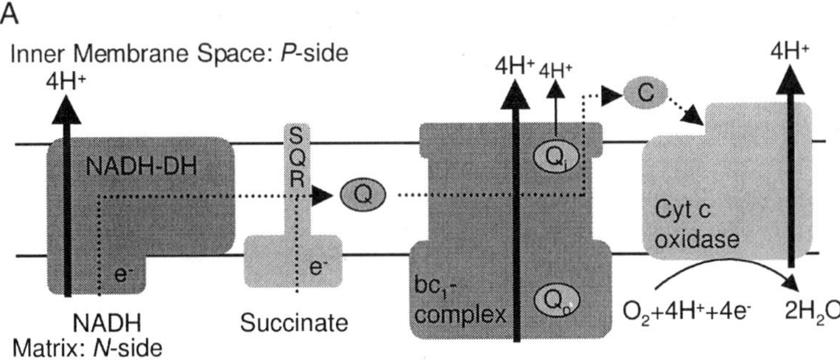

B

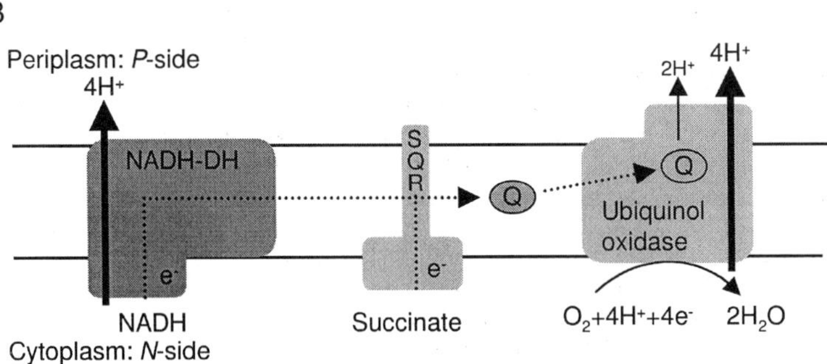

FIG. 1. A schematic illustration of the mitochondrial (A) and *Escherichia coli* (B) respiratory chains. Respiratory enzymes perform a series of oxidation–reduction reactions by transferring electrons (dashed lines) through mobile electron carriers. Electron transfer is coupled to the pumping of protons (thick black arrows) from the *N*-side (negative side) to the *P*-side (positive side) generating a proton gradient that ultimately drives the conversion of ADP to ATP.

of a proton gradient that can be used to drive the conversion of ADP to ATP.

Many prokaryotic organisms such as *Escherichia coli* have a simplified respiratory chain located in the inner cell membrane (Fig. 1B). The *E. coli* respiratory chain performs a function similar to as its mitochondrial counterpart but lacks Complex III. Instead, electrons are directly transferred from the ubiquinol molecule to Complex IV (ubiquinol oxidase in this case).

In both systems, membrane-bound ubiquinone plays crucial roles in the respiratory chain. Indeed, various quinones, including ubiquinone and menaquinone, are used to connect the redox reactions of various membrane proteins. In spite of the large amount of biochemical and biophysical data on quinone and quinone binding proteins, little structural information on the precise action mechanism of quinone in the respiratory chain is available. So far the structures of the following proteins have provided structural information on quinone binding sites: the photosynthetic reaction center (Deisenhofer *et al.*, 1985; Allen *et al.*, 1986), cytochrome bc_1 complex (Xia *et al.*, 1987; Zhang *et al.*, 1998; Iwata *et al.*, 1998; Hunte *et al.*, 2000), fumarate reductase (Iverson *et al.*, 1999; Lancaster *et al.*, 1999), ubiquinol oxidase (Abramson *et al.*, 2000), photosystem I (Jordan *et al.*, 2002), and formate dehydrogenase (Jormakka *et al.*, 2002). In this review, we will discuss the structure and function of the cytochrome bc_1 complex from bovine mitochondria and ubiquinol oxidase from *E. coli* focusing on the quinone binding sites within these proteins.

II. STRUCTURE OF CYTOCHROME bc_1 COMPLEX FROM BOVINE HEART MITOCHONDRIA

A. *Cytochrome* bc$_1$ *Complex*

The cytochrome bc_1 complex was isolated in the early 1960s (for review see Clejan and Beattie, 1986) but its redox centers had been already identified by Keilin in 1925 and Rieske in 1964 (for review see Edwards and Trumpower, 1986). The cytochrome bc_1 complex not only is part of the oxygen respiratory chain but is also involved in bacterial photosynthesis. Additionally, a related enzyme, the cytochrome b_6f complex, takes part in photosynthesis in chloroplasts. As mentioned above, the cytochrome bc_1 complex is an intermediate component of the oxygen respiratory chain which catalyzes the reduction of cytochrome *c* using membrane-bound ubiquinol. During the reaction, two protons are translocated across the membrane by a mechanism called "the proton motive Qcycle," discussed in the next section. Although mammalian cytochrome bc_1 complexes have 11 subunits, the number of the subunits varies between species from 3 to 11. The cytochrome bc_1 complex from bovine mitochondria exists as a dimer of the 11 subunits with a total molecular weight of about 480,000 (Schägger *et al.*, 1986, 1995). The subunit composition of the bc_1 complex is shown in Fig. 2.

All known cytochrome bc_1 complexes have three essential subunits, which contain all four redox centers in common. These subunits are cytochrome *b* (cyt. *b*), cytochrome c_1 (cyt. c_1), and the Rieske [2Fe-2S] protein (ISP). The redox centers are heme b_L and heme b_H in cytochrome *b* (L and H represent low and high potential, respectively), heme c_1 in cytochrome c_1, and Rieske [2Fe-2S] (FeS) cluster in ISP.

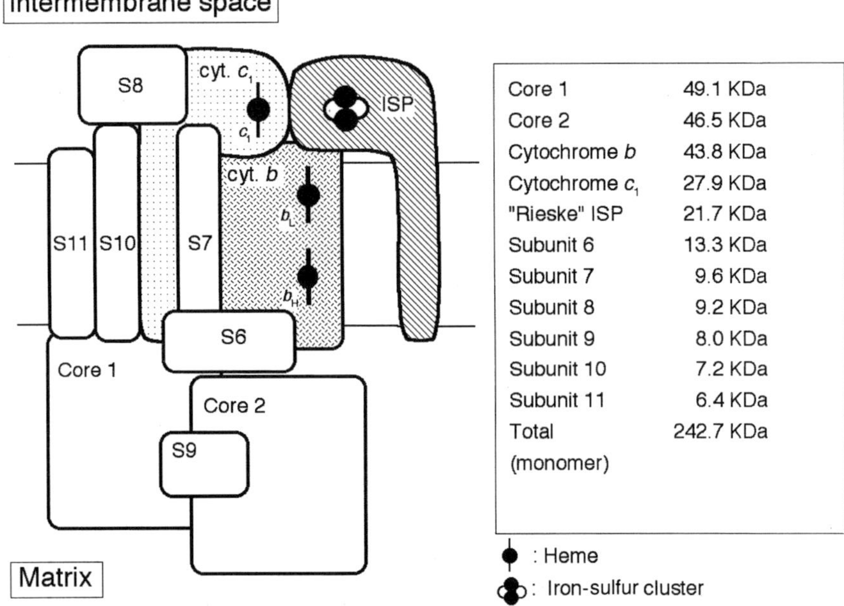

FIG. 2. Subunit composition of the bovine bc_1 complex.

B. *Redox-Active Subunits and the Proton Motive Qcycle*

The reaction mechanism of cytochrome bc_1 complex is known as "the proton motive Qcycle" originally proposed by Peter Mitchell (Mitchell, 1976). This mechanism is the basis of his chemiosmotic theory for which he was awarded the Nobel prize in 1978. Since then, the enzyme has been characterized extensively using various techniques. Redox centers have been characterized spectroscopically (for review, see Trumpower and Gennis, 1994), electron transfer pathways have been determined using kinetic experiments with specific inhibitors (De Vries 1986; Zhu *et al.*, 1984), and the positions of quinone binding sites and redox centers have been determined using biochemical and mutational analysis (for review, see Esposti *et al.*, 1993; Brasseur *et al.*, 1996). As a result of these efforts, the latest modified Qcycle has been widely accepted by researchers in the field (for reviews, see Crofts *et al.*, 1983; Trumpower, 1990; Berry *et al.*, 2000).

Figure 3 summarizes the subunits and redox centers involved in the Qcycle. As explained before only three subunits of the bc_1 complex, namely cytochrome b, cytochrome c_1, and ISP, are essential for the Qcycle. Cytochrome b has heme b molecules, heme b_L and heme b_H

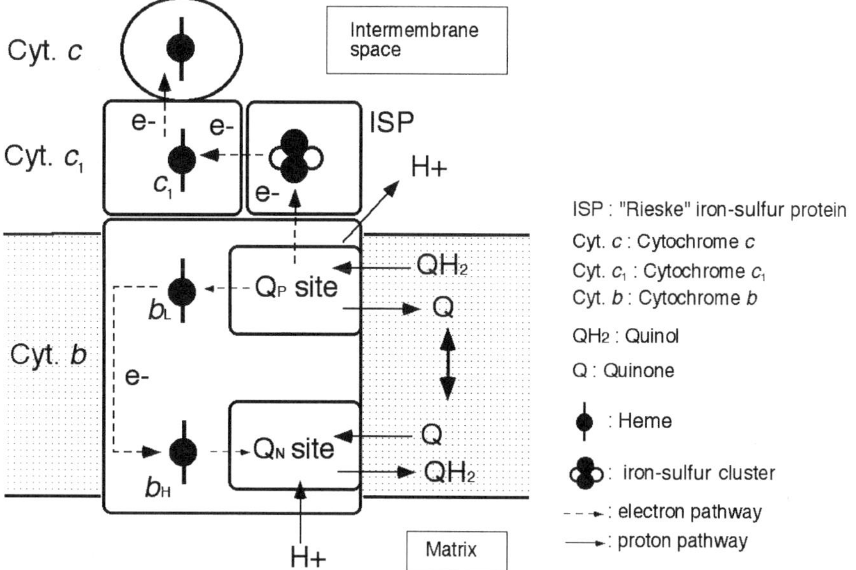

FIG. 3. Schematic view of the proton motive Qcycle.

associated with two quinone-binding sites Q_P and Q_N, respectively. P and N represent the positive and negative sides of the membrane. Q_P and Q_N sites are also known as Q_o and Q_i sites, where o and i represent proton output and input. The Q_P site is the ubiquinol oxidation site on the "out" side of the membrane and the Q_N site is the ubiquinone reduction site on the "in" side of the membrane. The Rieske FeS protein has the Rieske-type [2Fe-2S] iron sulfur cluster (FeS cluster), whose ligand is two histidine (His-141 and His-161) and two cysteine (Cys-138 and Cys-158) side chains instead of four cysteine side chains for ferredoxin type [2Fe-2S] cluster. Cytochrome c_1 has heme c_1, which transfers the electron to soluble cytochrome *c*.

The Qcycle can be summarized starting from Q_P site as follows. Membrane-bound ubiquinol binds to the Q_P site. During oxidation of ubiquinol the release of two protons to the mitochondrial intermembrane space is coupled to electron transfer. A ubiquinol molecule can donate two electrons when fully oxidized. The first electron is taken up by FeS cluster of ISP, then passed to heme c_1 of cytochrome c_1, where it is subsequently transferred to soluble cytochrome *c*. Interestingly, the second electron is transferred to bound ubiquinone in the Q_N site through heme b_L and b_H yielding a ubisemiquinone. At the Q_N site this ubisemiquinone, which is usually unstable, is stabilized by the protein environment. At the Q_P site another ubiquinol molecule replaces the oxidized ubiquinone and the oxidation reaction is repeated. During the second reaction, the second electron is transferred to the ubisemiquinone in the Q_N site yielding fully reduced ubiquinol. For the full reduction of ubiquinone, two protons are taken up from the matrix and the (re-) generated ubiquinol is released into the membrane. The released ubiquinol flips to the other side of the membrane and is, subsequently, used as a substrate in the Q_P site. Thus ubiquinone itself works as a proton carrier in the Qcycle and there are no proton pathways in the protein. In contrast, terminal oxidases, which use a "proton pump" for proton translocation, have specific proton pathways within the molecule.

The key step of the Qcycle is the "electron bifurcation" at the Q_P site. The FeS cluster has a far higher redox potential (+300 mV) than that of heme b_L (−90 mV). Considering the redox potential, both electrons could be transferred to the FeS cluster, although this never occurs in reality. The first electron is always transferred to the FeS cluster and the second electron to heme *b*. There was no clear answer how electrons could be bifurcated against the redox potential until the structures of the bc_1 complex were revealed.

In the Qcycle mechanism, two different types of quinone binding sites play critical roles. We will analyze different types of the quinone-binding sites in Section III,C.

C. *Overall Structure*

Crystals of the whole cytochrome bc_1 complex have been reported by different groups since 1991. Within them, Yu and Deisenhofer's group succeeded in providing the first structure of the complex in 1997 (Xia *et al.,* 1997). They have also shown the structures of various inhibitor complexes and revealed the positions of the quinone binding sites. Because of disorder, some of the subunits and domains, including the extrinsic domains of ISP and cytochrome c_1, were missing in the structure (a more complete structure was reported later (Kim *et al.,* 1998)). The structure was followed by the chicken bc_1 complex structure (Zhang *et al.,* 1998), which showed all 10 subunits of the complex. They built the structure using bovine sequence (75–90% similarity for known part) since, at the time, the sequence of the chicken complex subunit was unknown except for cytochrome *b* (Zhang *et al.,* 1998). Most importantly, they found the movement of the ISP extrinsic domain by comparison of the native and stigmatellin (a Q_P site inhibitor) complex structure. The structure of the bovine bc_1 complex from our group followed the structure of the chicken bc_1 complex (Iwata *et al.,* 1998).

This was the first complete structure of the bc_1 complex. The structure provided information about all 11 subunits and revealed that subunit 9, the mitochondrial targeting presequence of ISP, exists between two core subunits, which are most likely a mitochondrial targeting presequence peptidase. We have solved the structures of the bc_1 complex in two different crystal forms. Surprisingly, the conformation of the Rieske FeS protein was totally different between two crystal forms, and this provided a crucial insight of the electron bifurcation mechanism at the Q_P site (see Section II,F).

Figure 4 (see color insert) shows the overall structure of the bc_1 complex dimer form with a view parallel to the membrane and perpendicular to the noncrystallographic twofold axis. Each monomer is related by the twofold axis perpendicular to the membrane. The ISP extrinsic domain was found in different positions for different crystal forms as mentioned above. This difference was found to have a very important physiological implication in electron bifurcation at the Q_P site as discussed below. Interestingly, the transmembrane helix and the extrinsic domain of ISP are associated with different monomers within the dimer. This "domain swapping" strongly suggests that the physiological unit of the bc_1 complex is, indeed, dimer.

Figure 5 is a view of the dimer from the mitochondrial intermembrane space. Transmembrane helices, hemes b_L and b_H, inhibitors, and lipid molecules are shown. Site-specific inhibitors are superimposed in order to illustrate the positions of the quinone binding sites. The Q_P site, characterized by myxothiazol, was located by heme b_L toward the "out" side (the mitochondrial intermembrane space side) of the membrane, whereas the Q_N site, characterized by antimycin A, was found by heme b_H toward the "in" side (the matrix side) of the membrane.

Interestingly, the Q_P site from one monomer is close to the Q_N site of the other monomer rather than to the Q_N site of the same monomer. Moreover, this pair of the Q_P and Q_N sites from different monomers exists within the same hydrophobic cavity. The bc_1 complex dimer has four hydrophobic cavities: two large cavities between the monomers and one small cavity within each monomer. The small cavities are occupied with firmly bound phospholipid molecules. These lipid molecules are from the native mitochrondrial phospholipid membrane, maintained with the protein throughout the purification procedure. This means these lipids are not exchangeable. Each large cavity maintains only one firmly bound lipid molecule, while the rest of the cavity is filled with disordered electron densities. These densities could be interpreted as disordered detergent molecules, which indicates that the environment within the large cavities is highly mobile. This could allow fast exchange of the quinone and quinol molecules between the two sites and the

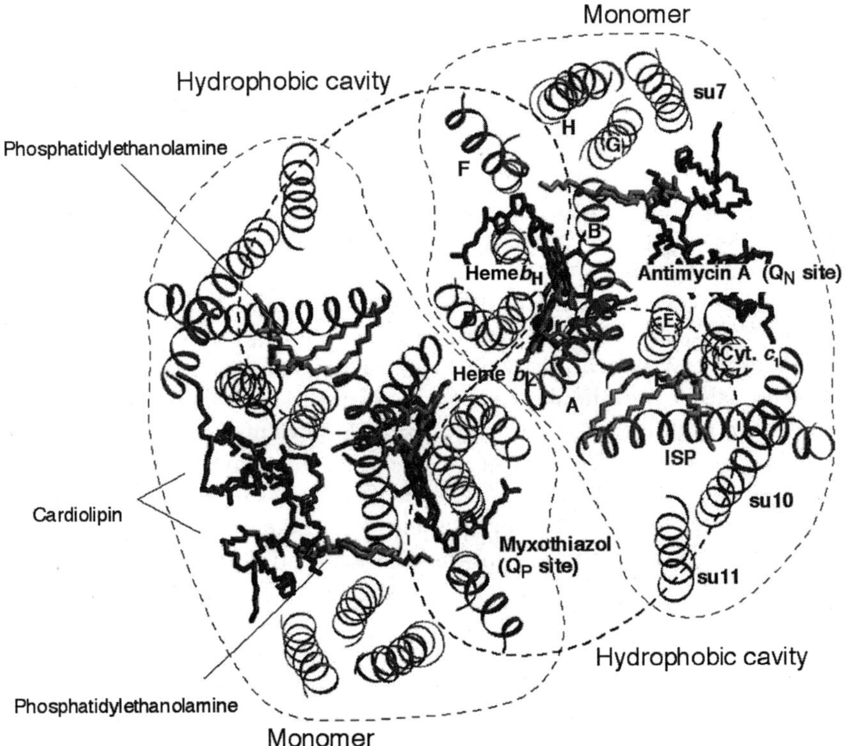

FIG. 5. Schematic view of the transmembrane section of the cytochrome bc_1 complex looking down from the intermembrane space. Transmembrane helices, hemes, bound inhibitors, and phospholipids are shown.

Q-pool outside. The minimum distance between myxothiazol and antimycin A molecules is only 18 Å; this indicates that the two sites can exchange a quinol/quinone molecule very quickly by simply flipping the head group.

D. *Cytochrome* b *and Quinone Binding Sites*

Cytochrome *b* is the only subunit of the complex encoded within the mitochondrial DNA. Cytochrome *b* is located in the center of the complex and mostly buried within the membrane. Both the NH_2- and COOH-termini are on the matrix side. Cytochrome *b* is composed of 8 transmembrane α-helices (αA to αH) and four horizontal α-helices (αab, αcd1, αcd2, and αef) on the mitochondrial intermembrane space side (Fig. 6). These helices are connected by four long (ab, cd, de, and ef)

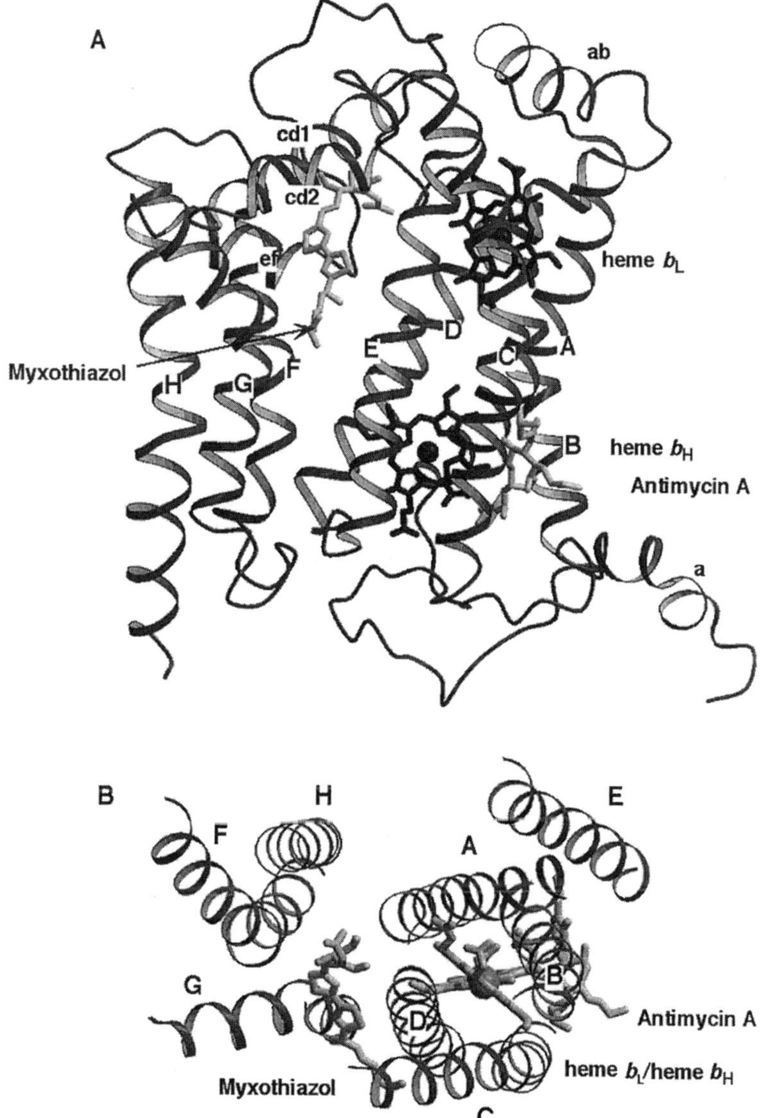

FIG. 6. Structure of cytochrome *b*. Transmembrane helices (A–H), horizontal helices (a, cd_1, cd_2, ef), hemes (b_L and b_H), and specific inhibitors (myxothiazol, antimycin A) are shown. (A) A view parallel to the membrane. (B) A view parallel to the membrane normal arrangement of the transmembrane helices is shown. Heme b_L is on the front and b_H is toward the back.

and 3 short (bc, fg and gh) loops. Hemes b_L and b_H are maintained within a bundle of four α-helices, αA, αB, αC and αD. Hemes b_L and b_H exist on the intermembrane space and the matrix sides, respectively. The ligands are His-83, His-182 for heme b_L and His-97, His-196 for heme b_H as predicted from biochemical studies (Yun *et al.*, 1991). These histidine residues are conserved in all known *b* type cytochromes (Widger *et al.*, 1984; Esposti *et al.*, 1993).

The Q_N site was found in a hydrophobic pocket near heme b_H, formed by the residues from αA, αD, αE, and αab, on the matrix side in cytochrome *b* (Fig. 6). In the native structure, the Q_N site seems empty and some water molecules are bond to His-201 and Asp-228. Antimycin A was found to be hydrogen bonded to Asp-228, Ser-35, Asn-32. It has also van der Waals contact with Phe-18, Tyr-224, Met-190 and a propionate group of the heme b_H. The hydrogen bond network between the inhibitor and the side chains in the binding pocket explain the high affinity and specificity of this inhibitor. UQ2 binds to a slightly different location in the Q_N site than the antimycin. Figure 7 shows a proposed quinone reduction mechanism at the Q_N site based on X-ray structures. First, an oxidized ubiquinone molecule from the membrane binds between His-201 and Asp-228. Both His-201 and Asp-228 should be protonated. The first electron transfer from heme b_H is coupled to the transfer of a proton from Asp-228 yielding a ubisemiquinone. To neutralize the negative charge, one proton is taken up to Asp-228 through the water channel. The formed ubisemiquinone should be stabilized until the second electron transfer occurs. The stabilization mechanism is still unclear, although the residues and groups in vicinity of the ubisemiquinone, including His-201 and heme b_H, which have direct contact to the quinone ring, could play crucial roles. Coupled to the second electron transfer, another proton is taken from His-201 yielding a fully reduced ubiquinol. To compensate for the negative charge, another proton is taken up to His-201 from the matrix space. Finally, the fully reduced ubiquinol is released into the membrane.

The Q_P site was found in a hydrophobic pocket near heme b_L, formed by residues from αC, αF, and αef, below αcd1, on the intermembrane space side of cytochrome *b*. No substrate structure in the Q_P site has been reported. The site has been characterized using various inhibitors by a number of different research groups. The Q_P site inhibitors are spectroscopically divided into two types: class I, inhibitors which affect heme b_L (myxothiazol and MOA type inhibitors); class II, inhibitors which affect both heme b_H and FeS cluster [stigmatellin, UBDBT [5-undecyl-6-hydroxy-4,7-dioxobenzothiazol).] We have succeeded in obtaining a class I inhibitor structure (myxothiazol); however, we were

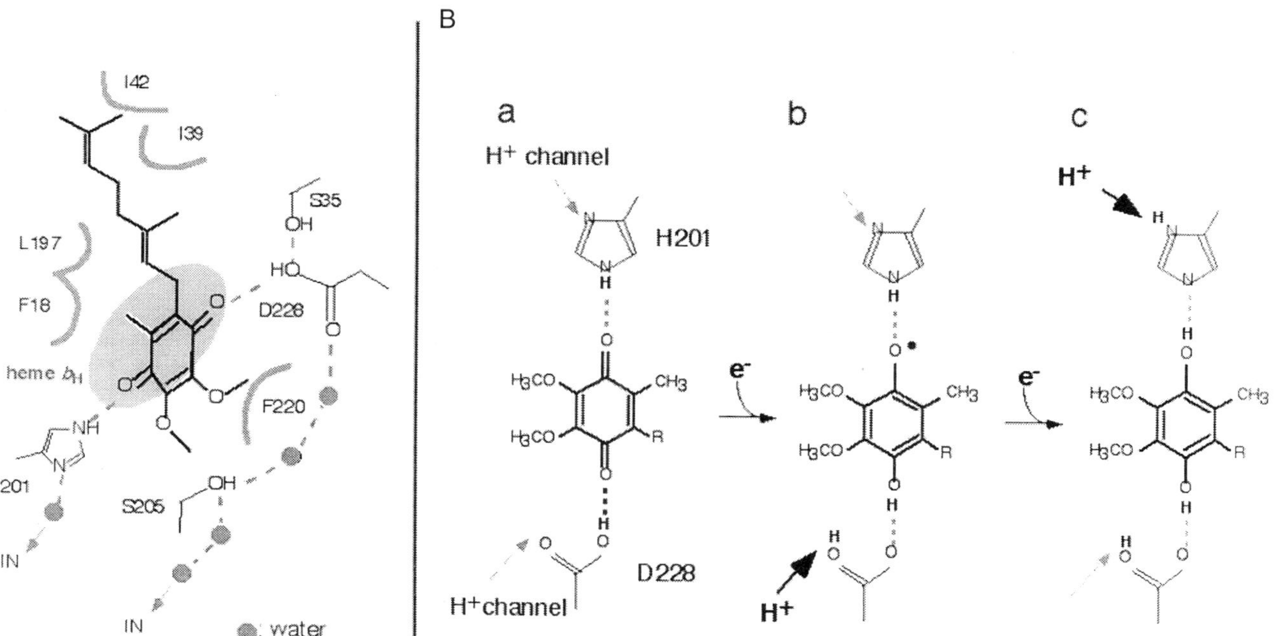

FIG. 7. (A) Quinone binding mode at the Q_N site. The residues involved in the binding are shown. (B) Possible quinone reduction mechanism in the Q_N site.

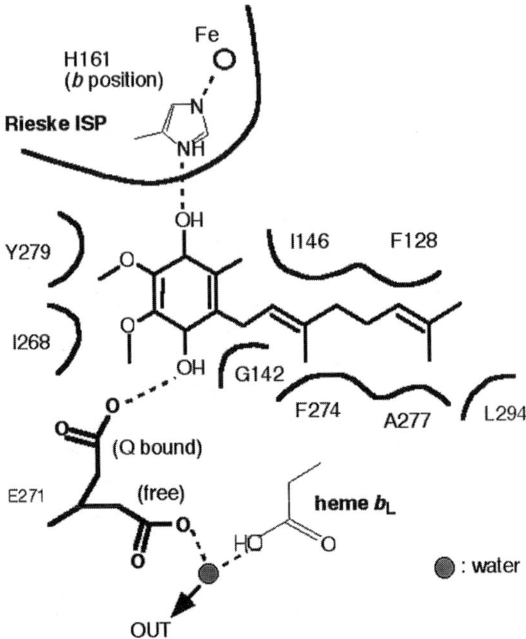

FIG. 8. Ubiquinol binding model in the Q_P site. Ubiquinol was modeled based on the inhibitor-bound bc_1 complex structures.

unable to obtain the class II inhibitor complex structure. Based on these results, a possible ubiquinol-bound structure was modeled as in Fig. 8. A ubiquinol occupies the same domain as myxothiazol with a small shift in such a position that it can bind to both His-161 (ISP) and flipped Glu-271 (cytochrome *b*) at the same time.

E. *ISP Extrinsic Domain Movement and Electron Bifurcation*

ISP has one transmembrane helix at the NH_2-terminal end the extrinsic domain at the COOH-terminal end, which are connected by a flexible linker with a short helix. A high resolution (1.5 Å) structure of the extrinsic domain (residues 71–196) has been reported previously (Iwata *et al.*, 1996). The domain is composed of 9 β-strands and one α-helix and has two subdomains called base and cluster-binding folds.

The extrinsic domain of ISP was found at different positions in the $P6_522$ and $P6_5$ forms (Iwata *et al.*, 1998). Various groups have also published different positions of the ISP extrinsic domain (Xia *et al.*, 1997; Kim *et al.*, 1998; Zhang *et al.*, 1998; Hunte *et al.*, 2000). These ISP positions are classified into three different groups using FeS–heme c_1 and FeS–heme b_L distances (Iwata *et al.*, 1998). The first group is composed of the positions observed in $P6_522$ and $P2_1$ crystals from bovine and $P6_522$ rabbit enzymes and native $P2_12_12_1$ from chicken enzyme (Zhang *et al.*, 1998). The ISP position for the group is called "c_1," because the ISP is located in close proximity to the cytochrome c_1 (Fig. 9). In the case of bovine $P6_522$, a direct hydrogen bond between a propionate of heme c_1 and His-161 of ISP was observed. The next group contains bovine $I4_122$, chicken, and yeast stigmatellin complex forms. The ISP position for this group is called "b" because ISP is in close proximity of cytochrome b. In the stigmatellin complex structures, a direct hydrogen bond between stigmatellin and His-161 of ISP was observed. Bovine $P6_5$ form is a member of the last group, where the ISP position is called "Intermediate (*Int*)." Similar positions were reported for the bovine $I4_122$ crystal forms with various inhibitors (Kim *et al.*, 1998). In this position, the extrinsic domain of ISP stays between "c_1" and "b" positions on loop ef of cytochrome b. Here, His-161 of ISP forms no hydrogen bond.

This extrinsic domain movement is explained using a combination of two rotations: (i) whole domain rotations against the rest of the complex; and (ii) the relative rotations within the domain between base and cluster binding folds. Between these, the rotation of the whole domain is more prominent than the rotation within the domain. Therefore, as a first approximation, the conformational change of ISP is well described by rotation of the whole extrinsic domain.

To study the flexible linker region of ISP, which seems to be important for the domain motion, various mutations of the bc_1 complexes from *Rhodobacter capsulatus* and yeast have been produced and characterized by research groups of Daldal and Trumpower, respectively (Darrouzet *et al.*, 2000 and Nett *et al.*, 2000). In these studies, both insertion and deletion mutants of the linker region caused some loss of the activity; however, the insertion mutants showed a more severe loss. The low activity of the mutants is probably due to restriction of domain movement, since the electron transfer rate from ubiquinol to ISP is not seriously affected in these mutants. Although the effects of the insertion or deletion could be more complex than just changing the length, the results clearly showed the extrinsic domain movement is a well-controlled process and essential for activity. If the movement of the ISP is too slow, the rate of enzymatic turnover could be reduced, and if the movement is too fast, short-circuit of electron transfer could result in dysfunction of the Qcycle (i.e., no electron bifurcation at the Q_P site). Darrouzet *et al.* also mentioned that the inhibitory effects of a linker insertion mutant

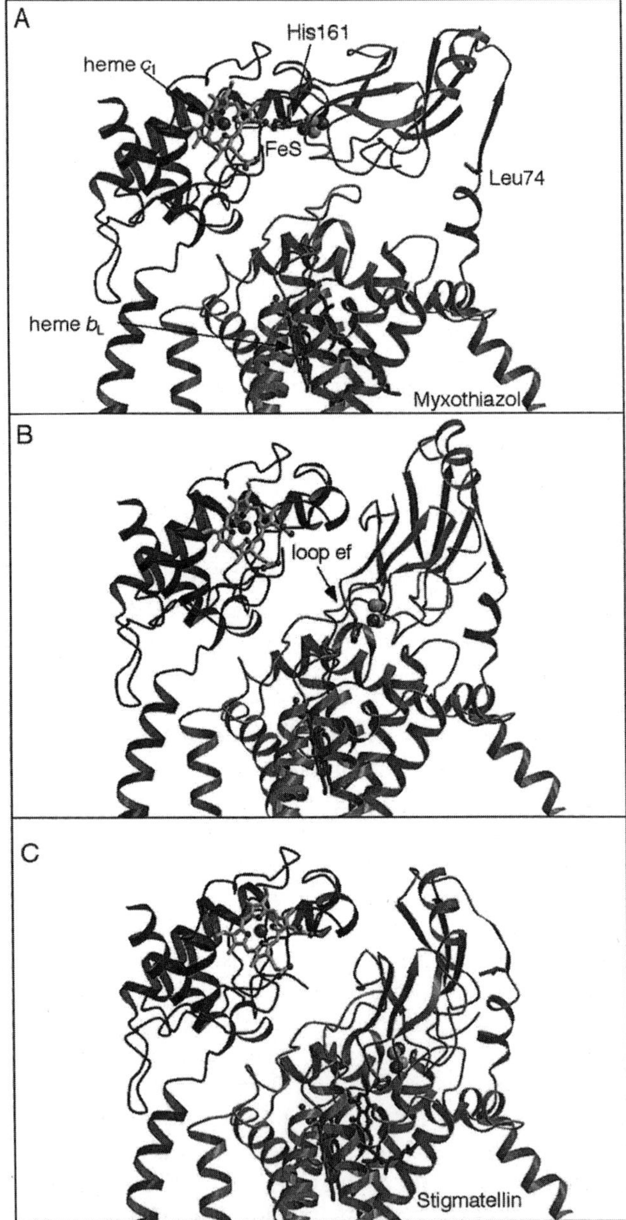

FIG. 9. Three different conformations of ISP in relation to its neighbors. (A) Structure in the "c_1" form. An inhibitor, myxothiazol, bound to the Q_P site is also shown. (B) Structure in "*Int*" form. (C) Structure in the "*b*" form. An inhibitor, stigmatellin, bound to the Q_P site is also shown.

could be reversed by a second mutation at Leu-286 (Leu-262 in bovine) in the loop ef. It is suggested that the conserved loop ef could be critical for rate control of ISP motion.

F. Electron Bifurcation Mechanism at the Q_P *Binding Pocket*

Based on the positions of metal centers and quinone binding sites and the different ISP positions, the electron bifurcation can be well explained (Zhang *et al.,* 1998; Iwata *et al.,* 1998, 1999; Crofts *et al.,* 1999).

Figure 10 shows the proposed ubiquinol oxidation and electron bifurcation mechanism at Q_P site. (A) In the absence of the ubiquinone, the side chain of Glu-271 is connected to the solvent in the mitochondrial intermembrane space via a water chain. (B) As a reduced ubiquinol molecule binds to the site, the side chain of Glu-271 flips to form a hydrogen bond to the bound ubiquinone. (C) Now, the ISP, which is moving around the intermediate position by thermal motion is trapped at the "*b*" position by a hydrogen bond to the bound ubiquinone. (D,E) Coupled to deprotonation, the first electron transfer occurs. Since the Rieske FeS cluster has a much higher redox potential (ca. +300 mV) than heme b_L (ca. 0 mV), the first electron is favorably transferred to ISP. This yields ubisemiquinone, (F,G). After ubisemiquinone formation, the hydrogen bond to the His-161 of ISP is destabilized. The ISP moves to the "c_1" position, where the electron is transferred from the Rieske FeS cluster to heme c_1. Now unstable ubisemiquinone is left in the Q_P pocket. The redox potential of the deprotonated ubisemiquinone is assumed to be several hundred millivolts. Now the electron transfer to the heme b_L is a "downhill" reaction. (H) Coupled to the second electron transfer, the second proton is transferred to Glu-271 and subsequently to the mitochondrial intermembrane space. The fully oxidized ubiquinone is released to the membrane.

III. The Structure of Cytochrome *bo*$_3$ Ubiquinol Oxidase from *Escherichia coli*

A. Cytochrome bo$_3$ *Ubiquinol Oxidase*

Cytochrome *bo*$_3$ ubiquinol oxidase from *E. coli* is a four-subunit heme-copper oxidase that catalyzes the four-electron reduction of O_2 to water and functions as a proton pump (Puustinen *et al.,* 1991; Fig. 11). All redox centers are located within the largest subunit (subunit I), with a low spin protoheme (heme *b*) acting as the electron donor to a binuclear center that is composed of an O-type heme (heme *o*$_3$) and a copper ion

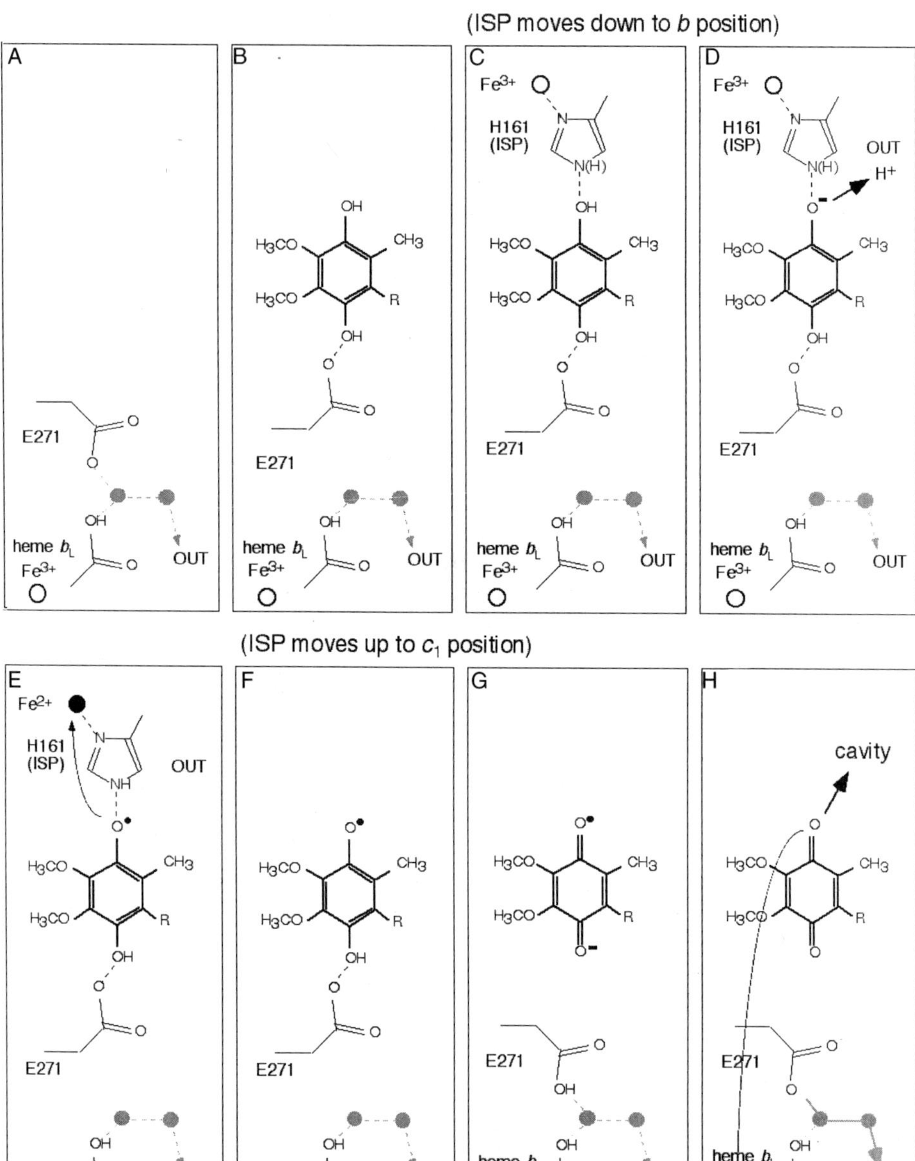

FIG. 10. Possible electron bifurcation mechanism at the Q_P site. See text for details.

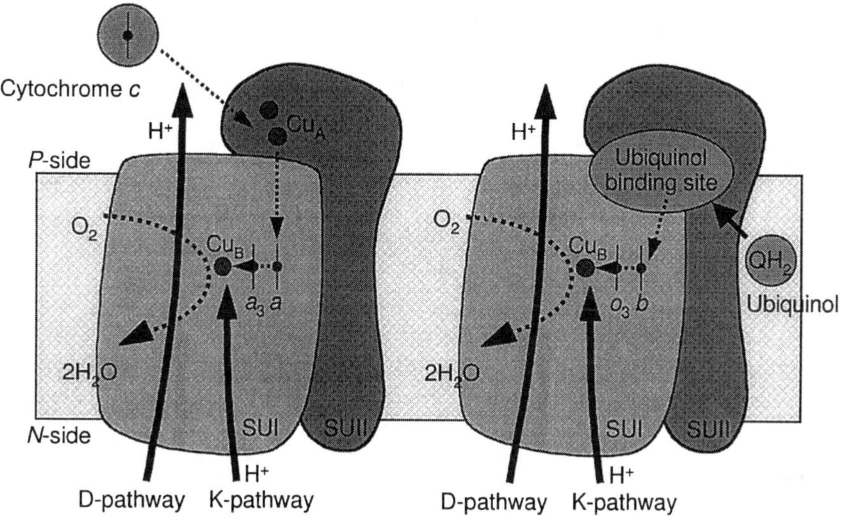

FIG. 11. A schematic illustration of subunits I (SUI) and II (SUII) of cytochrome *c* and ubiquinol oxidases. The proton pathways used for proton pumping and for the delivery of protons to the O_2 binding site (thick black lines) are called the D- and K-pathways (named after the conserved residues Asp for the D-pathway and Lys for the K-pathway). Dashed lines depict the electron path from the substrate to the O_2 binding site. In cytochrome *c* oxidases, electrons are delivered to the copper A center (Cu_A) from the substrate, cytochrome *c*, and then passed onto the low-spin heme (heme *a*) and finally to the binuclear center (heme a_3–Cu_B). In ubiquinol oxidases, electrons are delivered directly to the low-spin heme (heme *b*) by the substrate, ubiquinol, and then to the binuclear center (heme o_3–Cu_B).

(Cu_B). Subunits I, II, and III of ubiquinol oxidase are homologous to the corresponding subunits in the aa_3-type cytochrome *c* oxidases, and the ligands of the two heme groups and of Cu_B have been identified as the invariant His residues (Lemieux *et al.*, 1992). Subunit IV has little or no sequence homology with other oxidases and its function is unknown. In contrast to the cytochrome *c* oxidases, subunit II of ubiquinol oxidase has no Cu_A center, nor does it have a cytochrome *c* binding site. Instead, heme *b* receives electrons directly from a membrane solubilized ubiquinol molecule, and the protons produced are released on the "out" side of the membrane. The crystal structure of the ubiquinol oxidase from *E. coli* has been solved at 3.5 Å resolution (Abramson *et al.*, 2000). The structure confirms that the overall architecture of this complex is very similar to that of cytochrome *c* oxidase. Using site-directed mutagenesis and sequence comparisons with known ubiquinone binding sites from other membrane proteins, we have identified a novel ubiquinone binding site within this enzyme. These studies reveal how ubiquinol oxidase and cytochrome *c* oxidase can use a similar protein framework for oxygen reduction but with two entirely different electron donors.

B. *Overall Structure*

The cytochrome bo_3 structure is composed of four subunits containing 25 membrane-spanning helices with a total molecular mass of 130 kDa (Fig. 12, see color insert). The overall protein architecture is similar to that of the other cytochrome *c* oxidases solved so far. In fact, cytochrome bo_3 can be superimposed over *Paracoccus denitrificans* cytochrome *c* oxidase (*Paracoccus*-COX) with an rms difference of 1.6 Å for 781 C_α atoms using a distance cutoff of 3.8 Å (Iwata *et al.*, 1995). This is as expected since numerous biochemical studies have shown these enzymes to perform the same function in a similar manner. In the following section the amino acid numbering will be for cytochrome bo_3.

Subunit I of cytochrome bo_3 has 15 membrane-spanning helices. There are three additional helices in cytochrome bo_3 that are not present in subunit I of the other oxidase structures: one at the N-terminus and two at the C-terminus (Chepuri *et al.*, 1990). As expected, the barrel structure composed of the twelve membrane-spanning helices (helices I–XII) is almost identical to those found in *Paracoccus*-COX and bovine cytochrome *c* oxidase (bovine-COX, Tsukihara *et al.*, 1995). Both the low-spin heme and the binuclear center have identical coordination patterns and reside in similar positions in Pores C and B, respectively. The low-spin heme (heme *b*) is the initial electron acceptor coordinated by His-106 and His-421. From here, the electrons are passed to the binuclear center composed of the high-spin heme (heme o_3), coordinated by His-419, and the Cu_B, coordinated by His-284, His-333, and His-334. As expected from sequence analysis, there are no nonredox metal ions (*Paracoccus*-COX Mg^{2+}/Mn^{2+} or bovine-COX Ca^{2+}/Na^{+}) observed in the structure of cytochrome bo_3. The two polar channels (D- and K-pathways) that were found in bovine-COX and *Paracoccus*-COX are also present in subunit I of cytochrome bo_3 (Iwata *et al.*, 1995; Yoshiukawa *et al.*, 1998). Both channels maintain a similar arrangement of residues or moieties to facilitate proton movement from the cytoplasm to the binuclear center. The D-pathway again begins with an aspartic acid (Asp-135), proceeds through Asn-124, Thr-211, Asn-142, Asn-124, Tyr-61, Thr-204, Ser-145, Thr-201, and Thr-149, and terminates at Glu-286. Likewise, the cavity for the K-pathway is formed by Ser-315, Ser-299, Lys-362, Thr-359, the OH group of the hydroxyethylfarnesyl tail of heme o_3, and Tyr-288.

Subunit II is composed of two membrane-spanning helices and a C-terminal extrinsic domain that is situated above the membrane on the periplasmic side. In the second membrane-spanning helix there is a conserved glutamic acid residue (Glu-89) that has a similar conformation to its homologue in *Paracoccus*-COX. Glu-89 has been suggested as an alternative input site for regulating proton entry or transit through the K-pathway (Ma *et al.*, 1999). Subunit III of cytochrome bo_3 has five helices that can be superimposed onto helices III through VII of cytochrome *c* oxidase with an rms difference of 1.5 Å for 174 C_α atoms using a distance cutoff of 3.8 Å. A positional alignment places helices (0) and (XIV) from subunit I of cytochrome bo_3 over helices I and II from subunit III of *Paracoccus*-COX. Subunit IV of cytochrome bo_3 has three membrane-spanning helices that are in contact with both subunits I and II. The function of subunit IV is unknown, but it may be important in maintaining an intact structure at the binuclear center. The structure of cytochrome bo_3 revealed that the third helix from subunit IV interacts with helix VII from subunit I, near the binuclear center.

C. *Ubiquinone Binding Sites in Membrane Proteins*

Ubiquinones are energy transducers that are obligatory in many respiratory and photosynthetic electron transport chains. The ubiquinone enzymes involved in these reactions usually function in a manner that couples the electron transfer by the ubiquinone to proton translocation across the membrane.The structural makeup of the ubiquinone active site permits varying functional roles that influence the electron and proton chemistry.

Ubiquinone binding sites could be categorized into three different types termed donor, acceptor, and pair-splitting sites according to Fisher and Rich (2000). In both the donor and acceptor ubiquinone binding sites, there are two half-reactions which proceed with near equal potentials and pass through a stable semiquinone intermediate. In the case of the pair-splitting ubiquinone binding site, there is no stable semiquinone and two separate electron carriers perform the oxidation of the ubiquinol with large differences in redox potentials. It is this large difference in redox potential that allows the enzyme to bypass the semiquinone intermediate in this ubiquinone site. Although there have been more than 50 types of ubiquinone binding sites identified in electron transport chains, no real identifiable sequence motif has been observed except for a general triad assembly of residues on a single α-helix flanking the ubiquinone head group (Fisher and Rich, 2000). However, a structural motif is beginning to unfold based on the crystallographic data available from the available structures (see Section I).

In both the Q_B site of RC and the Q_N (Fig. 7) site of the bc_1 complex, a semiquinone intermediate is stabilized. This mechanism allows the ubiquinone to switch between two distinct one-electron transfers. Four contact points between the ubiquinone molecule and the enzyme were revealed in a brief structural analysis of these ubiquinone binding sites (Murray *et al.*, 1999). The ubiquinone molecule is stabilized in a hydrophobic pocket where the ring structure of the ubiquinone is sandwiched between two hydrophobic residues (contact points 1 and 2). The distal oxygen (distal from the hydrophobic tail) of the ubiquinone ring is always hydrogen bonded with a histidine residue (contact 3). The proximal oxygen is hydrogen bonded by various hydrogen bond partners (contact 4). The approximate dimensions of the binding site is 9.5 Å from the hydrogen bonding residues (contact point 3 and 4) and 7.5 Å between hydrophobic residues (contact points 1 and 2).

The cytochrome bc_1 complex maintains a pair-splitting ubiquinone binding site, Q_P. The Q_P site removes the need for the semiquinone intermediate through the bifurcation of electrons to the ISP and heme b_L. Unfortunately, there is no structure containing a ubiquinone at the Q_N site as mentioned, but complexes with inhibitors have been reported. In the presence of different inhibitors the hydrogen bonding pattern at the Q_P site and position of the ISP changes. Figure 8 shows four contact points positioned in a similar manner to the other ubiquinone binding sites presented. The binding pocket maintains the sandwich of hydrophobic residues (contacts 1 and 2) in the wild type as well as all inhibitor structures. Furthermore, a histidine residue from the ISP is hydrogen bonded to the distal oxygen (contact 3) and there is a glutamate residue hydrogen bonded to the proximal oxygen (contact 4).

D. *A Novel Ubiquinone Binding Site in Cytochrome* bo_3

Unlike the other ubiquinol binding sites presented above, the ubiquinol binding site in cytochrome bo_3 stabilizes a semiquinone during the oxidation of ubiquinol. A previous model placed the ubiquinone binding site of cytochrome bo_3 within the extrinsic domain of subunit II, replacing the Cu_A site (Murray *et al.*, 1999). However, several lines of evidence are available to refute this model.

To begin with, the proposed ubiquinone binding site was unlike any of the previous sites that have been described in membrane proteins that bind ubiquinone. As presented above the photosynthetic reaction center (Deisenhofer *et al.*, 1995), the bc_1 complex (Xia *et al.*, 1987; Zhang *et al.*, 1998; Iwata *et al.*, 1998; Hunte *et al.*, 2000), and fumarate reductase (Iverson *et al.*, 1999; Lancaster *et al.*, 1999) bind the ubiquinone molecule within membrane-spanning helices where the ring of the ubiquinone molecule is oriented near the phospholipid head group of the membrane. Second, a large mutational study of residues at or around this region in cytochrome bo_3 had little to no effect on the enzyme activity (Ma *et al.,* 1998). Third, ubiquinone is dissolved within the lipid bilayer and the proposed binding site is clearly outside the membrane-associated region. It is unlikely that the ubiquinone would leave the favorable environment of the membrane to release electrons. Finally, the structure of the soluble fragment of subunit II (Wilmanns *et al.,* 1995) is virtually identical to the same domain in the intact cytochrome bo_3 with the exception of a shift in a small loop composed of residues 209–217. This structural difference is small but important as it renders the previously proposed binding site inaccessible.

It has been proposed that cytochrome bo_3 has two ubiquinone binding sites, a high-affinity (Q_H) and a low-affinity (Q_L) site. The low-affinity site is said to function as a pair-splitting site donating one electron to heme *b* and the other to the high-affinity site (Sato-Watanabe *et al.,* 1994). The Q_H site maintains a tightly bound ubiquinone when the enzyme is solubilized in the detergent dodecyl maltoside (DDM) (Sato-Watanabe *et al.,* 1994; Puustinen *et al.,* 1996). Unfortunately, crystals could only be obtained using the detergent octyl glucoside (OG), under which conditions ubiquinone appears not to bind. Because of these difficulties, we have been unable to obtain a cytochrome bo_3 structure complexed with ubiquinone or any of its analogues.

Based on similarities with known structural motifs in membrane proteins which bind ubiquinone and biochemical data, a potential ubiquinone binding site was located in the membrane domain of subunit I of cytochrome bo_3. The proposed binding site is formed from a patch of conserved polar residues located near the surface of transmembrane helices I and II. To maintain such an energetically unfavorable structure in a hydrophobic environment suggests this feature may be functionally important. Sequence alignments show that the residues within this polar cluster, Arg-71 and Asp-75 in the C-terminal half of helix I, and His-98 and Gln-101 in the N-terminal half of helix II, are conserved in the ubiquinol oxidases but not in the cytochrome *c* oxidases (Fig. 13). Subunit I between cytochrome *c* and ubiquinol oxidases shows a high degree of homology but the sequence in this particular region is not conserved.

E. Analysis of Active Site Mutations

In order to ascertain the functional relevance of this polar cluster, site-specific mutagenesis was performed on the residues Arg-71, Asp-75, His-98, and Gln-101. The mutations of Arg-71, Asp-75, and His-98 all showed a block in enzymatic activity (Abramson *et al.,* 2000). Even the most conservative Q101N mutation inhibited activity by 75% and caused

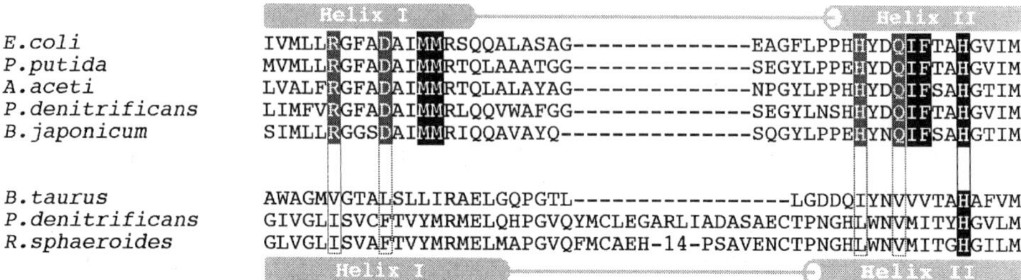

FIG. 13. Sequence alignment of membrane-spanning helices I and II in subunit I displaying the conserved polar residues of cytochrome bo_3. Five ubiquinol oxidase sequences (top) and three cytochrome *c* oxidase sequences (bottom) are aligned. Residues mutated in this work are boxed in light gray and other residues discussed in the text are boxed in dark gray. Helices I and II are marked. The histidine ligand of the low-spin heme, which is conserved in all heme–copper oxidases, is also shown in a black box in helix II. An interhelical loop in the *Rhodobacter sphaeroides* sequence has been omitted as marked. The following sequences were used: (top) *E. coli*, CYOB_ECOLI (Swissprot); *Pseudomonas putida*, BAA76357 (Genbank); *Acetobacter aceti*, QOX1_ACEAC (Swissprot); *Paracoccus denitrificans*, B54759 (Genbank); *Bradyrhizobium japonicum*, CO14_BRAJA (Swissprot); (bottom) *Bos taurus*, COX1_BOVIN (Swissprot); *P. denitrificans*, CX1B_PARDE (Swissprot); *R. sphaeroides*, COX1_RHOSH (Swissprot).

a 10-fold increase in the apparent K_M value for ubiquinol-1. Optical spectra of all these mutants showed no substantial perturbations, indicating no large structural changes near the heme groups.

Electrochemical studies of the wild-type as well as Q101N mutant enzyme reconstituted in proteoliposomes was monitored by Wikström's group and proved to be a powerful tool for following electron transfer (Abramson *et al.*, 2000). The electrometric response of the Q101N mutant showed that electron transfer from the bound ubiquinol molecule was lost. In the presence of bound ubiquinol, the initial fast electrometric response on oxidation of the reduced wild-type enzyme by O_2 can be assigned to the conversion of the enzyme–O_2 complex (**A** state) to the oxoferryl state (**F** state). The subsequent slower phase, with equal amplitude to the fast phase, is assigned to reduction of the oxoferryl state (**F** state) by the fourth electron from ubiquinol (**O** state, Jasaitis *et al.*, 1999). Without bound ubiquinol, the fully reduced enzyme has only three electrons available for oxygen reduction, so the reaction becomes truncated at the oxoferryl state (**F** state, Puustinen *et al.*, 1996). Oxygen reduction in the Q101N mutant enzyme stops at this state, indicating that normal fast electron transfer from bound ubiquinol is disrupted in this mutant. The validity of this method to assess normal binding of ubiquinone is confirmed by the control experiment in which bound ubiquinone had been removed from the wild-type enzyme with Triton X-100 (TX-100, Puustinen *et al.*, 1996).

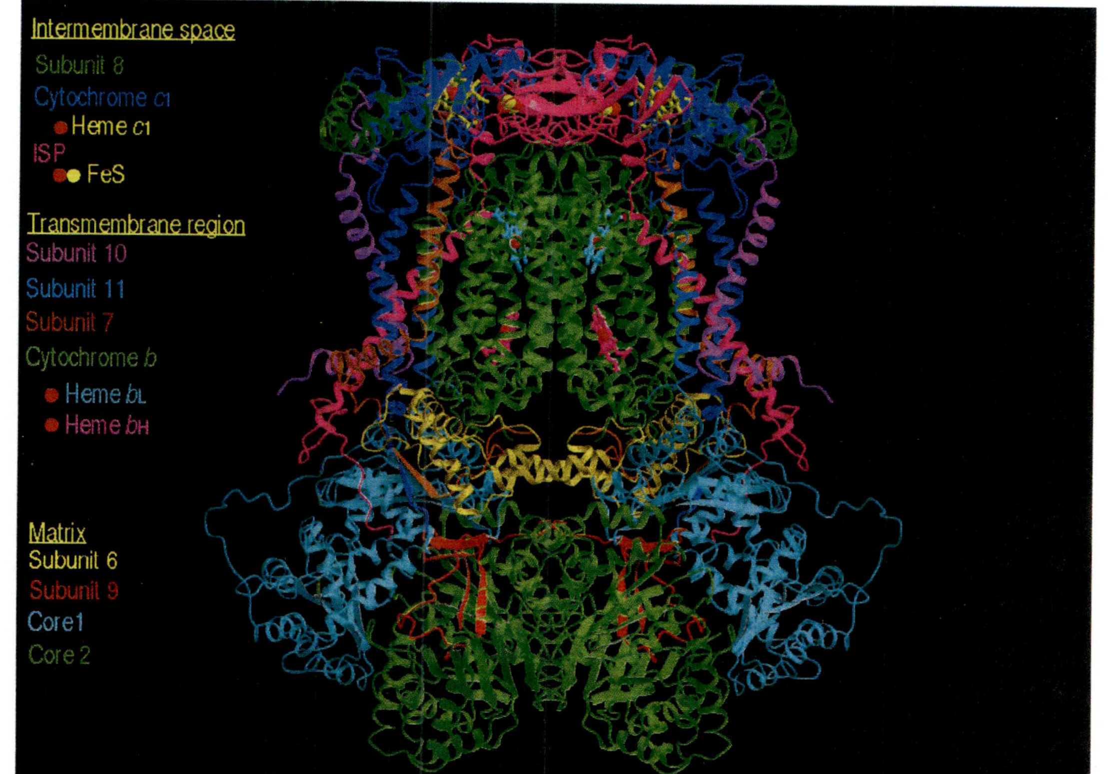

Iwata *et al.*, FIG. 4. Overall dimer structure of the cytochrome bc_1 complex from bovine heart mitochondria. Reprinted from Abramson, J., Riistama, S., Larsson, G., Jasaitis, A., Svensson-Ek, M., Laakkonen, L., Puustinen, A., Iwata, S., and Wikström, M. (2000). *Nat. Struct. Biol.* 7, 910–917, with permission from Nature Publishing Group.

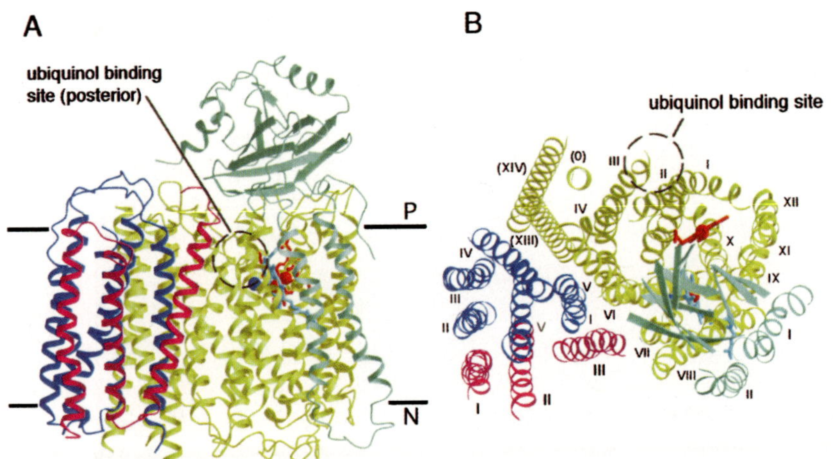

Iwata *et al.*, FIG. 12. Overall structure of cytochrome bo_3. Cytochrome bo_3 has four subunits that are colored yellow/green (subunit I), green (subunit II), blue (subunit III), and pink (subunit IV). The roman numerals represent the helix numbering for each subunit. Heme *b* (red), heme o_3 (cyan), and the Cu_B (blue) are all located in subunit I. The location of the proposed ubiquinol binding site is indicated by the dashed circle. (A) View parallel to the membrane. (B) View normal from the *P*-side of the membrane. Reprinted from Abramson, J., Riistama, S., Larsson, G., Jasaitis, A., Svensson-Ek, M., Laakkonen, L., Puustinen, A., Iwata, S., and Wikström, M. (2000). *Nat. Struct. Biol.* 7, 910–917, with permission from Nature Publishing Group.

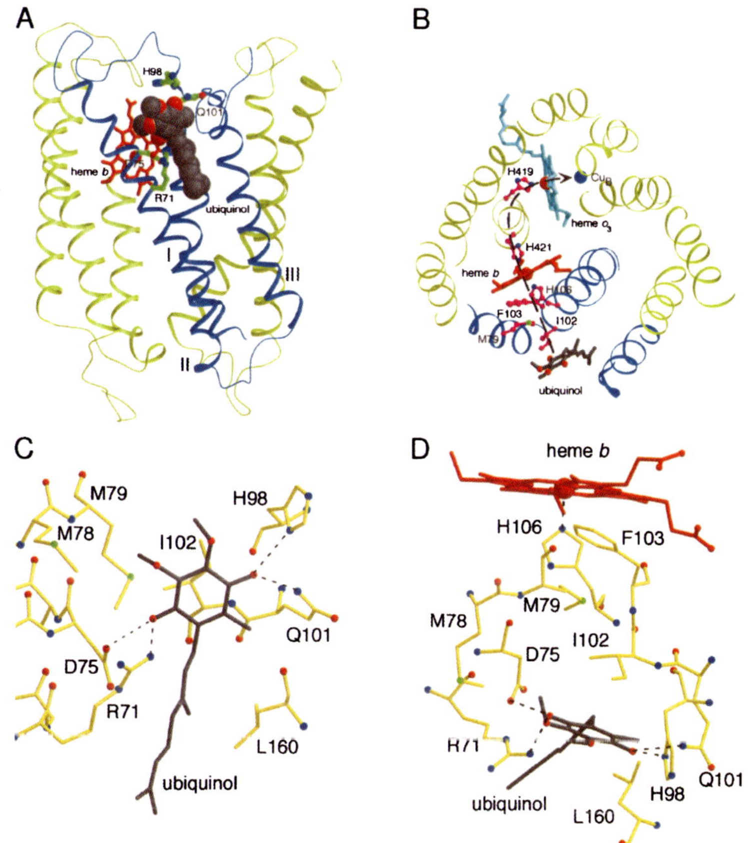

Iwata *et al.*, FIG. 14. The predicted ubiquinone binding site in cytochrome bo_3. (A) View parallel to the membrane of subunit I with the modeled ubiquinone-2. Helices I, II, and III of subunit I are colored blue; ubiquinol-2 and heme *b* are colored black and red, respectively. The polar residues (Arg-71, Asp-75, His-98, and Glu-101) forming the ubiquinol binding site are shown in green. (B) View normal from the *P*-side of the membrane of subunit I with the modeled ubiquinone-2. A possible electron path from the ubiquinone to the binuclear center (heme o_3 and Cu_B) via heme *b* is shown by a dotted line. The residues along the path (IIe-102, Met-79, Phe-103, and histidine ligands for hemes) are shown in pink. (C) and (D) Detailed view of the ubiquinone binding site parallel to the membrane (C) and along the membrane normal from the *P*-side (D). Reprinted from Abramson, J., Riistama, S., Larsson, G., Jasaitis, A., Svensson-Ek, M., Laakkonen, L., Puustinen, A., Iwata, S., and Wikström, M. (2000). *Nat. Struct. Biol.* 7, 910–917, with permission from Nature Publishing Group.

F. Modeling of the Ubiquinol Binding Site

From the structural and biochemical evidence presented above a molecule of ubiquinone-2 (UQ2) was modeled into the proposed site. The model was aligned in such a way as to allow interaction of the ubiquinone ring with the polar residues. Using the program CNS (Adams *et al.*, 1997) with energy minimization procedures the optimal position for UQ2 was obtained. The UQ2 molecule is positioned in a pocket formed from three membrane-spanning helices in subunit I (helices I, II and III). This site is exposed to the membrane bilayer providing direct access for the substrate molecule, ubiquinol. EPR measurement supports the modeling of UQ2, placing the ubiquinone ring near the surface of the phospholipid membrane at ~70° and ~15° from the plane of the membrane (Murray *et al.*, 1999). The distance from the face of the ubiquinone ring to heme *b* is ~13 Å permitting fast electron transfer from ubiquinol to heme *b* (Fig. 14, see color insert). The contact points that form the ubiquinone-binding site (discussed in the previous section) can also be seen in this binding site. His-98 (contact 3) and the carboxylate of Asp-75 (contact 4) stabilize the oxygen atoms of the ubiquinone ring. A similar arrangement has been seen in the ubiquinone binding sites of the cytochrome bc_1. In cytochrome bo_3, the distance between $N_{\varepsilon 2}$ of His-98 and $O_{\delta 1}$ of Asp-75 is 11 Å, which is similar to the distance observed in the Q_i site (10 Å) of the bovine cytochrome bc_1 complex (Iwata *et al.*, 1998). His-98 is located at the N-terminal end of transmembrane helix II and is exposed to the solvent in the periplasm, thus allowing direct proton release. The hydrophobic residues Ile-102 (contact 2) and also Leu-160 stabilize the ubiquinone ring.

The ubiquinol binding site modeled here appears to be that of the Q_H site. Optical spectra of the mutant enzymes show a slight redshift in the heme *b* spectrum which has also been observed in the TX-100-treated wild-type enzyme (no bound ubiquinol). This indicates that the loss of the bound ubiquinone-8 affects the spectrum of heme *b* and further suggests that the ubiquinone-binding site is relatively close to the heme, as also concluded by Sato-Watanabe *et al.* for the Q_H site (Adelroth *et al.*, 1998). Mutations in this site leads to the loss of activity that could be related to changes in the ubiquinone binding site and/or a disruption of electron transfer from the site. The steady-state level of reduction of heme *b* was decreased during turnover in all mutant enzymes, which shows that the inhibition of enzyme activity is, indeed, due to impaired electron transfer from ubiquinol to the heme group. Thus, all current mutagenesis data suggest that the site proposed here corresponds to the Q_H site.

One of the remaining questions about ubiquinone binding in cytochrome bo_3 is the existence of the low-affinity (Q_L) site. There appears to be no other reasonable site for the binding of a ubiquinone in this structure. However, it may be possible that the site found here is alone sufficient for the oxidation of ubiquinol by cytochrome bo_3. The binding site is directly exposed to the lipid bilayer, which makes the uptake of ubiquinol and the release of ubiquinone relatively easy. Following the binding of ubiquinol, an initial electron is transferred to heme *b* and two protons are released, forming the semiquinone anion. Multiple hydrogen bonds or electrostatic interaction from the positively charged arginine residue could stabilize the semiquinone anion. The stabilized semiquinone could then deliver the second electron to heme *b* at nearly the same redox potential (Ingledew *et al.,* 1995). Finally, the oxidized ubiquinone can exit the binding site and be replaced by the fully reduced ubiquinol. Further studies are required to determine whether one site is sufficient or whether two are essential for the function of this enzyme.

IV. Conclusion

We have summarized the structures of two quinone-binding membrane proteins in this chapter. Although there are some common motifs between the two quinone binding sites, they are quite different overall. This individuality of quinone binding sites has been further reinforced by the recent structures of fumarate reductase and formate dehydrogenase, which have different sites from any previously solved ones. It is interesting to know how many different types of quinone binding site in the membrane there are, and if it even makes sense to categorize them. This is a very challenging but important question for membrane biochemistry. In order to address this question, we are currently studying the quinone binding sites in a number of different proteins including Complex I and nitrate reductase.

References

Abramson, J., Riistama, S., Larsson, G., Jasaitis, A., Svensson-Ek, M., Laakkonen, L., Puustinen, A., Iwata, S., and Wikström, M. (2000). *Nat. Struct. Biol.* **7,** 910–917.

Adams, P. D., Pannu, N. S., Read, R. J., and Brunger, A. T. (1997). *Proc. Natl. Acad. Sci. USA* **94,** 5018–5023.

Adelroth, P., Gennis, R. B., and Brzezinski, P. (1998). *Biochemistry* **37,** 2470–2476.

Allen, J. P., Feher, G., Yeates, T. O., Rees, D. C., Deisenhofer, J., Michel, H., and Huber, R. (1986). *Proc. Natl. Acad. Sci. USA* **83,** 8589–8593.

Berry, E. A., Guergova-Kuras, M., Huang, L. S., and Crofts, A. R. (2000). *Annu. Rev. Biochem.* **69,** 1005–1075.

Brasseur, G., Saribas, A. S., and Daldal, F. (1996). *Biochim. Biophys. Acta* **1275,** 61–69.

Chepuri, V., and Gennis, R. B. (1990). *J. Biol. Chem.* **265,** 12978–12986.

Clejan, L., and Beattie, D. S. (1986). *Methods Enzymol.* **126,** 173–180.

Crofts, A. R., Meinhardt, S. W., Jones, K. R., and Snozzi, M. (1983). *Biochim. Biophys. Acta* **723,** 243–271.

Crofts, A. R., Hong, S., Ugulava, N., Barquera, B., Gennis, R., Guergova-Kuras, M., and Berry, E. A. (1999). *Proc. Natl. Acad. Sci. USA* **96,** 10021–10026.

Darrouzet, E., Valkova-Valchanova, M., Moser, C. C., Dutton, P. L., and Daldal, F. (2000). *Proc. Natl. Acad. Sci. USA* **97,** 4567–4572.

Deisenhofer, J., Epp, O., Miki, K., Huber, R., and Michel, H. (1985). *Nature* **318,** 618–624.

Deisenhofer, J., Epp, O., Sinning, I., and Michel, H. (1995). *J. Mol. Biol.* **246,** 429–457.

Edwards, C. A., and Trumpower, B. L. (1986). *Methods Enzymol.* **126,** 211–224.

Esposti, M. D., de Vries, S., Crimi, M., Ghelli, A., Patarnello, T., and Meyer, A. (1993). *Biochim. Biophys. Acta* **1143,** 243–271.

Fisher, N., and Rich, P. R. (2000). *J. Mol. Biol.* **296,** 1153–1162.

Hunte, C., Koepke, J., Lange, C., Rossmanith, T., and Michel, H. (2000). *Structure Fold. Des.* **8,** 669–684.

Ingledew, W. J., Ohnishi, T., and Salerno, J. C. (1995). *Eur. J. Biochem.* **227,** 903–908.

Iverson, T. M., Luna-Chavez, C., Cecchini, G., and Rees, D. C. (1999). *Science* **284,** 1961–1966.

Iwata, S., Ostermeier, C., Ludwig, B., and Michel, H. (1995). *Nature* **376,** 660–669.

Iwata, S., Saynovits, M., Link, T. A., and Michel, H. (1996). *Structure* **4,** 567–579.

Iwata, S., Lee, J. W., Okada, K., Lee, J. K., Iwata, M., Rasmussen, B., Link, T. A., Ramaswamy, S., and Jap, B. K. (1998). *Science* **281,** 64–71.

Jasaitis, A., Verkhovsky, M. I., Morgan, J. E., Verkhovskaya, M. L., and Wikström, M. (1999). *Biochemistry* **38,** 2697–2706.

Jordan, P., Fromme, P., Witt, H. T., Klukas, O., Saenger, W., and Krauss, N. (2001). *Nature* **411,** 909–917.

Jormakka, M., Tornroth, S., Byrne, B., and Iwata, S. (2002). *Science* **295,** 1863–1868.

Kim, H., Xia, D., Yu, C. A., Xia, J. Z., Kachurin, A. M., Zhang, L., Yu, L., and Deisenhofer, J. (1998). *Proc. Natl. Acad. Sci. USA* **95,** 8026–8033.

Lancaster, C. R., Kroger, A., Auer, M., and Michel, H. (1999). *Nature* **402,** 377–385.

Lemieux, L. J., Calhoun, M. W., Thomas, J. W., Ingledew, W. J., and Gennis, R. B. (1992). *J. Biol. Chem.* **267,** 2105–2113.

Ma, J., Puustinen, A., Wikstrom, M., and Gennis, R. B. (1998). *Biochemistry* **37,** 11806–11811.

Ma, J., Tsatsos, P. H., Zaslavsky, D., Barquera, B., Thomas, J. W., Katsonouri, A., Puustinen, A., Wikström, M., Brzezinski, P., Alben, J. O., and Gennis, R. B. (1999). *Biochemistry* **38,** 15150–15156.

Mitchell, P. (1976). *J. Theor. Biol.* **62,** 327–367.

Murray, L., Pires, R. H., Hastings, S. F., and Ingledew, W. J. (1999). *Biochem. Soc. Trans.* **27,** 581–585.

Nett, J. H., Hunte, C., and Trumpower, B. L. (2000). *Eur. J. Biochem.* **267,** 5777–5782.

Puustinen, A., Finel, M., Haltia, T., Gennis, R. B., and Wikström, M. (1991). *Biochemistry* **30,** 3936–3942.

Puustinen, A., Verkhovsky, M. I., Morgan, J. E., Belevich, N. P., and Wikstrom, M. (1996). *Proc. Natl. Acad. Sci. USA* **93,** 1545–1548.

Sato-Watanabe, M., Mogi, T., Ogura, T., Kitagawa, T., Miyoshi, H., Iwamura, H., and Anraku, Y. (1994). *J. Biol. Chem.* **269,** 28908–28912.

Schägger, H., Link, T. A., Engel, W. D., and von Jagow, G. (1986). *Methods Enzymol.* **126,** 224–237.

Schägger, H., Brandt, U., Gencic, S., and von Jagow, G. (1995). *Methods Enzymol.* **260,** 82–96.

Trumpower, B. L. (1990). *J. Biol. Chem.* **265,** 11409–11412.

Trumpower, B. L., and Gennis, R. B. (1994). *Annu. Rev. Biochem.* **63,** 675–716.

Tsukihara, T., Aoyama, H., Yamashita, E., Tomizaki, T., Yamaguchi, H., Shinzawa-Itoh, K., Nakashima, R., Yaono, R., and Yoshikawa, S. (1996). *Science* **272,** 1136–1144.

de Vries, S. (1986). *J. Bioenerg. Biomembr.* **18,** 195–224.

Widger, W. R., Cramer, W. A., Herrmann, R. G., and Trebst, A. (1984). *Proc. Natl. Acad. Sci. USA* **81,** 674–678.

Wilmanns, M., Lappalainen, P., Kelly, M., Sauer-Eriksson, E., and Saraste, M. (1995). *Proc. Natl. Acad. Sci. USA* **92,** 11955–11959.

Xia, D., Yu, C. A., Kim, H., Xia, J. Z., Kachurin, A. M., Zhang, L., Yu, L., and Deisenhofer, J. (1997). *Science* **277,** 60–66.

Yoshikawa, S., Shinzawa-Itoh, K., Nakashima, R., Yaono, R., Yamashita, E., Inoue, N., Yao, M., Fei, M. J., Libeu, C. P., Mizushima, T., Yamaguchi, H., Tomizaki, T., and Tsukihara, T. (1998). *Science* **280,** 1723–1729.

Yun, C. H., Crofts, A. R., and Gennis, R. B. (1991). *Biochemistry* **30,** 6747–6754.

Zhang, Z., Huang, L., Shulmeister, V. M., Chi, Y. I., Kim, K. K., Hung, L. W., Crofts, A. R., Berry, E. A., and Kim, S. H. (1998). *Nature* **392,** 677–684.

Zhu, Q. S., van der Wal, H. N., van Grondelle, R., and Berden, J. A. (1984). *Biochim. Biophys. Acta* **765,** 48–57.

PROKARYOTIC MECHANOSENSITIVE CHANNELS

By PAVEL STROP,* RANDAL BASS,† and DOUGLAS C. REES†,‡

*Biochemistry Option, California Institute of Technology, Pasadena, California 91125
†Division of Chemistry and Chemical Engineering 114-96, California Institute of Technology, Pasadena, California 91125, and ‡Howard Hughes Medical Institute, Pasadena, CA 91125

I. Introduction

Mechanosensitive ion channels are integral membrane proteins that open and close in response to mechanical stress applied either directly to the cell membrane (in the case of intrinsically mechanosensitive channels) or indirectly, through forces applied to other cytoskeletal components (Sachs, 1997; Gillespie and Walker, 2001; Hamill and Martinac, 2001). Cellular phenomena mediated by mechanosensitive channels include touch, hearing, cardiovascular tone, detection of gravity, pressure sensation, pain perception, and osmoregulation. Mechanosensitive channels are quite diverse both physiologically and structurally (Fig. 1) and have been discovered in all fundamental branches of the phylogenetic tree, Eubacteria, Eukarya, and Archaea. Eukaryotic mechanosensitive channels include the TRPV subclass of the Transient Receptor Potential channel family (Clapham *et al.*, 2001); the TREK-1 and TRAAK members of the two-pore domain potassium channel family K_{2P} (Maingret *et al.*, 1999a,b); and the DEG/ENaC superfamily (composed of degenerins, eptithelial sodium channels, and acid-sensing channels (see Alvarez de la Rosa *et al.*, 2000; Tavernarakis and Driscoll, 2001;

ADVANCES IN PROTEIN CHEMISTRY, Vol. 63

0065-3233/03 $35.00

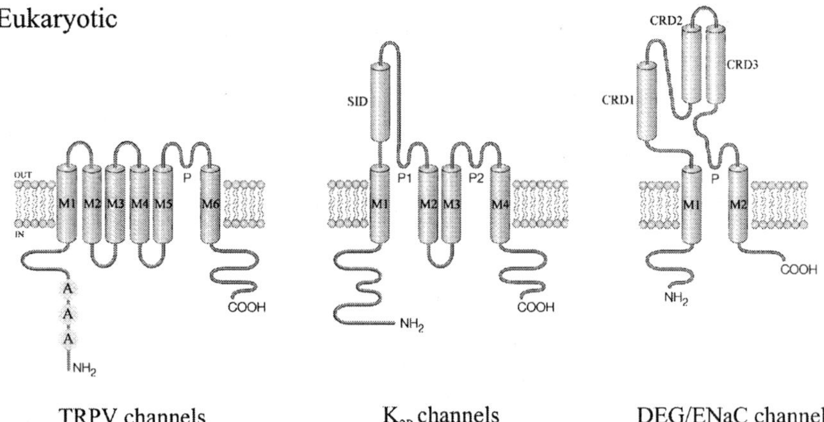

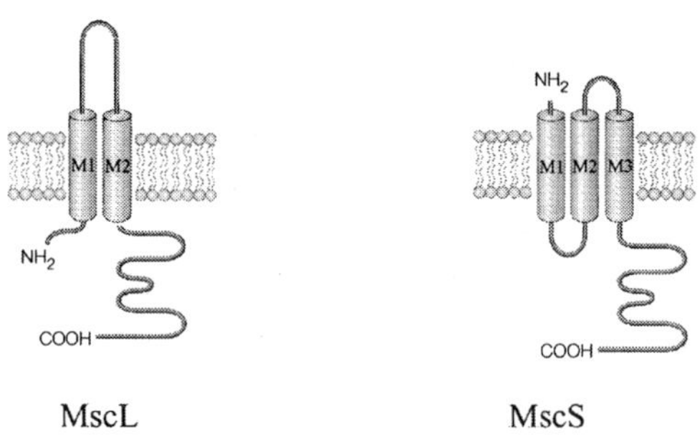

FIG. 1. Predicted membrane-spanning topology for mechanosensitive channels found in eukaryotes (TRPV, K_{2P}, and DEG/ENaC channels) and bacteria (MscL and MscS). In addition to the transmembrane helices (represented as cylinders), other motifs present in these channels are designated as follows. The TRPV channels contain several cytoplasmic ankyrin domains (A) at the N terminus, and one pore-forming loop (P). K_{2P} channels have two pore-forming loops and a self-interaction domain (SID) through which dimers are generated. DEG/ENaC sodium channels have a single pore-forming loop and three cysteine-rich domains (CRDs).

Welsh *et al.,* 2002), that have been implicated in the touch response of nematodes. Although many of these channels, including the degenerins and the yet-to-be identified channel receptors involved in hearing (Gillespie and Walker, 2001), appear to have an obligatory requirement for cytoskeletal coupling, others, such as the TREK-1 and TRAAK channels, exhibit intrisinic mechanosensitive activity. Unfortunately, eukaryotic mechanosensitive systems tend to be complex and have yet to be well defined biochemically. In contrast, prokaryotic mechanosensitive channels can be both relatively simple and intrinsically mechanosensitive. Consequently, bacterial channels have been much more amenable to biochemical, biophysical, and genetic characterization, and hence can serve as models for establishing basic features of mechanosensation that may also be relevant to the more complex eukaryotic mechanosensitive channels. Reviews of these prokaryotic channels from a variety of perspectives may be found in Sukharev *et al.* (1997), Batiza *et al.* (1999), Blount *et al.* (1999), Oakley *et al.* (1999), Spencer *et al.* (1999), Martinac (2001), and Kloda and Martinac (2002a).

Prokaryotic mechanosensitive channels have been proposed to function in the response of a microorganism to an abrupt transition from a high to low osmotic strength environment (Blount *et al.*, 1996a; Sukharev *et al.*, 1997; Booth and Louis, 1999; Poolman *et al.*, 2002). The response mechanisms of a bacteria to osmotic stress have been best characterized in *Escherichia coli* (see Csonka and Epstein, 1999). During downshock conditions, when bacteria are shifted from a high to a low osmolarity solution, water enters the cell and generates a large increase in turgor pressure. Mechanosensitive channels embedded in the plasma membranes of bacteria can respond to sudden increases in turgor pressure by opening under the most stressful conditions, thereby jettisoning water and solutes from the cytoplasm to prevent cell lysis during hypoosmotic shock. These channels are often nonselective and allow the passage of an astonishing variety of solutes across the membrane, ranging from monovalent ions and water to proteins such as thioredoxin (Cruickshank *et al.*, 1997; Ajouz *et al.*, 1998). Based on their distinct conductances and activation properties, three mechanosensitive channels have been identified in *E. coli* (Martinac *et al.*, 1987; Sukharev *et al.*, 1994; Le Dain *et al.*, 1998; Levina *et al.*, 1999): the mechanosensitive channels of large conductance (MscL), small conductance (MscS), and mini conductance (MscM). These channels are located in the inner membrane of the bacteria (Berrier *et al.*, 1989). As detailed below, the proteins responsible for the MscL and MscS activities have been identified (Sukharev *et al.*, 1994; Levina *et al.*, 1999). The presence of multiple mechanosensitive channels apparently confers a level of redundancy on the bacterial response to osmotic downshock; while knockout mutants in only one of these activities exhibit little phenotype, mutants lacking both MscL and MscS are severely compromised during osmotic downshock (Levina *et al.*, 1999).

In this review, we first introduce some general background considerations relevant to the description of mechanosensitive channels, followed by a discussion of the best characterized prokaryotic channels, MscL and MscS. An important objective of this analysis, still to be fully realized, is to identify the protein structural elements that are responsible for mechanosensitivity—we would like to be able to answer the question "What makes a mechanosensitive channel mechanosensitive?"

A. *Thermodynamic Aspects of Mechanosensitivity*

The consequences of applied membrane tension to an intrinsically mechanosensitive channel can be readily evaluated for a simple two-state system where the channel can exist in either closed (C) or open (O) conformations:

$$C \leftrightarrow O \tag{1}$$

Values for ΔG°, the standard free energy for channel opening in the absence of applied tension, for channels such as MscL are ~40 kJ/mol (see Sukharev *et al.,* 1997). If the difference in cross-sectional area, $\Delta A = A_{open} - A_{closed}$, between closed and open states is nonzero, then stretching the membrane will increasingly stabilize the state with the greater cross-sectional area. Typically, $\Delta A > 0$, so that stretching the membrane opens the channel. Following Howard *et al.* (1988), the contribution of membrane tension, σ, to the free-energy of channel opening, ΔG, may be included through the expression

$$\Delta G = \Delta G^\circ - \sigma \Delta A \tag{2}$$

This formalism is analogous to the treatments that can be used to incorporate the effects of membrane potential and ligand concentration on the free energy of opening for voltage gated and ligand gated channels, respectively.

For this two-state system, the ratio of channels in the closed to open states is given by a Boltzmann-type equilibrium expression

$$\frac{(O)}{(C)} = e^{-(\Delta G^\circ - \sigma \Delta A)/RT} \tag{3}$$

where R is the gas constant (8.3144 J mol^{-1} K^{-1}) and T is the absolute temperature. It should be noted that in the electrophysiology literature, the properties of single channels are typically under investigation, so that the corresponding expression addresses the energetics on a per molecule basis, with the result that $k_B T$ appears in place of RT, where k_B is the Boltzmann constant $= R/N_A$.

The applied tension, $\sigma_{1/2}$, required to open half the channels at equilibrium is equal to

$$\sigma_{1/2} = \frac{\Delta G^\circ}{\Delta A} \tag{4}$$

The steepness of the response of the channel to applied tension is determined by the magnitude of the change in cross-sectional area; for example, in terms of units typically employed in these calculations, if $\Delta A = 100$ Å^2, then an increase of 1 dyn cm^{-1} in membrane tension corresponds to a free energy change of -0.602 kJ/mol, and ΔG will scale proportionally to changes in ΔA. Consequently, for a channel with $\Delta G^\circ = 40$ kJ/mol and $\sigma_{1/2} = 10$ dyn cm^{-1} (typical for mechanosensitive channels such as MscL), $\Delta A \sim 660$ Å^2.

A convenient experimental system for manipulating membrane tension involves applying suction (negative pressure) in an electrophysiological patch clamp experiment, which leads to stretching and curvature of the membrane patch (see Sachs, 1997; Hamill and Martinac, 2001). For a film such as a membrane patch that separates two compartments, the condition of equilibrium reflects the balance between two competing effects: the surface energy of the film, $\sigma \delta A$, which will decrease as the surface area decreases, and a countering pressure differential, ΔP, across the film that will provide a work of expansion equivalent to $\Delta P \delta V$. Equating these terms for a spherical film of radius r, where $\delta A = 8\pi r dr$ and $\delta V = 4\pi r^2 dr$, gives the Laplace equation (Adamson, 1982) that relates these quantities.

$$\Delta P = \frac{2\sigma}{r} \tag{5}$$

Consequently, larger spheres require a smaller pressure differential to maintain a given membrane tension than smaller spheres, and vice versa. The magnitudes of the pressure differential required to generate specific levels of membrane tension can be estimated from the following considerations. For a membrane with a radius of curvature $= 3$ μm (typical of the dimensions in a patch clamp experiment), application of 0.1 atm pressure (where 1 atm $= 760$ mm Hg $= 1.013 \times 10^6$ dyn cm^{-2}) corresponds to $\sigma = 15$ dyn cm^{-1}. The maximum applied pressure depends on the limit at which the patches break ($\sim$20–30 dyn cm^{-1}) and the curvature of the membrane, but is typically $\sim$0.2 atm.

B. Osmotic Pressure

Although applying suction to a membrane patch is an experimentally convenient mechanism for generating pressure differentials and associated tension, it is not a very physiologically relevant mechanism for many mechanosensitive systems. For prokaryotes, a more likely situation involves experiencing changes in environmental conditions associated with changes in osmotic pressure between the outside and inside of the cell. These pressure differences can be substantial; for a concentration gradient across a membrane, ΔC, in mol/liter (or more precisely osmolarity, Osm), the osmotic pressure difference, $\Delta\Pi$, in atmospheres, at 25°C is given by the relationship

$$\Delta\Pi = 22.4\Delta C \tag{6}$$

so that a concentration difference of 1 *M* across a membrane gives an osmotic pressure difference of 22.4 atm. A concentration gradient of only ~0.01 *M* will produce a 0.2 atm osmotic pressure differential that was shown above to rupture membranes with radii ~3 μm.

The response of a bacteria to variations in the external osmolarity has been extensively characterized for *Escherichia coli* (Record *et al.,* 1998; Cayley *et al.,* 2000). While optimal growth rates occur in media of osmolarity ~0.3 *M*, cells can grow in medium varying by over 100-fold in external osmolarity (~0.02 *M* to ~3 *M*). A variety of osmoregulatory mechanisms are utilized by *E. coli* to minimize the consequences of variability in external osmolarity (Csonka and Epstein, 1999). The first line of defense is provided by the peptidoglycan outer membrane that provides the main physical barrier against osmotically induced swelling and shrinking. This cell wall can stretch like a balloon under outwardly directed turgor pressure (Doyle and Marquis, 1994; Cayley *et al.,* 2000), which is the osmotic pressure difference between the inside and outside of the cell. The turgor pressure has been measured as a function of external osmolarity for *E. coli* and has been found to decrease with increasing growth osmolarity, ranging from ~3.1 atm at 0.03 Osm, to less than 0.5 atm above 0.5 Osm. The inner membrane that separates the cytoplasmic and periplasmic volumes is considerably more fragile, and, as the preceding considerations indicate, can only withstand osmotic pressure differentials substantially below 1 atm. Consequently, the periplasm and cytoplasm are isoosmotic, and the turgor pressure is determined by the differences in osmolarities between the periplasm and surrounding solution. Since the outer membrane contains porins that are permeable to water and small solutes, while the inner membrane contains aquaporins that conduct only water and a very few other solutes, the cell can respond quickly to osmotic pressure differences by transport of water (Booth and Louis, 1999). In the case of osmotic upshock, *E. coli* responds by increasing the cytoplasmic concentrations of osmotically active solutes and decreasing the water concentration. Osmotic downshock will be associated with a strong driving force for water influx into the cell; if this driving force is sufficiently large, the mechanical integrity of the cell will be threatened (as in the use of osmotic shock for cell lysis), and it is in this capacity that mechanosensitive channels are believed to play critical roles.

C. *Helix Packing and Gating Transitions*

For channels such as MscL, the permeation pathway for solutes through the membrane is lined by helices. Consequently, changes in the permeation pathway associated with conformational gating between closed and open states will require changes in helix–helix packing arrangements. These arrangements may be characterized by a minimum pore radius, R_0, which will be relevant to the conductance of the pore. For symmetric, oligomeric proteins that have the permeation pathway generated by one helix per subunit, R_0 will depend on the number of helices, N, surrounding the pore, the tilt of the helix with respect to the membrane normal, η, and the interhelical crossing angle, α. With perfectly regular helices modeled as cylinders of diameter d and exact N-fold rotational symmetry, the variation in minimum pore radius with helix tilt may be shown to be (Spencer and Rees, 2002)

$$R_0 = \frac{d}{2}\left[\tan\eta\cot\left(\frac{\alpha}{2}\right) - 1\right] \tag{7a}$$

where

$$\cos\alpha = \cos^2\eta + \sin^2\eta\cos\theta \tag{7b}$$

and $\theta = 2\pi/N$. The dependence of R_0 on η is illustrated in Fig. 2A for different numbers of subunits in an oligomer. For perfectly regular helices and ideal oligomeric symmetry, an increase in pore radius may be generated by some combination of two mechanisms involving an increase in helical tilt η and/or or an increase in the number of helices N lining the permeation pathway. Although this model is idealized and cannot accommodate changes in the diameter of the permeation pathway due to bending or kinking of the pore forming helices

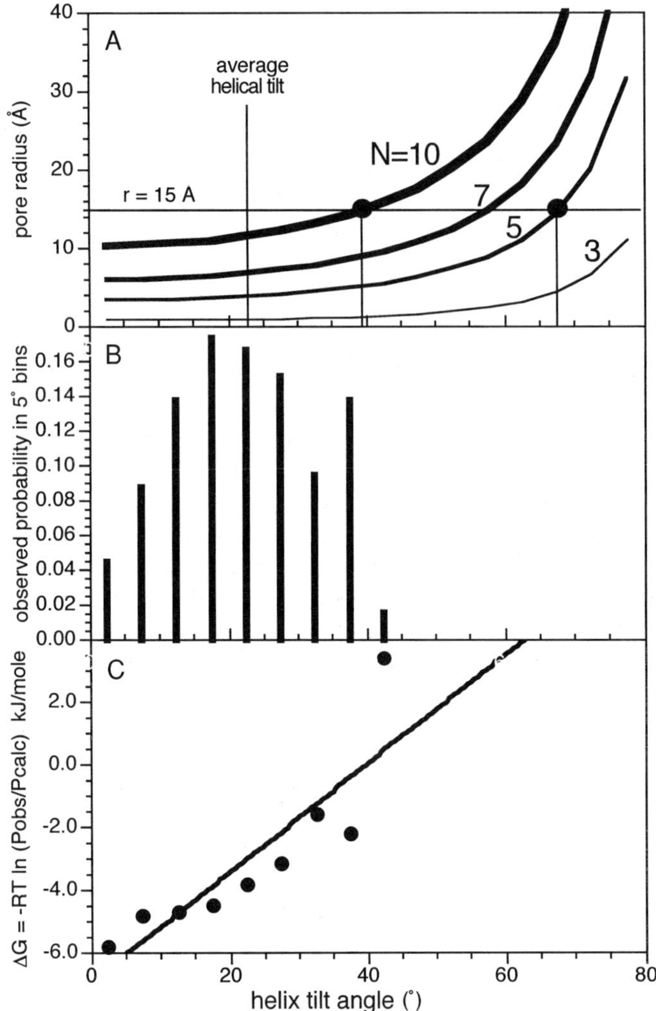

Fig. 2. (A) Dependence of the minimum pore radius on the helix tilt angle for ideal, oligomeric channels with one helix per subunit lining the permeation pathway, as calculated from Eq. (7) for $N = 3, 5, 7$, and 10 (thin to thick lines, respectively), with a helical diameter of 9 Å. The average helical tilt angle = 23° is indicated, as observed in a survey of membrane-spanning helices (Spencer and Rees, 2002). The points of intersections of the line corresponding to a pore radius = 15 Å with the curves for $N = 5$ and 10 are designated by filled circles. The vertical lines from the circle highlight the tilt angles associated with this pore size for these two oligomeric states. (B) Histogram of the observed probability (P_{obs}) distribution for the tilt angle between the axis of membrane-spanning helices and the bilayer normal, tabulated in 5° bins for proteins surveyed in Spencer and Rees (2002). The probability is calculated from the number of helices in each bin, divided

(see Unwin, 1995; Jiang *et al.*, 2002a,b), it will, it is hoped, capture the major features of the large-scale conformational transitions observed in mechanosensitive channels.

In the absence of applied tension, the axes of membrane-spanning helices tend to be oriented along the direction of the membrane normal (Bowie, 1997; Spencer and Rees, 2002), which, according to the analysis presented in Fig. 2A, will favor the closed state of a channel. The observed distribution of helical tilt angles clusters between 0 and 40° with an average value of ~23° (Fig. 2B); in this range, the calculated pore radius of a channel (Fig. 2A) is fairly insensitive to the value η of the helical tilt angle, so that channels should be stabilized in the closed state under these conditions. The energetics associated with changing the helical tilt can be estimated, in what we hope is not too fanciful a fashion, by using a mean-field approximation to extract the tilt energy of helices relative to the membrane normal. This involves correcting the observed helical tilt distribution, P_{obs}, by that expected for a completely random distribution P_{calc}, and then converting this ratio to an energy through the relationship $\Delta G = -RT \ln P_{\text{obs}}/P_{\text{calc}}$ (Fig. 2C). A linear fit to these data suggests that the free energy change required to increase the tilt of a helix from $\eta = 20°$ to $60°$, which would be associated with an increase in pore radius to ~15 Å for $N = 5$, would be ~8 kJ/mol/helix, or a total of ~40 kJ/mol/pentameric channel. For reference, the average thermal energy at room temperature is ~2.5 kJ/mol.

II. MscL: Structure and Mechanism

Owing to the pioneering efforts of C. Kung, the best characterized mechanosensitive channel is the prokaryotic MscL, the mechanosensitive channel of large conductance. MscL was originally identified, isolated, and characterized by Kung and co-workers using a biochemical approach in combination with a patch-clamp assay to isolate an intrinsically stretch activated channel from *E. coli* membranes (Sukharev *et al.*,

by the total number of helices in the sample set (139). (C) Estimation of ΔG for helix tilting evaluated from a mean-field analysis through the relationship $\Delta G = -RT \ln(P_{\text{obs}}/P_{\text{calc}})$, where P_{obs} is taken from (B), and P_{calc} is the expected probability [$= \cos(\theta) - \cos(\theta + \Delta\theta)$] that a randomly chosen line will make an angle with the membrane normal between θ and $\theta + \Delta\theta$, with $\Delta\theta = 5°$ in this calculation. The line through these points is of no obvious theoretical significance, but represents an empirical, linear fit with the equation $\Delta G(\eta) - 7.0 + 0.17\eta$, where ΔG and η are in kJ/mol and degrees, respectively. This relationship is used in the text to estimate free energy changes associated with helix tilting.

1994, 1997). Homologues of MscL have subsequently been identified in other bacteria (Moe *et al.,* 1998) and have been reported in Archaea (Kloda and Martinac, 2001a,b,c). MscL is a nonselective ion channel of ~3.5 nS conductance activated *in vitro* by the application of membrane tension; with a potential difference of 100 mV, this conductance is equivalent to the flow of $\sim 10^9$ ions/s across the membrane. For reference, this conductance level is ~100 times greater than more extensively characterized channels, such as the potassium channel and acetylcholine receptor that have conductances of ~30 pS. Kung and co-workers established that the MscL channel consists of a single type of subunit of molecular weight ~16,000 (Fig. 3A), with hydropathy analyses (Fig. 2B) indicating that each subunit contains two transmembrane helices. Both the NH_2 and COOH termini have been shown to be cytoplasmic (Blount *et al.,* 1996b, 1999; Häse *et al.,* 1997b). In view of the large conductance and the small subunit size, MscL was proposed to be organized as a homooligomer, now known to be a pentamer (Chang *et al.,* 1998; Sukharev *et al.,* 1999a).

The initial characterizations of the dependence of channel conductance on membrane tension for MscL were interpreted in terms of a two-state model analogous to Eq. (1) (Sukharev *et al.,* 1997). In a more detailed analysis, Sukharev, Sachs, and co-workers (Sukharev *et al.,* 1999c) established the energetic parameters for the gating transition in the *E. coli* MscL (Ec_MscL) and obtained evidence for three subconductance states (S2, S3, and S4) between the fully closed (C1) and fully open (O5) states:

$$\text{C1} \leftrightarrow \text{S2} \leftrightarrow \text{S3} \leftrightarrow \text{S4} \leftrightarrow \text{O5} \tag{8}$$

In an unstressed membrane, the C1 state is favored by ~40 kJ/mol relative to the O5 state. The open probability exhibits a strong sigmoidal dependence on the tension applied to the membrane, with a midpoint of ~ 12 dyn cm^{-1}. Assuming that tension favors the state with the largest cross-sectional area in the membrane plane, the increase in area between C1 and O5 was calculated to be ~ 650 Å^2. From the conductance and biochemical studies, pore diameters of ~30 Å have been estimated (Cruickshank *et al.,* 1997; Ajouz *et al.,* 1998), which suggests that the bulk of the increased cross-sectional area of MscL in the open state corresponds to formation of the pore. The functional properties of MscL suggest a physiological role in the regulation of osmotic pressure in the cell (Sukharev *et al.,* 1997; Levina *et al.,* 1999); since tension near the breaking point of the membrane is required to open the channel, it is believed that this channel represents a last-ditch safety valve that transiently places sufficiently large holes in the membrane to rapidly

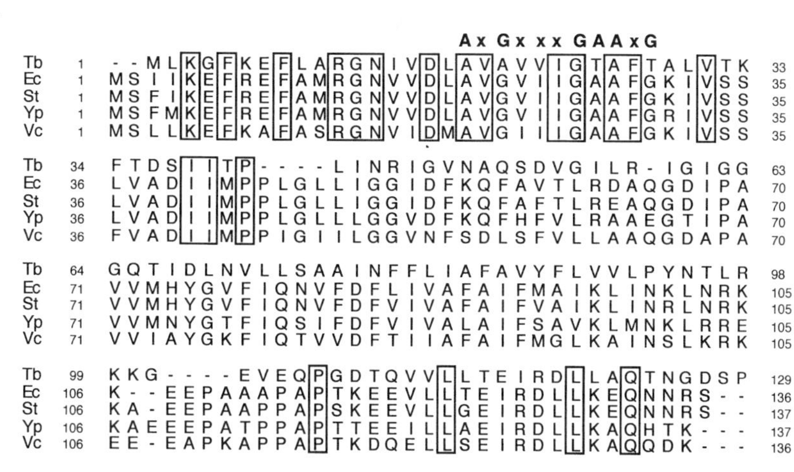

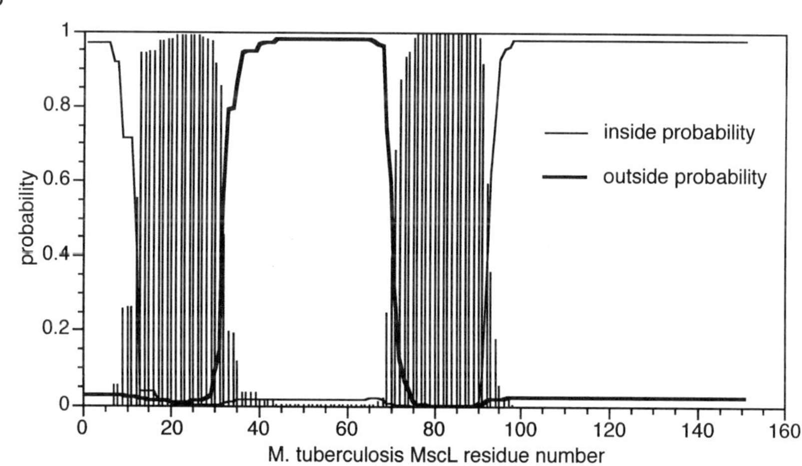

FIG. 3. (A) Sequence alignment of five MscL homologues prepared by the program CLUSTAL-W (Thompson *et al.*, 1997). The sequences were selected to be representative of a more complete sequence alignment (see Spencer *et al.*, 1999) and are designated as Tb, Ec, St, Yp, and Vc for the MscL homologues from *M. tuberculosis, E. coli, Salmonella typhimurium, Yersinia pestis,* and *Vibrio cholerae,* respectively. Identical residues in these sequences are enclosed by boxes. Residues 130–151 of Tb_MscL have no equivalents in the other sequences and have been omitted from this alignment. The pattern of conserved Ala and Gly between residues 18–28 is indicated. (B) Hydropathy analysis of the *M. tuberculosis* MscL as calculated with the program TMHMM (Krogh *et al.*, 2001). The narrowly spaced vertical lines represent the probability that a given residue is found in a membrane spanning helix. The helical segments are predicted to include residues 13–32 and 69–91, compared to residues 15–43 and 69–89 observed in the crystal structure (Chang *et al.*, 1998). The predicted topology of the interhelical loops, either inside or outside the cell, is in agreement with experimental observations (Blount *et al.*, 1996b).

equilibrate osmotic pressure differences. Although other mechanosensitive channels with lower thresholds are present in *E. coli,* principally MscS, evidence has been obtained that demonstrates MscL does open under downshock conditions (Batiza *et al.*, 2002) even when these other channels are present.

A. *Structure of MscL*

MscL offers many advantages for structural studies of a gated channel system—it has a relatively small subunit size and simple architecture, and prokaryotic membrane proteins are generally more easily overexpressed. Although the probability of crystallizing any particular membrane protein may be low (at present), the probability of success can be greatly increased by screening homologous proteins with similar, but distinct, amino acid sequences to find one that might be more readily compatible with the formation of a three-dimensional crystal lattice. The identification of MscL homologues in numerous prokaryotes, as originally reported by Moe *et al.*, (1998), provided an opportunity to search for appropriate crystallization conditions using naturally occurring sequence variants that might be more optimal for crystal formation. This is the same strategy utilized by Kendrew *et al.* (1954) for the structure determination of myoglobin, except that rather than obtaining different types of whale meat from around the world, we now obtain the appropriate DNA. To facilitate this screening process, standardized expression, purification, and crystallization protocols were employed to streamline the preparation and testing of the various homologues (Spencer *et al.*, 2002), which culminated in the structure determination of the MscL homologue from *Mycobacterium tuberculosis* (Tb_MscL) at 3.5 Å resolution (Chang *et al.*, 1998).

The crystal structure of Tb_MscL established that this protein assembles as a homopentamer that is organized into two domains, the transmembrane domain and the cytoplasmic domain. The transmembrane domain consists of 10 helices (2 per subunit) connected by an extracellular loop, while the cytoplasmic domain contains 5 helices that form a left-handed pentameric bundle. The sequence of the Tb_MscL subunit has 151 amino acids and can be further subdivided into five segments: the N terminus, the first transmembrane helix (TM1), an extracellular loop, the second transmembrane helix (TM2), and a cytoplasmic domain (Fig. 4, see color insert). Each of the segments is discussed in more detail below. The pore is aligned along the fivefold symmetry axis and is formed by the first transmembrane helix (TM1) and an extracellular loop from each subunit. The channel has overall dimensions of approximately $85 \times 50 \times 50$ Å and both the N and C termini reside on the cytoplasmic side of the membrane (Blount *et al.*, 1996b). Unless otherwise indicated, the residue numbering used here is that of Tb_MscL.

1. Amino Terminus

Despite the fact that the first 10 residues (and the His tag) are disordered in the crystal, the high sequence conservation of the N terminus, together with mutagenesis studies, suggests the importance of the N terminus in channel function. Deletion of the first two residues of Ec_MscL or addition of residues at the N terminus makes little difference in channel function; however, the deletion of the first nine residues is not tolerated (Blount *et al.*, 1996c; Häse *et al.*, 1997a). More recent N-terminus proteolysis experiments have shown that limited N-terminal truncations result in functional channels with increased sensitivity to pressure (Ajouz *et al.*, 2000). Sukharev and Guy, in their studies of the gating transition in Ec_MscL, have proposed that the N-terminal residues of each subunit are organized into a five helix bundle that functions as the actual gate (Sukharev *et al.*, 2001a,b) (see below for further discussion).

2. TM1 Helix

The first transmembrane helix (residues 15–43) crosses the membrane from the cytoplasm to the periplasm and, together with the four TM1 helices from the other subunits, forms the permeation pathway. The interior of the channel on the periplasmic side is mostly hydrophilic and is formed by residues Thr-25, Thr-28, Thr-32, Thr-35, Lys-33, and Asp-36 (Fig. 4C); the abundance of polar residues toward the C-terminal end of the helix is responsible for the shorter predicted helix in the hydropathy analysis (Fig. 3B). The TM1 helix packs against the adjacent TM1 helix with right-handed crossing angles of −43° and makes little contact with the lipid bilayer. Each TM1 helix is tilted approximately 35° with respect to the membrane normal, with the channel constriction occurring near the cytoplasmic side of the membrane. The structure of Tb_MscL is believed to be in the closed or nearly closed state, since the channel constriction formed by the hydrophobic residues Ile-14 and Val-21 is about 2 Å in diameter (Fig. 4D). The equivalent residues of the Ec_MscL also form the constriction site, based on numerous mutagenesis studies (Ou *et al.*, 1998; Yoshimura *et al.*, 1999, 2001; Moe *et al.*, 2000). A sequence comparison of MscL homologues reveals that many of the most highly conserved residues in this channel are found in TM1 (Fig. 3A), particularly at positions involved in helix–helix contacts and in forming the pore constriction (Fig. 5).

Perozo, Martinac, and co-workers (Perozo *et al.*, 2001) have reported a site-directed spin-labeling (SDSL) analysis of Ec_MscL in lipid vesicles by EPR spectroscopy under physiological conditions. In a SDSL

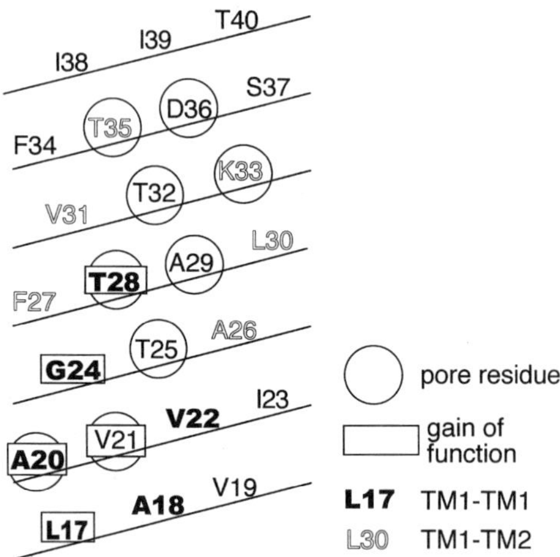

FIG. 5. A helical net representation for the inner TM1 helix of Tb_MscL illustrating the positions of helix interface residues, pore residues, and gain-of-function mutations. The side chain of Val-21 serves to constrict the channel at the narrowest position. The correspondence between residues at the interface between the inner helices and the gain-of-function mutations is evident (Ou *et al.,* 1998), suggesting this interface is critical for the gating process. Residues 17, 20, 24, and 28 participate in the helix interface with one neighboring TM1 helix, while 18, 22, and 26 participate in the interface with the other neighboring TM1 helix; if these residues are found in subunit A, then the former group interacts with subunit E, and the latter with subunit B.

experiment, as pioneered by W. Hubbell (Hubbell *et al.,* 1998), a spin label is covalently attached to a series of single cysteine mutants introduced throughout the region of interest in a protein (such as a membrane-spanning helix). From the accessibility of the spin label to various reagents, and from the presence or absence of spin–spin interaction between probes, the environment and relative location of the spin label can be deduced. Specifically, a high accessibility of the spin label to the nonpolar dioxygen molecule is indicative of a hydrophobic environment, while exposure of the spin label to the polar species NiEdda is indicative of exposure to an aqueous environment. In addition, analysis of the spectral line shape gives information about spin label mobility. From the variation in these properties with sequence position, information on the protein structure and environment can be obtained. In the case of Ec_MscL, cysteine mutants were generated for residues in transmembrane helices TM1 and TM2, and subsequently modified with a methanethiosulfonate spin label. In general, the structure and environment of labeled residues established by this study correlated well with the Tb_MscL structure. In particular, severely restricted accessibility to

NiEdda was observed for residues 19–26, which in the crystal structure create the channel constriction, while strong periodicities in the mobility and accessibility of the label were seen for residues 19–38 that indicate a helical structure. The periodicity of NiEdda and O_2 accessibilities are 180° out of phase, suggesting that one face of the helix is water accessible, while the opposite face of the helix is in a hydrophobic environment. This is in good agreement with the crystal structure, where one face of the TM1 helix forms the water-accessible pore and the opposite face packs against surrounding helices and therefore resides in a hydrophobic environment.

3. Extracellular Loop

The connection between the TM1 and TM2 helices is mediated by the extracellular loop formed from residues 44–68, which is poorly ordered in the crystal structure. This extracellular loop dips into the channel and lines this region of the pore to form an outer-lip of diameter ~20 Å. The high glycine content of the extracellular loop (Gly-47, Gly-55, Gly-60, Gly-62, Gly-63, and Gly-64 in Tb_MscL) likely contributes to the flexibility seen in this region of the crystal structure (Fig. 4B). The high sequence variability and structural flexibility of this region might argue against an important functional role; however, proteolysis experiments have demonstrated that cleavage of the extracellular loops makes it easier to open the channel (Ajouz *et al.*, 2000), while mutagenesis studies have shown that substitutions in the extracellular loop can lead to a gain-of-function phenotype (Maurer *et al.*, 2000)

4. TM2 Helix

The second transmembrane helix (residues 69–89) traverses the lipid bilayer from the periplasm back to the cytoplasm along the outside of the TM1 helix bundle, and is consequently responsible for most of the contact with the lipid bilayer. The lipid exposed surface of MscL is composed of approximately 35% TM1 and 65% TM2. The face of TM2 that comes in contact with the lipid bilayer is lined with many hydrophobic residues (Leu-69, Leu-72, Leu-73, Ile-77, Phe-79, Phe-80, Leu-81, Phe-84, and Phe-88) and is more hydrophobic than an average protein core (Rees *et al.*, 1989; Wallin *et al.*, 1997; Spencer and Rees, 2002). The TM2 helix, like the TM1 helix, is tilted about 35° with respect to the membrane normal; however, unlike the TM1 helices, the TM2 helices from different subunits are separated by ~20 Å and do not contact each other.

SDSL studies of the TM2 helix (Perozo *et al.*, 2001) revealed that it is slightly less motionally restricted, relative to TM1. This is to be expected, since TM2 is less tightly packed than TM1, because of the accessibility to the solvating lipid bilayer. As with TM1, the periodicity of oxygen accessibility is in general agreement with the crystal structure, where one side of the TM2 faces the lipid bilayer. NiEdda accessibility decreases as the TM2 helix enters the membrane, and sharply increases at residue 100 where it presumably exits the membrane. The spin labeling data deviate from the crystal structure at the C terminus of the TM2 helix, where the local environment derived from EPR studies does not match the environment derived from the crystal structure. These differences may reflect some combination of the presence of partially ordered electron density near the C terminus of the TM2 helix (corresponding to ordered detergent molecules or part of the disordered polypeptide chain), structural differences between *E. coli* and *M. tuberculosis* homologues or differences arising from the use of bilayers and detergent to solubilize Eco_MscL and Tb_MscL, respectively.

5. *Cytoplasmic Domain*

After the TM2 helix returns to the cytoplasm, a second loop connects the TM2 and cytoplasmic helices. The cytoplasmic helices from all five subunits (residues 102–115) associate together to form a left-handed helix bundle. Mutagenesis and proteolysis data have shown that the C terminus of Ec_MscL can be removed up to residue 104 without any change in gating properties of MscL (Blount *et al.*, 1996c; Häse *et al.*, 1997a; Ajouz *et al.*, 1998). A cluster of charged residues (Arg-98, Lys-99, Lys-100, Glu-102, and Glu-104) is present in the loop connecting TM2 to the cytoplasmic helix (Fig. 4E) and has been shown to be important for channel gating (Sukharev *et al.*, 1994, 1997; Hamill and Martinac, 2001). The overall negative charge of the C-terminal domain and the relatively low pH (3.7) used for crystal growth have led to the hypothesis that the cytoplasmic helix bundle is not present at physiological pH (Chang *et al.*, 1998); however, there is no experimental evidence for the loss of helicity at high pH (Strop, 2002), suggesting that the bundle remains intact, or that the helices partition into the membrane–water interface region under these conditions. The rest of the C terminus (residues 116–151) does not show any ordered structure in the crystal and corresponds to the least conserved region of the protein.

B. *Gating Mechanism: General Considerations*

In view of the substantial conductance and solute nonselectivity of MscL, the general picture that emerges for the open state of this channel is of a large, water-filled pore with a likely diameter of up to 30 Å. Assuming that the fivefold symmetry of MscL is maintained in both states, generation of a permeation pathway with these properties in the open state would require a significant rearrangement of all subunits in response to application of membrane tension (Blount *et al.*, 1996a). Given that the membrane-spanning region of the channel is constructed from α-helices, a change in subunit–subunit interactions will necessarily require changes in the packing of helices against each other. Following the analysis presented at the beginning of this chapter, there are two basic mechanisms for increasing the minimum pore radius created by a set of symmetrically interacting helices: either the helical tilt must increase, or the number of helices surrounding the permeation pathway must increase. For the specific case of MscL with $\eta \sim 35^\circ$ in the closed state, the transition to an open state with a minimum diameter of 30 Å (radius = 15 Å) could be achieved either by increasing the tilt of the five TM1 helices to $\sim 70^\circ$, or by incorporating the TM2 helices to create a permeation pathway surrounded by 10 helices, each with a tilt of $\sim 40^\circ$ (Fig. 2A).

An important development in identifying residues critical to the gating transition has been the isolation of "gain-of-function" mutants that display a slow or no-growth phenotype as a result of the leakage of solutes out of the cell under low osmotic strength growth conditions (Ou *et al.*, 1998). *In vitro* characterization of these mutant channels demonstrated that they generally exhibit a reduction in the tension required for channel gating, suggesting that the closed state in these mutants is destabilized relative to the wild-type channel [i.e., the equilibrium is shifted to the right in Eq. (1)]. Many of the mutations associated with severe phenotypes are located at the interface between adjacent TM1 helices (Fig. 5), in the region of the membrane-spanning domain where the pore is most restricted. Three of the residues with the most severe phenotypes, corresponding to positions 20, 24, and 28 of Tb_MscL, are typically found to be glycine in the different MscL homologs (Fig. 3A). Intriguingly, glycines are frequently found at the interface between transmembrane helices (Javadpour *et al.*, 1999; Eilers *et al.*, 2000; Russ and Engelman, 2000), and these residues of MscL are among the most highly conserved positions. An extensive mutagenesis study of residue Gly-22 in *E. coli* MscL (equivalent to Ala-20 in *M. tuberculosis* MscL), which is at the interface between adjacent inner helices, revealed that hydrophobic and hydrophilic substitutions stabilized the closed and open states of the channel, respectively, suggesting that this residue becomes exposed to a water-filled channel in the open state (Yoshimura *et al.*, 1999). In experiments where cysteine residues were substituted for channel-constricting residues (Yoshimura *et al.*, 2001), reversible modification of hydrophilicity was possible under patch clamp. Chemical modifications were consistent with the mutagenesis studies, resulting in functional channels when hydrophobic modifying agents were used, and resulting in spontaneous gating when hydrophilic modifying agents were used. These observations suggest that contacts between TM1 helices play a crucial role in the gating mechanism, and, particularly, changes made to residues at this interface tend to destabilize the closed state and make it easier to open the channel.

Although the most severe phenotypes identified in the initial "gain-of-function" screen were found in the membrane-spanning helices (Ou *et al.,* 1998), residues in other regions throughout the protein were also identified that influenced the gating behavior of MscL. Proteolysis of the extracellular loop between the membrane-spanning helices in Ec_MscL has been found to significantly increase the mechanosensitivity of the channel, without changing conductance (Ajouz *et al.,* 2000), and mutagenesis studies in this region of Tb_MscL have also picked up gain-of-function mutants (Maurer *et al.,* 2000). Variants at position 56 of Ec_MscL, also in the extracellular loop, lead to significant alterations in open time (Blount *et al.,* 1996c). On the other side of the membrane, deletion of either the first 12 or last 33 residues of Ec_MscL results in loss of channel activity, although deletions of the first 3 or last 27 residues can be tolerated (Blount *et al.,* 1996c). Intriguingly, these sensitive regions include relatively well-conserved clusters of charged residues that have been suggested to be important for the gating mechanism (Gu *et al.,* 1998).

At present, most of the mutational studies have been conducted in *E. coli,* whereas the structure has only been determined for the *M. tuberculosis* MscL. The emphasis on functional studies of Ec_MscL reflects not only that this was the first MscL to be characterized, but also that opening Tb_MscL requires tensions near the breaking point of the patch (Moe *et al.,* 2000), and hence Tb_MscL is technically more difficult to study than Ec_MscL. Although the structures of these two homologues are undoubtedly similar, particularly in the membrane-spanning regions with the highest degree of sequence conservation, differences in the properties of corresponding mutations constructed in Ec_MscL and Tb_MscL have been reported and interpreted as indicating there may be mechanistic distinctions between these two homologues (Maurer *et al.,* 2000; Moe *et al.,* 2000). Clearly, more extensive functional and structural studies are required to establish the extent of the mechanistic similarities in the gating of these two channels.

C. MscL Open State

The earliest structural models for the open state of MscL emphasized the involvement of all the membrane-spanning helices of MscL to create the permeation pathway (Cruickshank *et al.,* 1997; Spencer *et al.,* 1999). In particular, an alteration of the TM1 and TM2 helices was suggested to create a 10-helix "barrel-stave" type channel with the α-helices nearly parallel to the membrane normal in the open state (Spencer *et al.,* 1999). The conformational rearrangement associated with this transition could plausibly involve the coordinated movement of adjacent pairs of TM1 and TM2 helices that are nearly antiparallel, since these types of helix–helix interfaces tend to be more stable (Gerstein *et al.,* 1994). From the extent of buried surface area, the most stable TM1–TM2 interface would involve TM1 from subunit A with TM2 of the neighboring subunit B (and the symmetry equivalents) that have a antiparallel crossing angle of $-170°$ and a total buried surface area of 700 Å^2. This interface exceeds the 540 Å^2 surface area buried between TM1 and TM2 in the same subunit with a crossing angle of 138°. Attempts to generate explicit models for a "barrel-stave" type channel have indicated, however, that a substantial number of hydrophobic residues would be exposed to water in the open state, which would be energetically unfavorable (Sukharev *et al.,* 2001b). An alternative mechanism for increasing the number of helices surrounding the permeation pathway in the open state, by increasing the number of subunits in the oligomer, appears unlikely in view of Sukharev's observations that the MscL oligomer is not in dynamic equilibrium with monomers (Sukharev *et al.,* 1999b).

A conceptual breakthrough in the modeling of the open state of MscL has been reported by Sukharev and Guy, using modeling studies combined with disulfide trapping and electrophysiological characterizations (Sukharev *et al.,* 2001a,b; Betanzos *et al.,* 2002). This model is based on the second general mechanism for generating the open state from the closed state, namely through a substantial increase in helical tilt (Fig. 6). The symmetry of MscL is assumed to be conserved in both closed and open states. The driving force for this rearrangement is likely to maintain the lipid solvation of the membrane-spanning helices in response to a stretch-induced thinning of the bilayer (Hamill and Martinac, 2001; see below for a more extensive discussion of the role of hydrophobic mismatch in MscL gating). In this model, the structural rearrangements for forming a large, ~30 Å pore in the open state involve increasing the helical tilt from 35° (closed) to 70° (open); the TM1 helices line the permeation pathway in both states. In the open state model, the interhelical crossing angle between adjacent TM1 helices changes from $-43°$ to $-68°$. Although both the tilt and the helix crossing angle are essentially unprecedented in structurally characterized membrane proteins (Fig. 2B), they are proposed to exist in a form that is significantly less stable than the closed state under resting conditions; the estimate from the mean-field approximation described in Fig. 2C suggests an energetic

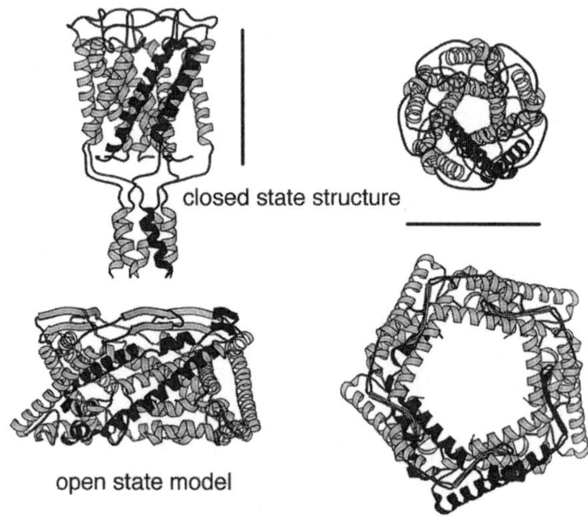

FIG. 6. Top: Comparisons of the structures of MscL as observed crystallographically in the closed state (Chang *et al.*, 1998), and bottom: as modeled in the open state by Sukharev and Guy (Sukharev *et al.*, 2001a,b). Views down the membrane normal (right) and in the plane of the membrane (left) are provided. One subunit is highlighted in each structure for clarity. The increased tilt of the membrane-spanning helices in the open state model, along with the increased helical length necessary to span the bilayer in the open state model, is evident. The dark lines indicate 40 Å.

cost of ~6 kJ/mol/helix to increase the helical tilt by 35°. Rather than the side chains of Val-21 in the inner helices serving as the gate, Sukharev and Guy propose that the real gate is a five-helix bundle containing the highly conserved amino-terminal residues of each subunit. These residues are not observed in the crystal structure, however, presumably because of disorder or other types of conformational heterogeneity. Experimental support for the Sukharev and Guy model is provided by the ability to stabilize an open state of MscL through disulfide bond formation between specifically incorporated cysteines that should be spatially adjacent only in the open state. Disulfide cross-linking studies have demonstrated that by cross-linking the N terminus to the C-terminal end of the TM2 helix (residues 1 and 91), the channel rarely reaches the fully closed state; instead, it flickers mostly between the open state and an intermediately open substate (Sukharev *et al.*, 2001a).

As implemented by Sukharev and Guy, the combination of patch-clamp measurements with disulfide trapping provides a sensitive approach for testing models of the open state by characterizing the consequences of disulfide bond formation between specific cysteines on the closed-to-open transition. The ability to monitor the conductance of single channels is a great advantage over bulk phase measurements, since the effect of the presence or absence of a disulfide bond on channel function can be directly analyzed, even if the overall equilibrium strongly favors one state or the other. While the ability to form a disulfide bridge between two cysteines clearly requires that these residues be in spatial proximity at the time of bond formation, the restrictions on the time-average structure imposed by this technique can be relatively weak. Since disulfide bond formation is an effectively irreversible trapping technique, disulfides can form between pairs of distant cysteines if they are brought into proximity by even infrequently occurring conformational fluctuations. This behavior has been detailed in studies on water-soluble proteins (Careaga and Falke, 1992; Butler and Falke, 1996), where cysteines separated by up to 15 Å in the time average structure can still be disulfide trapped.

In view of these considerations, a detailed, experimentally based model of the open state of MscL will likely require crystallographic and spectroscopic approaches. The main barrier to studying the open state of MscL using these techniques is that they require a large number of molecules to obtain an experimentally recordable signal; as the equilibrium strongly favors the closed state in the absence of applied tension, these observations will consequently reflect the properties of the closed state. The challenge of stabilizing the open state of MscL is shared with other protein systems, such as allosteric enzymes, that exist in multiple conformations and must be dealt with by establishing conditions where the alternate conformation dominates. Stabilization of a state disfavored at equilibrium may be achieved, for example, through the addition of ligands that bind preferentially to that state (allosteric effectors), through the use of mutants that stabilize the desired state (or equivalently destabilize the undesired states), or by covalently trapping the desired state. Unfortunately, in contrast to many medically relevant channels, no state-specific ligands have been identified that could be used to trap MscL in an open state. Mutants that favor the open state of MscL are known; while the equilibrium is shifted toward the open state in the gain-of-function mutants isolated by Kung and co-workers (Ou *et al.*, 1998), these mutants are not locked in the open state in the absence of applied tension. Hence, these mutants, as isolated, are not suitable for structural studies of the open state. Finally, for reasons noted above, while trapping methods could be potentially used to stabilize an open state, it is difficult to find conditions that will generate sufficient quantities of homogenous material that captures the ground-state structure of the open state.

Perozo and Martinac (Perozo *et al.*, 2002a,b) have reported a major experimental breakthrough in the structural characterization of the MscL open state by using SDSL methods in conjunction with solution conditions that stabilize the open state at equilibrium. This advance was based on an analysis of potential coupling mechanisms between applied membrane tension and the conformational state of MscL. Because of differences in the relative cross-sectional areas of the head-group and hydrocarbon-chain components of phospholipids, substantial lateral pressures are generated within a membrane bilayer (Gruner, 1985; Cantor, 1997, 1999; de Kruijff, 1997) that will act on integral membrane proteins. When a membrane is deformed, the lateral pressure profile will necessarily be altered, with consequent changes in the conformational energetics of any embedded membrane protein, such as MscL, that possesses multiple states of differing cross-sectional area. Perozo and Martinac evaluated two possible mechanisms for the transduction of mechanical energy derived from membrane stretching to shifting the conformational equilibrium toward the open state of MscL: a decrease in bilayer thickness and an increase in bilayer curvature.

Changes in bilayer thickness could alter the equilibrium between conformational states of a channel through hydrophobic mismatch effects, where the ability of a membrane to solvate the nonpolar region of a protein will change when the membrane thickness is varied. To assess the contribution of bilayer thinning when the membrane is stretched to driving the open state transition, Ec_MscL was reconstituted into vesicles formed from phosphatidylcholine of varying acyl chain lengths (Perozo *et al.*, 2002b). For chain lengths of 16, 18, and 20 carbons (designated PC16, PC18, and PC20, respectively), reconstituted MscL required greater applied tension for opening with increasing chain length, with ΔG° values measured to be 10, 24, and 59 kJ/mol, respectively. Despite the differences in pressure dependence, the conductances of the channel were essentially constant under all conditions. However, under no condition could the equilibrium be shifted to predominantly favor the open state, a result corroborated by SDSL studies with MscL containing spin labels positioned near the constriction in TM1 reconstituted into vesicles composed of PC10 to PC20. Thus, while bilayer thinning/hydrophobic mismatch effects do influence the energetics of MscL gating, they are insufficient to preferentially stabilize the open state under equilibrium conditions. However, evidence for a distinct, closed state form of MscL was observed in vesicles reconstituted with PC14.

The possibility that changes in bilayer curvature might be implicated in the gating of mechanosensitive channels was motivated by the observation that addition of amphiphiles to one side of the membrane can lower the activation tension required for channel opening (Martinac *et al.*, 1990). The basic idea is that if non-cylindrically shaped molecules, such as the cone-shaped amphiphile lysophosphatidylcholine (LPC) that contains a "big head" relative to the small acyl chain, are incorporated asymmetrically into membranes, geometric considerations will lead to mechanical deformation (curvature) of the bilayer to optimize packing throughout the membrane leaflet. This induced curvature will necessarily alter the lateral pressure profile throughout the membrane, which will consequently perturb the conformational equilibrium of proteins exhibiting multiple states of different cross-sectional areas. Using SDSL probes incorporated into residues near the constricted region of TM1, Perozo and Martinac found that an open state conformation of MscL could be stabilized under equilibrium conditions by the addition of ~10–25 mol% LPC from solution to the externally facing leaflet of the bilayer.

By establishing conditions that stabilize the open state of MscL, a systematic analysis of the conformation and environment of residues in the TM1 and TM2 helices could be conducted using the SDSL approach (Perozo *et al.*, 2002a). The results of this study clearly establish that (1) the permeation pathway is lined primarily, if not exclusively, by residues in the TM1 helix; (2) the TM2 helix remains in contact with the lipid bilayer, but the interactions between TM2 and TM1 shield the latter from the extensive interactions with the membrane; and (3) equivalent residues in the TM1 helices of different subunits move sufficiently far apart that they no longer exhibit spin–spin coupling that is apparent for residues near the channel constriction in the closed state. A structural model for the MscL open state has been generated that is in good agreement with the constraints established by the SDSL analysis, which indicates that the pore diameter is ≥25 Å. As anticipated by the Sukharev and Guy model (Fig. 6), the open state is derived from the closed state by a substantial increase in tilt of the TM1 and TM2 helices that maintains the right-handed packing interactions between TM1s. Further developments in the structural characterization of this system, both within and external to the membrane bilayer, will be very informative in defining the gating mechanism of MscL.

D. *Computational Studies*

In addition to experimental studies, the combination of relative simplicity and structural characterization have made MscL an attractive target for computational molecular dynamic simulations of channel gating (Elmore and Dougherty, 2001; Gullingsrud *et al.*, 2001; Bilston and

Mylvaganam, 2002; Kong *et al.,* 2002). These studies have highlighted regions of relative structural stability found in the membrane, particularly near the channel constriction, as well as more poorly defined regions involving residues in the extracellular loop and carboxy terminus. Calculations on the closed state of wild-type MscL have been extended to include evaluation of the sensitivity of the channel structure and dynamics to mutations, and, of particular mechanistic interest, the response of the channel to applied tension (Gullingsrud *et al.,* 2001; Bilston and Mylvaganam, 2002). In one study, the trajectory of closed to open state developed by Sukharev and Guy has been modeled by targeted molecular dynamics (Kong *et al.,* 2002), which suggests that the intermediate open states are asymmetric. Computational analyses would seem to hold particular potential for establishing the coupling between membrane and protein structures as a function of applied tension, and in structurally defining the progression of intermediates between closed and open states.

III. MscS and Other Prokaryotic Mechanosensitive Channels

A. *MscS*

Following the identification of *E. coli* MscS in 1987 (Martinac *et al.,* 1987), this channel has been electrophysiologically characterized (Sukharev *et al.,* 1997; Blount *et al.,* 1999) and shown to have a conductance of approximately 1 nS, roughly one-third that of MscL, with a pressure threshold for channel opening approximately 60% that of MscL. From a fit to the Boltzmann equation [Eq. (3)] the closed-to-open transition for MscS may be characterized by $\Delta G^\circ = 28$ kJ/mol, $\Delta A = 840$ Å^2, and the midpoint tension is 5–6 dyne cm^{-1} (Sukharev, 2002). From the observed conductance, a pore diameter of approximately 18 Å is estimated for the open state of MscS. Unlike the MscL channel, MscS exhibits a voltage-dependent gating. On reaching its pressure threshold, there is an *e*-fold increase in open probability per 15 mV of membrane depolarization (positive voltage) (Martinac *et al.,* 1987). At moderate voltages (± 20 mV), MscS has slow kinetics, dwelling in an open or closed state for up to seconds (Sukharev *et al.,* 1997). MscS is mostly nonselective with a slight anion selectivity (Kloda and Martinac, 2002a; Sukharev, 2002).

Booth and co-workers established that the yggB gene was associated with MscS activity in membrane patches of *E. coli* (Levina *et al.,* 1999). As shown by a series of yggB knockout strains of *E. coli,* the yggB gene was required for MscS channel activity. Whereas deletion of either the mscL or yggB gene appears to be without a phenotype, mutants lacking both genes are hypersensitive to osmotic downshock. This finding suggested that each channel can compensate for the other in responding to osmotic stress. KefA [subsequently renamed MscK (Li *et al.,* 2002)], a protein found by homology to contain an MscS-like domain, also displays MscS-like activity, but is dispensable. Deletion of the kefA gene in concert with either the mscS gene or the MscL gene was found not to have any phenotype. Although it appears that kefA is not required for the response to hypoosmotic shock, it contains a domain highly homologous to yggB at the carboxy-terminal end of the protein and does encode a channel of similar conductance to MscS (~1.0 nS). In contrast to the slight anionic preference observed for the MscS activity associated with yggB, a mutational analysis has suggested that KefA is a cation-specific channel involved in osmotic adaptation in *E. coli* (McLaggan *et al.,* 2002).

Whereas homologues of MscL range from around 130 to 150 amino acids, MscS homologues are typically at least twice that size (Fig. 7A). By separating *E. coli* MscS and MscL channels on a size exclusion column and subsequently analyzing the resulting fractions by patch clamp, it was demonstrated that MscS and MscL activities can be resolved (Sukharev *et al.,* 1993). MscS elutes from the column first, indicating that the functional channel is larger than MscL. Based on the size exclusion chromatography, the fully assembled MscS channel was estimated to have a molecular weight of 200,000–400,000, in comparison to 70,000 for MscL. From these studies, MscS is expected to be a homomultimer (Sukharev *et al.,* 1993), with cross-linking data suggestive of a hexamer (Sukharev, 2002).

Hydropathy plots of MscS indicate that the N-terminal half of the protein is highly hydrophobic (Fig. 7B). Transmembrane helix predictions programs such as TMHMM (Krogh *et al.,* 2001) predict three transmembrane helices for *E. coli* MscS (Fig. 7B), followed by a large (160 amino acids) soluble domain that is predicted to be cytoplasmic on the basis of PhoA fusion studies on KefA (McLaggan *et al.,* 2002) and from the sequence properties identified in the hydropathy analysis (Fig. 7B). Predictions for the larger MscS proteins from *Bacillus subtilis* and *Methanococcus jannaschii* suggest two additional transmembrane helices are present beyond those in *E. coli* MscS. From sequence comparisons (Fig. 7A), regions with a high degree of sequence conservation are found throughout the entire yggB protein. It is striking, however, that residues in the region corresponding to the third predicted transmembrane helix are especially well conserved. It has not been established which membrane-spanning helix forms the permeation pathway, although mutants that alter the

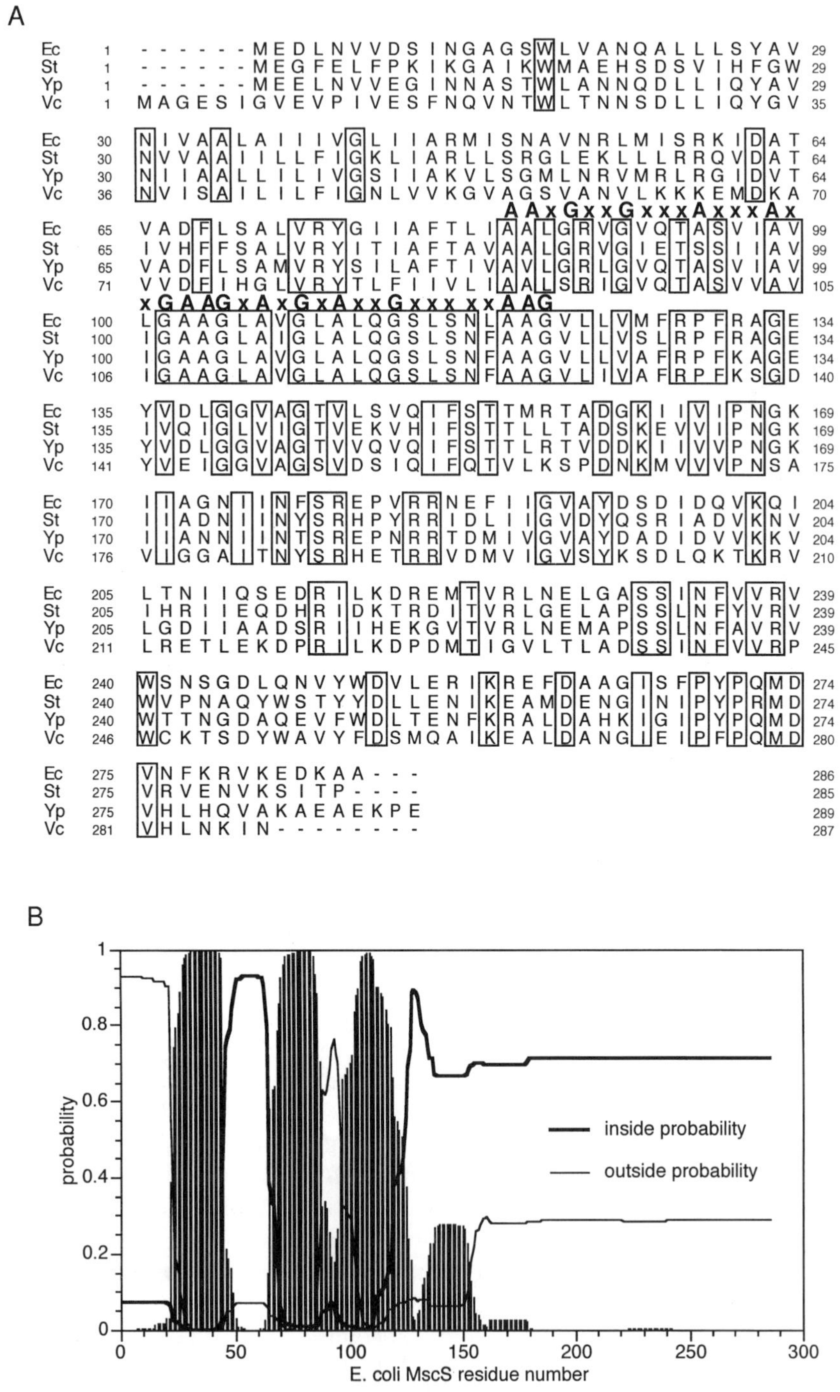

```
                                     A A x G x x G x x x A x x x A x
Ec   1   - - - - - - M E D L N V V D S I N G A G S W L V A N Q A L L L S Y A V   29
St   1   - - - - - - M E G F E L F P K I K G A I K W M A E H S D S V I H F G W   29
Yp   1   - - - - - - M E E L N V V E G I N N A S T W L A N N Q D L L I Q Y A V   29
Vc   1   M A G E S I G V E V P I V E S F N Q V N T W L T N N S D L L I Q Y G V   35

Ec  30   N I V A A L A I I I V G L I I A R M I S N A V N R L M I S R K I D A T   64
St  30   N V V A A I I L L F I G K L I A R L L S R G L E K L L L R R Q V D A T   64
Yp  30   N I I A A L L I L I V G S I I A K V L S G M L N R V M R L R G I D V T   64
Vc  36   N V I S A I L I L F I G N L V V K G V A G S V A N V L K K K E M D K A   70

Ec  65   V A D F L S A L V R Y G I I A F T L I A A L G R V G V Q T A S V I A V   99
St  65   I V H F F S A L V R Y I T I A F T A V A A L G R V G I E T S S I I A V   99
Yp  65   V A D F L S A M V R Y S I L A F T I V A V L G R L G V Q T A S V I A V   99
Vc  71   V V D F I H G L V R Y T L F I I V L I A A L S R I G V Q T A S V V A V  105
         x G A A G x A x G x A x x G x x x x x A A G
Ec 100   L G A A G L A V G L A L Q G S L S N L A A G V L L V M F R P F R A G E  134
St 100   I G A A G L A I G L A L Q G S L S N F A A G V L L V S L R P F R A G E  134
Yp 100   I G A A G L A V G L A L Q G S L S N F A A G V L L V A F R P F K A G E  134
Vc 106   I G A A G L A V G L A L Q G S L S N F A A G V L I V A F R P F K S G D  140

Ec 135   Y V D L G G V A G T V L S V Q I F S T T M R T A D G K I I V I P N G K  169
St 135   I V Q I G L V I G T V E K V H I F S T T L L T A D S K E V V I P N G K  169
Yp 135   Y V D L G G V A G T V V Q V Q I F S T T L R T V D D K I I V V P N G K  169
Vc 141   Y V E I G G V A G S V D S I Q I F Q T V L K S P D N K M V V V P N S A  175

Ec 170   I I A G N I I N F S R E P V R R N E F I I G V A Y D S D I D Q V K Q I  204
St 170   I I A D N I I N Y S R H P Y R R I D L I I G V D Y Q S R I A D V K N V  204
Yp 170   I I A N N I I N T S R E P N R R T D M I V G V A Y D A D I D V V K K V  204
Vc 176   V I G G A I T N Y S R H E T R R V D M V I G V S Y K S D L Q K T K R V  210

Ec 205   L T N I I Q S E D R I L K D R E M T V R L N E L G A S S I N F V V R V  239
St 205   I H R I I E Q D H R I D K T R D I T V R L G E L A P S S L N F Y V R V  239
Yp 205   L G D I I A A D S R I I H E K G V T V R L N E M A P S S L N F A V R V  239
Vc 211   L R E T L E K D P R I L K D P D M T I G V L T L A D S S I N F V V R P  245

Ec 240   W S N S G D L Q N V Y W D V L E R I K R E F D A A G I S F P Y P Q M D  274
St 240   W V P N A Q Y W S T Y Y D L L E N I K E A M D E N G I N I P Y P R M D  274
Yp 240   W T T N G D A Q E V F W D L T E N F K R A L D A H K I G I P Y P Q M D  274
Vc 246   W C K T S D Y W A V Y F D S M Q A I K E A L D A N G I E I P F P Q M D  280

Ec 275   V N F K R V K E D K A A - - -                                         286
St 275   V R V E N V K S I T P - - - -                                         285
Yp 275   V H L H Q V A K A E A E K P E                                         289
Vc 281   V H L N K I N - - - - - - - -                                         287
```

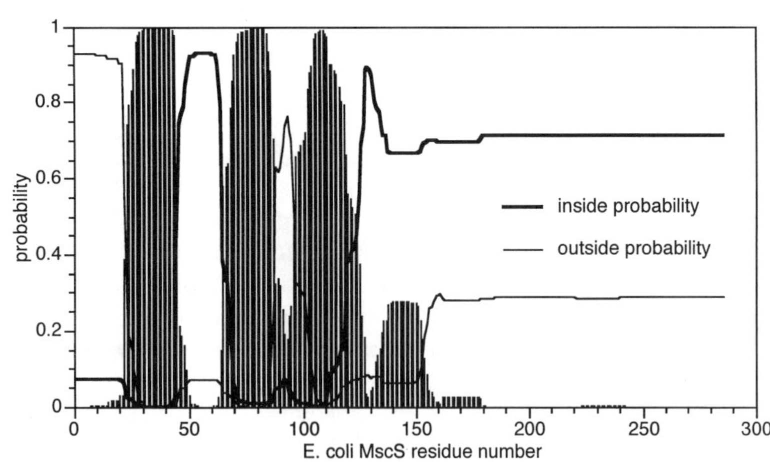

channel properties have been selected in the first (Okada *et al.,* 2002) and last (McLaggan *et al.,* 2002) membrane-spanning helices of YggB and KefA, respectively.

B. MscM

A third mechanosensitive channel of mini conductance (MscM) has been functionally identified in *E. coli* (Berrier *et al.,* 1989, 1996), although the encoding gene has not yet been reported. This channel has a conductance of less than half that of MscS (0.1–0.4 nS), with slow opening/closing kinetics relative to MscS, and especially MscL (Berrier *et al.,* 1993). In contrast to MscL and MscS, MscM has pronounced cation selectivity.

C. Archaeal MscS-like Proteins

Once the genes encoding MscL and MscS were identified, it became possible to search genomic databases for homologous channels in a wide array of prokaryotes. The results of these sequence searches suggest that MscS may be more widely distributed in prokaryotes than MscL. Unlike MscL, MscS homologues appear to be prevalent in Archaea as well as in at least some eukaryotes. Kloda and Martinac (2001b,c, 2002a) identified and characterized two archaeal mechanosensitive channels that were reported to be similar to both MscS and MscL. Using the first transmembrane helix of MscL as a sequence probe, they identified an apparently homologous channel in the archaea *Methanococcus jannaschii,* termed MscMJ. Using this channel as a search criterion, they subsequently found a second mechanosensitive channel MscMJLR. Although the original search was performed using TM1 of MscL, both channels are homologous to MscS. On the basis of these sequence similarities, prokaryotic mechanosensitive channels have been proposed to be evolutionarily related (Kloda and Martinac, 2002b), although this

FIG. 7. (A) Sequence alignment of four MscS homologues prepared by the program CLUSTAL-W (Thompson *et al.,* 1997). The sequences correspond to MscS homologues from the organisms included in Fig. 3A for MscL, with the exception of *M. tuberculosis,* where a MscS homologue has not been identified. Identical residues in these sequences are enclosed by boxes. The pattern of highly conserved Ala and Gly between residues 84 and 121 is indicated. (B) Hydropathy analysis of the *E. coli* MscS as calculated with the program TMHMM (Krogh *et al.,* 2001). The narrowly spaced vertical lines represent the probability that a given residues is found in a membrane-spanning helix. The helical segments are predicted to include residues 22–44, 65–87, and 102–124, with the N and C termini positioned in the periplasm and cytoplasm, respectively.

interpretation has been questioned (Okada *et al.,* 2002). MscMJ was found to have a conductance of approximately 0.3 nS and was activated over a negative pressure range of 25–80 mmHg. In contrast, MscMJLR was found to have a much larger conductance of 2.0 nS, more similar to MscL than to MscS, but which opened at a lower and narrower negative pressure range similar to MscS. This indicates that MscMJ and MscMJLR are not strict functional homologues of MscL and MscS, but instead exhibit characteristics of both channels.

IV. What Makes a Mechanosensitive Channel Mechanosensitive?

As evident from the functional and structural diversity of mechanosensitive channels (Fig. 1), there will be no unique answer to this question. Part of the challenge in characterizing mechanosensitive channels is that there have been no obvious clues in the protein sequences to indicate whether or not a particular protein may be mechanosensitive. In contrast to voltage gated potassium channels, which contain characteristic sequence motifs such as "GYG" in the selectivity filter or the pattern of positively charged residues in the S4 voltage sensor, comparable signature sequences have not yet been identified in mechanosensitive channels. Consequently, the mechanosensitivity of a channel can only be identified through functional characterization, at least for the first member of a family. In the case of prokaryotic channels such as MscL and MscS that are intrisincally mechanosensitive, however, basic features of this process are emerging that may reveal clues hidden in the amino acid sequences of these proteins relevant for mechanosensation. The substantial conductance of these channels requires a large pore diameter in the open state that in turn will require substantial conformational rearrangements during the transition from the closed state. As envisioned in general models for mechanogated channels (Howard *et al.,* 1988), the open state of MscL has an increased cross-sectional area relative to the closed state. Experimental studies have clearly established that the transition from the closed to the open state is associated with increased tilts for the membrane-spanning helix that require significant rearrangements in the helix–helix interfaces (Sukharev *et al.,* 2001a; Perozo *et al.,* 2002a). Consequently, an essential property of MscL must be the ability to stably adopt at least two distinct conformations with different helix packing interactions.

As discussed earlier in this chapter, residues at the interfaces between adjacent TM1 helices of MscL are relatively well conserved. These interface residues tend to occur along two stripes that run along the i, i + 4 grooves on opposite sides of the helices (Fig. 5), which is consistent with the packing of TM1 helices in a right-handed bundle with a crossing angle of $\sim -40^\circ$ (Chothia *et al.,* 1981). Residues found occurring in these positions of TM1 are generally small, specifically Gly and Ala, as is typical for the interface between membrane-spanning helices, and are relatively well conserved in MscL. In the model for the transition between the closed to open states developed by Sukharev and Guy (Sukharev *et al.,* 2001b), adjacent helices slide against one another through contacts mediated by residues positioned along the i, i + 4 direction. Consequently, residues at these positions in the TM1-TM1 interface must be compatible with helix packing interactions in both closed and open states. The use of small amino acids, particularly Gly and Ala, at these positions may facilitate the interconversion between conductance states. Examination of the MscL sequence alignments (Fig. 3A) shows some tendency for Gly and Ala to alternate at every second residue between positions 18 and 28 to give the pattern AxGxxxGxAxG. The spacing is such that the Glys form an interface with one neighboring TM1 helix, while the Alas participate in the interface with the other neighboring TM1. Although the pattern GxxxG tends to be relatively common in membrane-spanning helices (Russ and Engelman, 2000), with the glycines positioned at the interface between helices, and although Ala is often commonly found in membrane-spanning helices, the combination GxAxG does not appear to be especially frequent (Senes *et al.,* 2000), suggesting the presence of this motif in MscL may be functionally relevant.

Intriguingly, the amino acid sequence of the third predicted transmembrane helix of MscS exhibits similar patterns to those present in TM1 of MscL (Levina *et al.,* 1999). In particular, there is a striking pattern of conserved Gly and Ala residues in the extended region between residues 84–121 of the *E. coli* yggB gene product, with sequence AAxGxxGxxxAxxxAxxGAAGxAxGxAxxGxxxxxAAG. The tendency of Alas to appear every fourth residue, particularly between positions 94–113, is quite noticeable, while the pattern involving Gly is more variable. These residues are present in one of the most strongly conserved stretches of sequence in MscS, and the properties are suggestive of participation in helix–helix interactions that may be relevant to mechanosensitive channel function.

V. Concluding Remarks

The past few years have witnessed significant advances in establishing the gating mechanism of prokaryotic mechanosensitive channels, particularly the recent characterization of the open state of MscL (Sukharev *et al.,* 2001a; Perozo *et al.,* 2002a) and the identification and initial characterization of the protein responsible for MscS activity (Levina *et al.,* 1999). Although biochemically more challenging, steady and increasing progress is being made in the identification and characterization of eukaryotic mechanosensitive channels (see Gillespie and Walker, 2001), including those that function through coupling to the cytoskeleton. Although an understanding in molecular detail of a complex mechosensory process such as hearing is still rather far off in the future, substantial progress has been made in deciphering basic features of mechanosensitive channels involved in simpler systems that will, it is hoped, reveal shortcuts towards this goal.

Acknowledgments

Discussions with Sidney Wang, Meg Barclay, Robert Spencer, Kaspar Locher, Geoffrey Chang, Dennis Dougherty, Henry Lester, Rob Phillips, and Eduardo Perozo are greatly appreciated. This work was supported in part by NIH grants GM62532 to D.C.R. and GM20705 to R.B.B.

Note Added in Proof

The crystal structure of the *Escherichia coli* MscS has been determined at 3.9 Å resolution [Bass, R. B., Strop, P., Barclay, M. and Rees, D. C. (2002). *Science* **298,** 1582–1587.] and demonstrates that this channel folds as a membrane-spanning heptamer with a large, cytoplasmic region. Each subunit contains three transmembrane (TM) helices, with the TM3 helices lining the pore, while TM1 and TM2 are likely candidates for the tension and voltage sensors. The transmembrane pore, apparently captured in an open state, leads into a large chamber within the cytoplasmic region, that connects to the cytoplasm through openings that may serve to filter out impermeant species.

References

Adamson, A. W. (1982). *Physical Chemistry of Surfaces.* Wiley, New York.

Ajouz, B., Berrier, C., Garrigues, A., Besnard, M., and Ghazi, A. (1998). *J. Biol. Chem.* **273,** 26670–26674.

Ajouz, B., Berrier, C., Besnard, M., Martinac, B., and Ghazi, A. (2000). *J. Biol. Chem.* **275,** 1015–1022.

Alvarez de la Rosa, D., Canessa, C. M., Fyfe, G. K., and Zhang, P.-C. (2000). *Annu. Rev. Physiol.* **62,** 573–594.

Batiza, A. F., Rayment, I., and Kung, C. (1999). *Structure* **7,** R99–R103.

Batiza, A. F., Kuo, M. M.-C., Yoshimura, K., and Kung, C. (2002). *Proc. Natl. Acad. Sci. USA* **99,** 5643–5648.

Berrier, C., Coulombe, A., Houssin, C., and Ghazi, A. (1989). *FEBS Lett.* **259,** 27–32.

Berrier, C., Coulombe, A., Houssin, C., and Ghazi, A. (1993). *J. Memb. Biol.* **133,** 119–127.

Berrier, C., Besnard, M., Ajouz, B., Coulombe, A., and Ghazi, A. (1996). *J. Memb. Biol.* **151,** 175–187.

Betanzos, M., Chiang, C.-S., Guy, H. R., and Sukharev, S. (2002). *Nature Struct. Biol.* **9,** 704–710.

Bilston, L. E., and Mylvaganam, K. (2002). *FEBS Lett.* **512,** 185–190.

Blount, P., Sukharev, S. I., Moe, P. C., Nagle, S. K., and Kung, C. (1996a). *Biol. Cell* **87,** 1–8.
Blount, P., Sukharev, S. I., Moe, P. C., Schroeder, M. J., Guy, H. R., and Kung, C. (1996b). *EMBO J.* **15,** 4798–4805.
Blount, P., Sukharev, S. I., Schroeder, M. J., Nagle, S. K., and Kung, C. (1996c). *Proc. Natl. Acad. Sci. USA* **93,** 11652–11657.
Blount, P., Sukharev, S. I., Moe, P. C., Martinac, B., and Kung, C. (1999). *Methods Enzymol.* **294,** 458–482.
Booth, I. R., and Louis, P. (1999). *Curr. Opin. Microbiol.* **2,** 166–169.
Bowie, J. U. (1997). *J. Mol. Biol.* **272,** 780–789.
Butler, S. L., and Falke, J. J. (1996). *Biochemistry* **35,** 10595–10600.
Cantor, R. S. (1997). *J. Phys. Chem. B* **101,** 1723–1725.
Cantor, R. S. (1999). *Biophys. J.* **76,** 2625–2639.
Careaga, C. L., and Falke, J. J. (1992). *J. Mol. Biol.* **226,** 1219–1235.
Cayley, D. S., Guttman, H. J., and Record, M. T., Jr. (2000). *Biophys. J.* **78,** 1748–1764.
Chang, G., Spencer, R. H., Lee, A. T., Barclay, M. T., and Rees, D. C. (1998). *Science* **282,** 2220–2226.
Chothia, C., Levitt, M., and Richardson, D. (1981). *J. Mol. Biol.* **145,** 215–250.
Clapham, D. E., Runnels, L. W., and Strubing, C. (2001). *Nat. Rev. Neurosci.* **2,** 387–396.
Cruickshank, C. C., Minchin, R. F., LeDain, A. C., and Martinac, B. (1997). *Biophys. J.* **73,** 1925–1931.
Csonka, L. N., and Epstein, W. (1999). In: Escherichia coli *and* Salmonella: *Cellular and Molecular Biology* (F. C. Neidhardt, Ed.), pp. 1210–1223. Am. Soc. for Microbiology, Washington, DC.
de Kruijff, B. (1997). *Curr. Opin. Chem. Biol.* **1,** 564–569.
Doyle, R. J., and Marquis, R. E. (1994). *Trends Microbiol.* **2,** 57–60.
Eilers, M., Shekar, S. C., Shieh, T., Smith, S. O., and Fleming, P. J. (2000). *Proc. Natl. Acad. Sci. USA* **97,** 5796–5801.
Elmore, D. E., and Dougherty, D. A. (2001). *Biophys. J.* **81,** 1345–1359.
Gerstein, M., Lesk, A. M., and Chothia, C. (1994). *Biochemistry* **33,** 6739–6749.
Gillespie, P. G., and Walker, R. G. (2001). *Nature* **413,** 194–202.
Gruner, S. M. (1985). *Proc. Natl. Acad. Sci. USA* **82,** 613–619.
Gu, L. Q., Liu, W. H., and Martinac, B. (1998). *Biophys. J.* **74,** 2889–2902.
Gullingsrud, J., Kosztin, D., and Schulten, K. (2001). *Biophys. J.* **80,** 2074–2081.
Hamill, O. P., and Martinac, B. (2001). *Physiological Rev.* **81,** 685–740.
Häse, C. C., LeDain, A. C., and Martinac, B. (1997a). *J. Membr. Biol.* **157,** 17–25.
Häse, C. C., Minchin, R. F., Kloda, A., and Martinac, B. (1997b). *Biochem. Biophys. Res. Commun.* **232,** 777–782.
Howard, J., Roberts, W. M., and Hudspeth, A. J. (1988). *Annu. Rev. Biophys. Biophys. Chem.* **17,** 99–124.
Hubbell, W. L., Gross, A., Langen, R., and Lietzow, M. A. (1998). *Curr. Opin. Struct. Biol.* **8,** 649–656.
Javadpour, M. M., Eilers, M., Groesbeek, M., and Smith, S. O. (1999). *Biophys. J.* **77,** 1609–1618.
Jiang, Y., Lee, A., Chen, J., Cadene, M., Chait, B. T., and MacKinnon, R. (2002a). *Nature* **417,** 515–522.
Jiang, Y., Lee, A., Chen, J., Cadene, M., Chait, B. T., and MacKinnon, R. (2002b). *Nature* **417,** 523–526.
Kendrew, J. C., Parrish, R. G., Marrack, J. R., and Orlans, E. S. (1954). *Nature* **174,** 946–949.
Kloda, A., and Martinac, B. (2001a). *Cell Biochem. Biophys.* **34,** 349–381.
Kloda, A., and Martinac, B. (2001b). *Biophys J* **80,** 229–240.

Kloda, A., and Martinac, B. (2001c). *EMBO J.* **20,** 1888–1896.

Kloda, A., and Martinac, B. (2002a). *Archaea* **1,** 35–44.

Kloda, A., and Martinac, B. (2002b). *Eur. Biophys. J.* **31,** 14–25.

Kong, Y., Shen, Y., Warth, T. E., and Ma, J. (2002). *Proc. Natl. Acad. Sci. USA* **99,** 5999–6004.

Kraulis, P. J. (1991). *J. Appl. Cryst.* **24,** 946–950.

Krogh, A., Larsson, B., von Heijne, G., and Sonnhammer, E. L. (2001). *J. Mol. Biol.* **305,** 567–580.

Le Dain, A. C., Saint, N., Kloda, A., Ghazi, A., and Martinac, B. (1998). *J. Biol. Chem.* **273,** 12116–12119.

Levina, N., Tötemeyer, S., Stokes, N. R., Louis, P., Jones, M. A., and Booth, I. A. (1999). *EMBO J.* **18,** 1730–1737.

Li, Y., Moe, P. C., Chandrasekaran, S., Booth, I. R., and Blount, P. (2002). *EMBO J.* **21,** 5323–5330.

Maingret, F., Fosset, M., Lesage, F., Lazdunski, M., and Honore, E. (1999a). *J. Biol. Chem.* **274,** 1381–1387.

Maingret, F., Patel, A. J., Lesage, F., Lazdunski, M., and Honore, E. (1999b). *J. Biol. Chem.* **274,** 26691–26696.

Martinac, B., Buechner, M., Delcour, A. H., Adler, J., and Kung, C. (1987). *Proc. Natl. Acad. Sci. USA* **84,** 2297–2301.

Martinac, B., Adler, J., and Kung, C. (1990). *Nature* **348,** 261–263.

Martinac, B. (2001). *Cell. Physiol. Biochem.* **11,** 61–76.

Maurer, J. A., Elmore, D. E., Lester, H. A., and Dougherty, D. A. (2000). *J. Biol. Chem.* **275,** 22238–22244.

McLaggan, D., Jones, M. A., Gouesbet, G., Levina, N., Lindey, S., Epstein, W., and Booth, I. R. (2002). *Mol. Microbiol.* **43,** 521–536.

Merritt, E. A., and Bacon, D. J. (1997). *Methods Enzymol.* **277,** 505–524.

Moe, P. C., Blount, P., and Kung, C. (1998). *Mol. Microbiol.* **28,** 583–592.

Moe, P. C., Levin, G., and Blount, P. (2000). *J. Biol. Chem.* **275,** 31121–31127.

Oakley, A. J., Martinac, B., and Wilce, M. C. J. (1999). *Prot. Sci.* **8,** 1915–1921.

Okada, K., Moe, P. C., and Blount, P. (2002). *J. Biol. Chem.* **277,** 27682–27688.

Ou, X. R., Blount, P., Hoffman, R. J., and Kung, C. (1998). *Proc. Nat. Acad. Sci. USA* **95,** 11471–11475.

Perozo, E., Kloda, A., Cortes, D. M., and Martinac, B. (2001). *J. Gen. Physiol.* **118,** 193–206.

Perozo, E., Cortes, D. M., Sompornpisut, P., Kloda, A., and Martinac, B. (2002a). *Nature* **418,** 942–948.

Perozo, E., Kloda, A., Cortes, D. M., and Martinac, B. (2002b). *Nature Struct. Biol.* **9,** 696–703.

Poolman, B., Blount, P., Folgering, J. H. A., Friesen, R. H. E., Moe, P. C., and van der Heide, T. (2002). *Mol. Microbiol.* **44,** 889–902.

Record, M. T. Jr., Courtenay, E. S., Cayley, D. S., and Guttman, H. J. (1998). *Trends Biochem. Sci.* **23,** 143–148.

Rees, D. C., DeAntonio, L., and Eisenberg, D. (1989). *Science* **245,** 510–513.

Russ, W. P., and Engelman, D. M. (2000). *J. Mol. Biol.* **296,** 911–919.

Sachs, F. (1997). In: *Cytoskeletal Regulation of Membrane Function* (S. C. Froehner, Ed.), vol. 52, pp. 209–218. Rockefeller Univ. Press, New York.

Senes, A., Gerstein, M., and Engelman, D. M. (2000). *J. Mol. Biol.* **296,** 921–936.

Spencer, R., Chang, G., and Rees, D. C. (1999). *Curr. Opin. Struct. Biol.* **9,** 448–454.

Spencer, R. H., Chang, G., Bass, R. B., and Rees, D. C. (2002). In: *Methods and Results in Membrane Protein Crystallization* (S. Iwata, Ed.). International Univ. Line, La Jolla, CA.

Spencer, R. H., and Rees, D. C. (2002). *Annu. Rev. Biophys. Biomol. Struct.* **31,** 207–233.

Strop, P. (2002). Ph.D. Thesis, California Institute of Technology, Pasadena.

Sukharev, S., Schroeder, M. J., and McCaslin, D. R. (1999a). *Biophys. J.* **76,** A138.

Sukharev, S., Betanzos, M., Chiang, C.-S., and Guy, H. R. (2001a). *Nature* **409,** 720–724.

Sukharev, S., Durell, S. R., and Guy, H. R. (2001b). *Biophys. J.* **81,** 917–936.

Sukharev, S. (2002). *Biophys. J.* **83,** 290–298.

Sukharev, S. I., Martinac, B., Arshavsky, V. Y., and Kung, C. (1993). *Biophys. J.* **65,** 177–183.

Sukharev, S. I., Blount, P., Martinac, B., Blattner, F. R., and Kung, C. (1994). *Nature* **368,** 265–268.

Sukharev, S. I., Blount, P., Martinac, B., and Kung, C. (1997). *Annu. Rev. Physiol.* **59,** 633–657.

Sukharev, S. I., Schroeder, M. J., and McCaslin, D. R. (1999b). *J. Memb. Biol.* **171,** 183–193.

Sukharev, S. I., Sigurdson, W. J., Kung, C., and Sachs, F. (1999c). *J. Gen. Physiol.* **113,** 525–539.

Tavernarakis, N., and Driscoll, M. (2001). *Cell. Biochem. Biophys.* **35,** 1–18.

Thompson, J. D., Gibson, T. J., Plewniak, F., Jeanmougin, F., and Higgins, D. G. (1997). *Nucleic Acids Res.* **25,** 4876–4882.

Unwin, N. (1995). *Nature* **373,** 37–43.

Wallin, E., Tsukihara, T., Yoshikawa, S., von Heijne, G., and Elofsson, A. (1997). *Protein Sci.* **6,** 808–815.

Welsh, M. J., Price, M. P., and Xie, J. (2002). *J. Biol. Chem.* **277,** 2369–2372.

Yoshimura, K., Batiza, A., Schroeder, M., Blount, P., and Kung, C. (1999). *Biophys. J.* **77,** 1960–1972.

Yoshimura, K., Batiza, A., and Kung, C. (2001). *Biophys. J.* **80,** 2198–2206.

THE VOLTAGE SENSOR AND THE GATE IN ION CHANNELS

By FRANCISCO BEZANILLA* and EDUARDO PEROZO†

*Department of Physiology and Department of Anesthesiology, David Geffen School of Medicine at UCLA, Los Angeles, California, 90095, and †Department of Molecular Physiology and Biological Physics, School of Medicine, University of Virginia, Charlottesville, Virginia 22908

I. Introduction

Ion channels are specialized membrane proteins that reduce significantly the energy barrier of ion conduction across membranes. A region called the pore is where ion conduction occurs and the flux is proportional to the electrochemical gradient between the two sides of the membrane. The easiness of ion permeation or *conductance* of the channel and the selection of the preferred types of ions that can go through the channel or *selectivity* are two properties of the pore region of the ion channel. The pore may be open in several different conductance states or altogether closed and the process that controls these conductance states is called *gating*.

In this chapter we will not address ion conduction proper but we will concentrate on two aspects of ion channels: the gating process and the transduction from the initiating stimulus to the operation of the gate. The gating process is the operation of the gate proper while the transducer may be considered the *sensor* of the stimulus that ultimately opens and closes the channel. Specifically, we will address one type of sensor: the membrane potential or voltage sensor.

Our understanding of voltage sensing and gating has been the result of a large number of physiological and biophysical observations that were initiated by the pioneering work of Hodgkin and Huxley (1952) in

ADVANCES IN PROTEIN CHEMISTRY, Vol. 63

0065-3233/03 $35.00

a large variety of ion channels from squid axon sodium and potassium channels to potassium channels from bacteria. The correlation between structure and function has benefited especially from the possibility of expressing ion channels in heterologous expression systems that allows the study of specially engineered channels where specific sites can be tested with EPR or fluorescent probes.

The description of structure–function relations of the voltage sensor will be based on experiments performed in the Shaker K channel, the first cloned potassium channel from *Drosophila melanogaster,* or the rat muscle sodium channel. At present, there is no three-dimensional structure of voltage-dependent channels; therefore the experiments are a combination of site-directed mutagenesis, cysteine scanning, and probing that, studied simultaneously with the function, helps in inferring the major conformational changes of the channel protein. Although some of these measurements can be very precise using optical techniques, most of these inferences are indirect and they allow only a hypothetical picture of channel operation that will be confirmed only when a high-resolution structure becomes available.

The pore and its gates have also been studied using approaches similar to those outlined for the voltage sensor. However, the availability of the high-resolution crystal structures from prokaryotic K^+ channels, KcsA and MthK (Doyle *et al.,* 1998a; Jiang *et al.,* 2002a,b), has provided a solid basis for understanding conformational changes with structural dynamic techniques such as EPR. The movements of the gates, described in the second part of this chapter, can then be considered as real conformational changes in the atomic structure of the channel pore.

II. The Voltage Sensor

Voltage-dependent channels, such as the classical sodium or potassium channels in nerve tissue, change their conductance with membrane potential. The changes in conductance are a very steep function of membrane voltage: conductance values can increase as much as 150 times for an increment of 10 mV in membrane potential (Hodgkin and Huxley, 1952).

The ion current through the channel is the result of flow through an open pore under the influence of the electrochemical gradient. Thus the effective conductance of the channel depends on the open channel conductance and the fractional time the channel spends in the open state, also called open probability (P_o). By recording the currents through single channels it is possible to measure the voltage dependence of the conductance of the open pore, and it has been found to be weakly voltage dependent. This means that the high voltage dependence of the conductance originates from the voltage dependence of the open probability. As the gating of the channel controls the open probability our understanding of the voltage dependence will involve the description of the mechanisms that sense the membrane potential and couple the sensor to the operation of the gate.

If voltage changes regulate gating the basic question is how voltage is detected by the channel. The two obvious candidates are electric or magnetic charges or dipoles that can reorient in the electric field. We will concentrate on electric charges because there is no evidence for their magnetic counterparts in channel proteins. The inescapable conclusion is that a change of position of charges or reorientation of dipoles will be a required mechanism to sense voltage changes (Hodgkin and Huxley, 1952).

Early evidence of the presence of the gating charge was obtained by recording the current produced by their displacement as a result of a change in membrane potential. These currents, which are transient in nature, were called *gating currents* (see Fig. 1). The kinetics and steady-state properties of these currents gave information on channel states which were not directly measured by the ionic current (Fig. 1). This is because in the process of channel opening there are multiple states which are nonconducting preceding the conducting state(s) and the transition of the voltage sensor between these states moved charge, thus contributing to the gating currents. This is summarized in Fig. 1B. The gating charge Q, which corresponds to the time integral of the gating currents, has a voltage dependence different from the open probability (P_o in Fig. 1B) such that there is a region of voltage where there is sizable movement of the charge with little or undetectable open probability (Fig. 1B). This basic experimental observation has drawn much attention and many kinetic models have been presented that can explain the basic features of Q and P_o. However, the focus in this review will be on the structural basis of these observations with the long-term goal of formulating physical models of channel operation.

A. *Charge Required to Open the Channel*

The large voltage dependence of these voltage dependent channels dictate that a large amount of charge move in the membrane electric field. Typically, at negative (inside) membrane potential, the channel is mostly closed and at positive potentials is mostly open. Then, the open

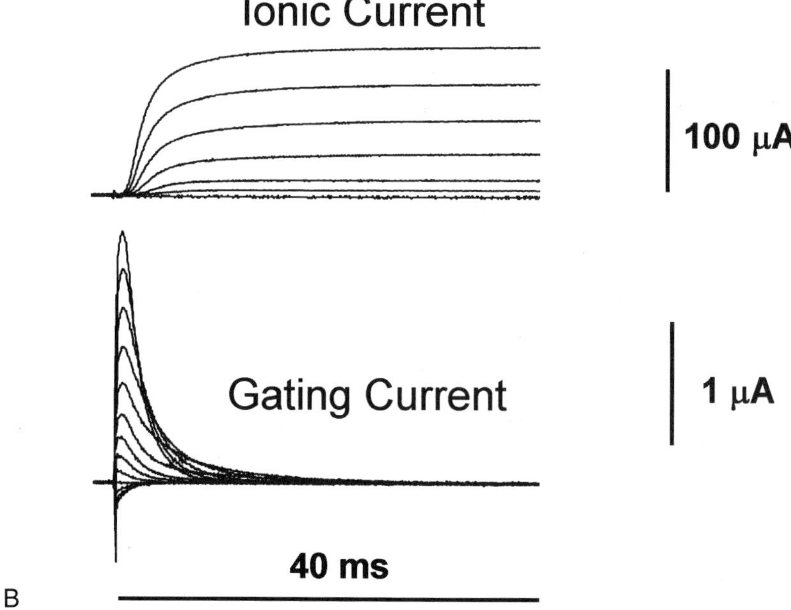

B

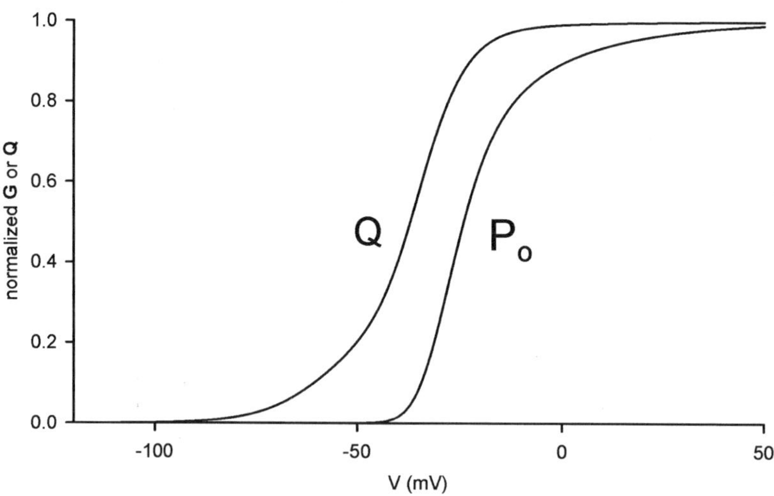

FIG. 1. (A) Recordings of ionic currents and gating currents of Shaker K channel with fast inactivation removed. The holding potential was −90 mV and the steps range from −120 to 0 mV in increments of 10 mV. The gating currents were recorded from a mutant that does not have pore conduction (W434F, Perozo *et al.*, 1993). Notice that ionic currents are only visible with pulses that are more positive than −60 mV. (B) The voltage dependence of the integral of the gating current or gating charge (Q) and the open probability of the channel (P_o).

probability P_o is a sigmoid function of membrane potential that can be roughly approximated by a Boltzmann distribution. One can ask how much charge is necessary to account for the voltage dependence. This question can be answered by first defining the activation potential (W_a) which is a measure of the electrical energy necessary to open the channel: $W_a = -kT \ln P_o$. The mean activation charge $\langle q_a \rangle$ is next defined as the negative gradient of W_a or

$$\langle q_a \rangle = kT \frac{d \ln P_o}{dV}$$

(Almers, 1978; Sigg and Bezanilla, 1997). Now we need to relate this mean activation charge with the gating charge displacement. We first find the distribution of open probability P_o, knowing the energies F_i of each one of the states i along the activation pathway where the reaction coordinate is the gating charge displacement q. Then, by assuming that the energy $F_i = G - qV$, where G represents the energy that depends on thermodynamic variables other than voltage (V), one obtains the derivative of ln P_o with respect to V and thus obtains a relation between $\langle q_a \rangle$ and q and V (Sigg and Bezanilla, 1997). If the channel has no charge movement between the open states, the total charge per channel is equal to

$$Q_{max} = \lim_{V \to -\infty} kT \frac{d \ln P_o}{dV}$$

The total gating charge Q_{max} can also be estimated by measuring the total gating charge Q and the total number of channels N and then dividing Q by N. However, in this case the estimation of Q_{max} may also contain contributions of charge that is not thermodynamically coupled to the opening of the channel. These two types of measurements have been done in the Shaker K channel, and remarkable agreement has been found between the two methods, indicating that there is very little or no peripheral charge (Fig. 2). It is important to notice that the value of Q_{max} corresponds to the charge that traverses the whole electric field of the membrane. If, for example, the movement is only half of the total electric field then the actual charge must be double to account for the same amount of Q_{max}. We should then consider Q_{max} as the total electric charge times the fraction of the field it traverses.

B. The Residues That Make Up the Gating Charge

In Shaker the value of Q_{max} is about 13 e per channel (Schoppa *et al.*, 1992). In the skeletal muscle Na channel the value of Q_{max} was 12 e

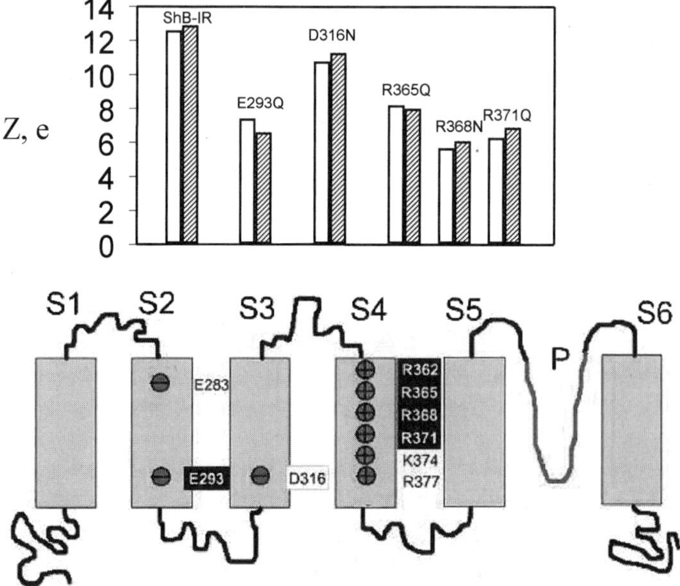

FIG. 2. Top: The neutralization of some residues in the S2, S3, and S4 segments decrease the gating charge. Each bar represents the charge measured after neutralization by the limiting slope method (empty bars) or the *Q/N* method (filled bars). ShB-IR represents the wild-type channel. Data from Seoh *et al.* (1996) and Aggarwal and MacKinnon (1996). Bottom: Topology of the transmembrane segments of Shaker K channel. Four subunits are required for channel function. The highlighted residues have been proved to be part of the gating charge. In sodium and calcium channels the four subunits are strung together in one polypeptide with four domains that are homologous among them but not identical to each other.

as estimated with limiting slope using single-channel measurements (Hirschberg *et al.,* 1996). What structure within the protein can account for such large values of charge displacement? The simple tilting of alpha helices may not be enough because each helix of the length of the membrane thickness is equivalent to a dipole separating 0.5 e, and as there are a total of 24 transmembrane segments they all would have to make a substantial conformational change. It is more likely that either acidic residues or basic residues bearing a permanent charge may be oriented and translocated by the change in the electric field.

When the first voltage-dependent ion channel was cloned and sequenced, it was recognized that the fourth transmembrane segment had a string of basic amino acids organized as one every three residues was separated by hydrophobic residues. This pattern was found in voltage-dependent sodium, potassium, and calcium channels. In addition, the second and third transmembrane segments had conserved acidic residues. These charges are shown in the lower half of Fig. 2 that depicts the transmembrane segments of the Shaker potassium channel. Four subunits are necessary to assemble a functional channel. In the case of the sodium and calcium channels a long polypeptide codes for the whole pore-forming subunit, but it has four repeats or domains that are similar to the four subunits of the potassium channel. The proof that these basic and acidic amino acids were responsible for the gating charge was obtained when the neutralization of a particular residue was shown to decrease the total charge per channel (Aggarwal and MacKinnon, 1996; Seoh *et al.*, 1996). Thus, the first four most extracellular basic residues of the S4 segment and the most intracellular acidic residue of the S2 segment of Shaker were found to contribute to the gating currents (see Fig. 2). If these five charges move across the whole electric field, the total charge per channel would be more than satisfied. The simplest interpretation is that the charges traverse a fraction of the electric field giving a total contribution of about 13 e for the total movement.

C. *Movement of Charged Residues with Membrane Voltage*

From the structural point of view, the movement of the voltage sensor is quite challenging because it needs to transport a large amount of charge across the membrane field. The first problem is the stability of the S4 segment with its basic residues, presumably all positively charged and embedded in the protein matrix. Initial hypotheses were that these basic residues were forming salt bridges with acidic residues, thus lowering the conformational energy to reasonable values. In this hypothesis, the change in the electric field would break these salt bridges allowing the basic residues to translocate to the other side of the membrane, producing the charge movement that could be recorded as gating current.

Papazian *et al.* (1995) found that in fact the basic residue K374 (the fifth charge of the S4 segment in Shaker) forms a network of charges with acidic residues E293 from S2 and D316 from S3. Disruption of this network prevents normal protein folding. It is important to remember that K374 does not contribute much charge to the gating currents. In addition, Tiwari-Woodruff *et al.* (2000) found another network that operates between R368 and R371 (the third and fourth charge of the S4 segment) and the acidic residue E283 of the S2 segment. However, there is no evidence that R362 and R365 (first and second charges of the S4 segments) pair with any acidic group.

The translocation of the basic residues of the S4 segment in response to voltage changes has been a matter of intense study. The techniques used have all been indirect because no three-dimensional structure information is available. We could classify the techniques into three broad categories. The first is the use of reactive agents against an engineered residue in the S4 segment; the second is the use of site-directed fluorescent probes to test conformational changes (Mannuzzu *et al.*, 1996; Cha and Bezanilla, 1997, 1998; Loots and Isacoff, 1998); and the third is the use of fluorescence resonance energy transfer (FRET) to measure actual distances and changes in distances between fluorophores inserted in the structure of the voltage sensor. The three types of techniques have given complementary information that has started to delineate a possible mechanism of operation of the voltage sensor.

1. Testing the Exposure of Residues with Probes

The use of cysteine reactive probes allows the testing of exposure of residues replaced by cysteine from either the inside or the outside of the membrane and as a function of the state of the sensor. The technique requires a mutation of the residue under study to a cysteine. Then, by applying a cysteine modifying reagent that on reaction changes the electrophysiological properties it is possible to test whether the reaction is occurring from the inside or the outside by using nonpermeant modifying reagents. The test is done in the hyperpolarized or depolarized membrane, thus probing the accessibility of the residue to the reagent mainly in two different conformations of the channel molecule. These experiments have been carried out in the S4 segment of the fourth domain of the skeletal muscle sodium channel and in the S4 segment of Shaker K channel (Larsson *et al.* 1996; Baker *et al.*, 1998; Yang and Horn, 1995; Yusaf *et al.*, 1996). The results show that indeed several charged residues of the S4 segment could be exposed to the inside when the membrane is hyperpolarized or to the outside when it is depolarized. However, some residues may be exposed to one side in one state and buried in the other state, indicating that they do not change sides when the membrane potential is changed.

The other method consists of a replacement of the basic residues by histidine (Starace *et al.*, 1997; Starace and Bezanilla, 2001). The rationale is that histidine can be titrated by pH and the pK_a is close to the physiological pH allowing a range of H^+ concentrations that can span from full loading of the histidine—thus bearing a positive charge—to full unloading making it a neutral residue. Histidine scanning has several advantages over cysteine scanning when dealing with the charges of the sensor. First, it is possible to conserve or eliminate the charge of the basic residue under study, and second, the probe is a proton, which is much smaller than cysteine reagents, allowing the probing of exposure to restricted aqueous crevices. These experiments have been carried out in a nonconducting mutant of Shaker (mutation in the pore region, W434F; Perozo *et al.*, 1993) because it requires recording gating currents to detect the proton accessibility under a proton gradient across the membrane.

If the histidine residue alternates exposure sides in response to membrane potential changes, one can observe an asymmetry in the gating current between the On and Off response. This is due to the fact that histidine is positively charged in the low pH solution but is uncharged in the high pH solution, introducing an asymmetry in the transported charge depending on the direction of movement of the sensor. In addition, if the membrane potential is in the range where the P_o is near 0.5, the sensor makes frequent transitions between the two extreme positions; therefore there will be a *net current* because each excursion of the voltage sensor from the low pH side to the high pH side will transport a proton, but none will be transported in the opposite direction. This current has a bell-shaped voltage dependence because it tends to decrease toward the extreme potentials at which the voltage sensor makes infrequent excursions between the two states. This voltage sensor-driven proton current was observed for the second, third, and fourth charges (R365H, R368H, R371H) of the S4 segment of Shaker showing that these three residues can be alternately exposed to either side depending on the state of the channel (see Fig. 3).

For the first charge (R362H), the observation was that there is a proton current that is not bell-shaped but increases monotonically as the membrane is hyperpolarized. This proton current has all the characteristics

362H	⊕	**Accessible from BOTH sides (H^+ channel)**
365H	⊕	**Accessible from BOTH sides (H^+ transport)**
368H	⊕	**Accessible from BOTH sides (H^+ transport)**
371H	⊕	**Accessible from BOTH sides (H^+ transport)** **(H^+ channel at positive potentials)**
374H	⊕	**NOT titratable**
377H	⊕	**NOT titratable**

FIG. 3. Summary of the results of histidine scanning in Shaker K channel. When replaced by histidine, the first four basic residues (R362 through R371) translocate completely from intracellular to extracellular while the next two residues (K374 and R377) are not titratable.

of an ionic current through a channel, as opposed to the other case that represents a typical transporter. The interpretation of this result is that in the hyperpolarized condition the histidine in position 362 acts like a bridge between the inside and outside allowing protons to circulate according to their electrochemical gradient using the histidine as a selectivity filter. The implication of this interpretation is that in the hyperpolarized state the electric field is concentrated in an extremely narrow region in the vicinity of residue 362 (Starace and Bezanilla, 1999). When the fourth charge is replaced by histidine (R371H) it forms a proton transporter; however at large depolarizations it seems to also form a proton channel as is the case for R362H (Starace and Bezanilla, 2001). This may be interpreted as the formation of a narrow constricted region with the consequent concentration of the electric field near R371 at high depolarizations, delineating certain symmetry in the movement of the voltage sensor.

When replaced by histidine, the fifth and sixth charges (K374H and R377H) are not titratable from either side of the membrane at any membrane potential. This may imply that these residues are always inaccessible or they do not move in the electric field. As the neutralization studies indicate that K374 does not contribute much charge, most likely this residue does not move in the field. As residue R377 could not be studied in the neutralized case because the protein did not express, we do not know whether it contributes to the gating charge. However, as it is even further along the sequence than any of the other charges and when replaced by histidine is not titratable we might consider it not moving in the electric field (see Fig. 3).

The results of cysteine and histidine scanning show that the gating charge residues reside in a solvent-accessible region of the protein. This is in contrast to the idea that the charges would be buried in the protein and protected by salt bridges. It is interesting to compare the results of cysteine scanning and histidine scanning because the probe differs drastically in size. Thus, positions that are not accessible with the cysteine reagent are accessible with protons, and this normally happens for positions that are far from the side where the test is made at potentials that take the residue away from the testing side. This result implies that the charges reside in water-filled crevices that may narrow down significantly away from the protein surface. At hyperpolarized potentials the internal crevice will contain at least the first four charges, but the first charge will be in the narrowest part that will be extremely close to the external face. When the first charge is replaced by a histidine it makes a bridge to the outside, but in the normal situation, arginine has a longer chain and volume; therefore is unlikely that it will span as a bridge.

In addition, the pK_a of argininine is so high that it would rarely lose its charge.

These experiments have given a first glimpse of how the charges may reside in the protein and in relation to the bulk solution, and also show us that the S4 is clearly optimized as a voltage sensor. We must remember that the charge movement is the product of the actual charge times the fraction of the field it moves. Then, to minimize large movements and consequently large conformational changes, the field should be concentrated in a small distance as seems to be the case for the narrow crevices. The other aspect, already mentioned, that the charges normally reside in the aqueous medium, has a stabilizing effect on the hyperpolarized and depolarized conformations. It is then clear that to change the conformation from closed to open or vice versa it will be necessary to overcome a large energy barrier as the charge is translocated through the hydrophobic part of the protein. Once the barrier is overcome, the charges will jump across in a rapid motion giving origin to the gating shots that have been inferred from noise analysis of the gating currents (Conti and Stuhmer, 1989; Sigg *et al.*, 1994). The waiting time to overcome this barrier is proportional to the time constant of the gating current and is shortened by a favoring electric field that by subtracting to the total energy of the barrier, effectively lowers it, facilitating the jump.

2. *The Electric Field in the Channel Protein*

The discussion of the previous paragraph emphasizes the importance of knowing the potential profile in the protein structure to account for the charge movement and its voltage dependence. Some progress has been made in measuring the local electric field using fluorescent probes (Asamoah *et al.*, 2002). There is a class of fluorophores that contain an electrochromic group with an intrinsic dipole moment that can be modified by the absorption of a photon. This differential dipole moment couples to the local electric field and perturbs the electronic energy levels between the ground and excited state. Consequently, the electrochromic effect is manifested as a spectral shift. By measuring the fluorescence with a filter that spans the region of maximum shift of the spectrum, it is possible to detect the shifts as changes in the fluorescence intensity. These electrochromic probes have been used extensively to measure membrane potential because they partition in the lipid bilayer, and their spectral shift is proportional to the electric field.

One of these probes was modified so as to have a cysteine reactive site, and at the same time it was made less hydrophobic to decrease its partition in the lipid bilayer with the idea of inserting the probe in predetermined sites of a channel. For this purpose the Shaker K channel was engineered to have cysteines in specific sites and after staining with dye the fluorescence was measured as a function of imposed voltage and time. In most cases, the time course of the fluorescence change was found to have two components. The fast component followed the charging of the membrane capacitance and the slow component varied widely depending on the position of the dye, voltage, and pulse duration. The fast component was interpreted as the electrochromic response of the dye, while the slow component was interpreted as a change in environment around the probe produced by a conformational change of the channel protein. Starting from a fixed potential, the electrochromic component had a linear dependence on the voltage of the pulse, and on return to the original potential, the electrochromic signal reversed its direction and frequently had a different magnitude. This means that the probe responded very quickly to the imposed potential and that the subsequent conformational changes in the channel made the response different, presumably because the field or the environment around the probe was modified by the voltage.

The slope of the fluorescence change as a function of membrane potential was measured in several sites of the Shaker K channel, and it was found that the strength of the electric field was maximum in the S4 segment. The magnitude of these changes were calibrated in terms of the electric field using the probe that partitions in the bilayer and comparing the $\Delta F/F$ of this probe for a change in V with the equivalent value from each one of the signals recorded from the different sites in the channel. In the S4 the strength of the field was found to be 3 times the field strength in the bilayer. This means that most of the field is concentrated in a thickness of about 10 Å. This result is in agreement with the conclusion drawn from histidine scanning where the first charge replaced by histidine generates a proton pore (Starace and Bezanilla, 1999). This requires a very concentrated field where the selectivity filter of the proton corresponds to the histidine group bridging the gap between the inside and outside solutions.

3. Direct Measurements of Distance Changes during Activation

Two groups have used FRET to estimate distance changes as a result of changing the membrane potential (Cha *et al.*, 1999; Glauner *et al.*, 1999). In FRET a donor fluorescent molecule can transfer energy to another probe (most generally another fluorophore) through dipole–dipole interactions. The efficiency of energy transfer depends on several known or computable factors such as the spectral overlap of the probes, the quantum yield of the donor, the index of refraction, and a much less known factor: the κ^2 or relative orientation of the dipoles of the donor and acceptor probes. It is generally assumed that the fluorophores are randomly oriented giving a $\kappa^2 = 2/3$, allowing the computation of the distance at 50% energy transfer efficiency or R_o. Knowing R_o, the distance between the fluorophores can be computed with extreme precision based on the energy transfer and it has a power dependence of 6 with respect to distance. This makes the selection of the fluorophore pair quite critical to optimize the high precision obtained when the measured distance is close to the R_o value.

Distance measurements have been done on the Shaker K channel taking advantage that it is a homotetramer. In this way, by labeling one subunit the channel acquires four labels. By adjusting the ratios of the donor and acceptor fluorophores, it is possible to label channels with an excess of donors with respect to acceptors and estimate distances between subunits. The logic of these experiments is that if there is movement of the voltage sensor (S4 and its neighborhood) with respect to the center of the channel then it should be possible to detect a distance change between subunits when the membrane potential is changed.

FRET measurements were done using the transfer of energy between fluorescein and tetramethylrhodamine and it was found that several sites in the S3–S4 linker changed positions during activation (Glauner *et al.*, 1999). The estimation of the energy transfer was done by recording the change in the photobleaching time constant when the acceptor was present. This required measurement on several oocytes to allow for different membrane potentials and the presence and absence of acceptor. Their results were interpreted as a rotation with a possible transmembrane translation of the S4 segment.

Cha *et al.* (1999) used a variant of FRET called LRET for lanthanide-based fluorescence energy transfer. In this technique (Selvin, 1996) the donor is terbium or europium which, in fact, is luminescent. There are several advantages of this technique over regular FRET. It has been found that terbium emits isotropically, which means that the uncertainty due to the dipole orientation is decreased to a maximum error of ±10%. This error can be decreased even further if the anisotropy of the acceptor is also known. The second advantage is that the fluorescence decay has a time constant of about 1.5 ms, making it easily measurable with conventional recording techniques. The third advantage is that the emission of terbium is peaked and one can find fluorophores that emit in between peaks. This means that the fluorescence of the acceptor can be measured with little or no contamination from the donor. In addition, as the acceptor has a fast decay, any measurement of the acceptor fluorescence with decays comparable to the donor will exclude any possible direct excitation of the acceptor. For these reasons, the measurement of the decay of the acceptor fluorescence, also called sensitized emission, is the cleanest way to detect transfer. The efficiency of transfer is computed from the ratio of the donor decay without acceptor (τ_D) and the sensitized emission decay (τ_{AD}), and the distance R is computed from Forster theory (Cantor and Schimmel, 1980) with

$$R = R_o\left(\frac{\tau_{DA}}{\tau_D - \tau_{DA}}\right)^{1/6}$$

Selvin (1996) has designed a cage that holds the terbium and excludes most of the water that normally acts as a quencher. The chelate has a carbostyril group that is excited at 334 nm by a nitrogen laser and transfer energy to the terbium that in turn emits. This improves enormously the absorption of light by terbium that has a very low extinction coefficient. The chelate has a maleimide group that is used to react with cysteine to attach the terbium chelate to the engineered site in the channel. The experiments of Cha *et al.* (1999) used this Tb chelate as donor and fluorescein as acceptor. Oocytes expressing Shaker K channel with the engineered cysteine site were stained with a mixture of Tb chelate and fluorescein in a 3:1 molar ratio. This ensured that most of the channels that contained donor and acceptor had three donors per each acceptor (see Fig. 4A). The channels containing only donors were invisible when

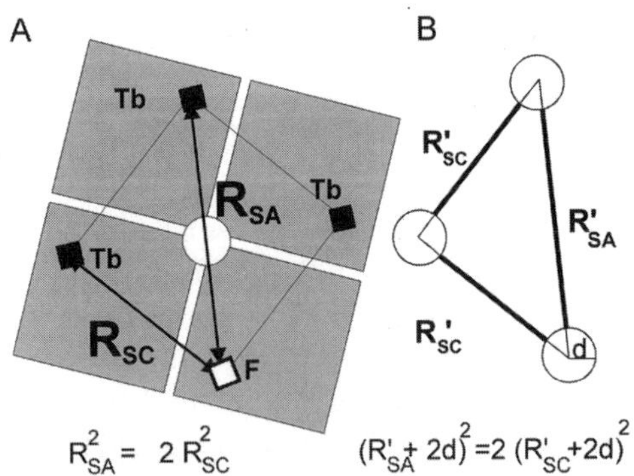

FIG. 4. (A) Measured distances using LRET in the Shaker K channel. Tb represents the attached terbium chelate and F represents the attached fluorescein maleimide. The distances between contiguous subunits (R_{SC}) and the distances across the pore (R_{SA}) are related by the Pythagorean theorem as shown by the equation at bottom, left. (B) When the fluorophores have rotational freedom with radius d, R_{SA} and R_{SC} are related by the equation and bottom, right that differs from the equation in A.

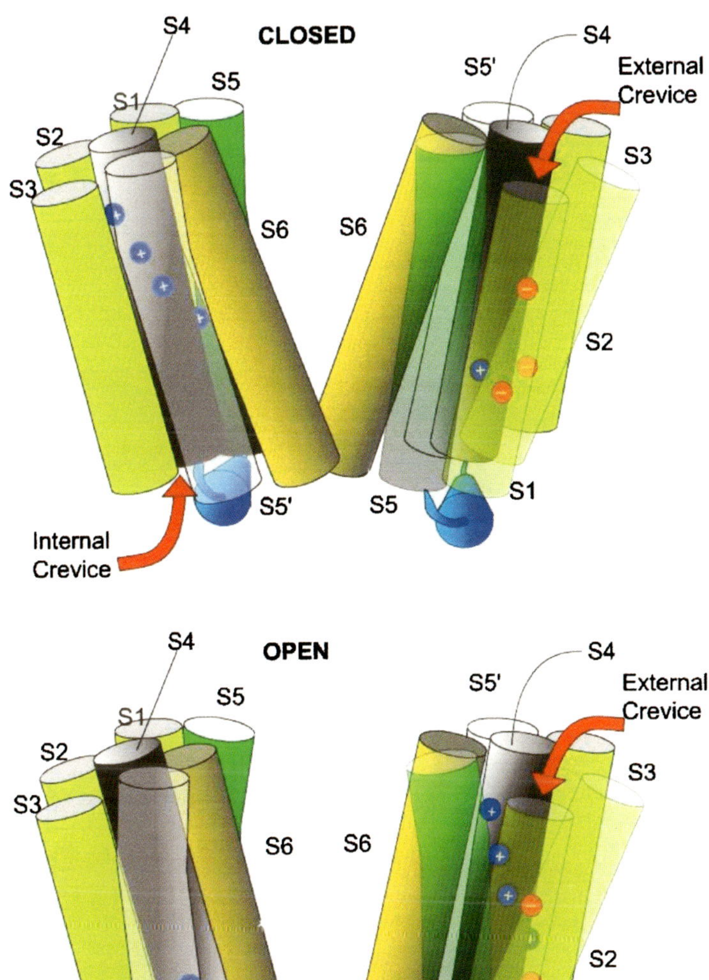

Bezanilla and Perozo, FIG. 5. A model of the voltage sensor that accounts for most of the biophysical observations. Two opposing subunits are represented together with the S5 segment of the contiguous subunits (shown as S5'). Some of the segments are semitransparent to show the details of the charged residues in the S2, S3, and S4 segments. The internal crevice is formed between S3, S6, S4, and S5 of the contiguous subunit (S5'). The external crevice is formed by S1, S2, S5, and S4. In the closed state the first four basic residues of S4 lie in the internal crevice (see CLOSED state, left subunit). The fifth charge faces the opposite side and is in proximity to D316 (on S3) and E293 (in S2) (see CLOSED state, right subunit). On depolarization, a rotation of the S4 segment moves the first four charges in the external crevice and the fifth to the opposite side (see OPEN state). In the depolarized state, E283 (on S2) is in proximity to R368 and R371 of S4. The rotation of S4 pulls the S4–S5 linker (represented by the horizontal cylinder), which retracts S5, carrying with it S6 and opening the conductive pathway.

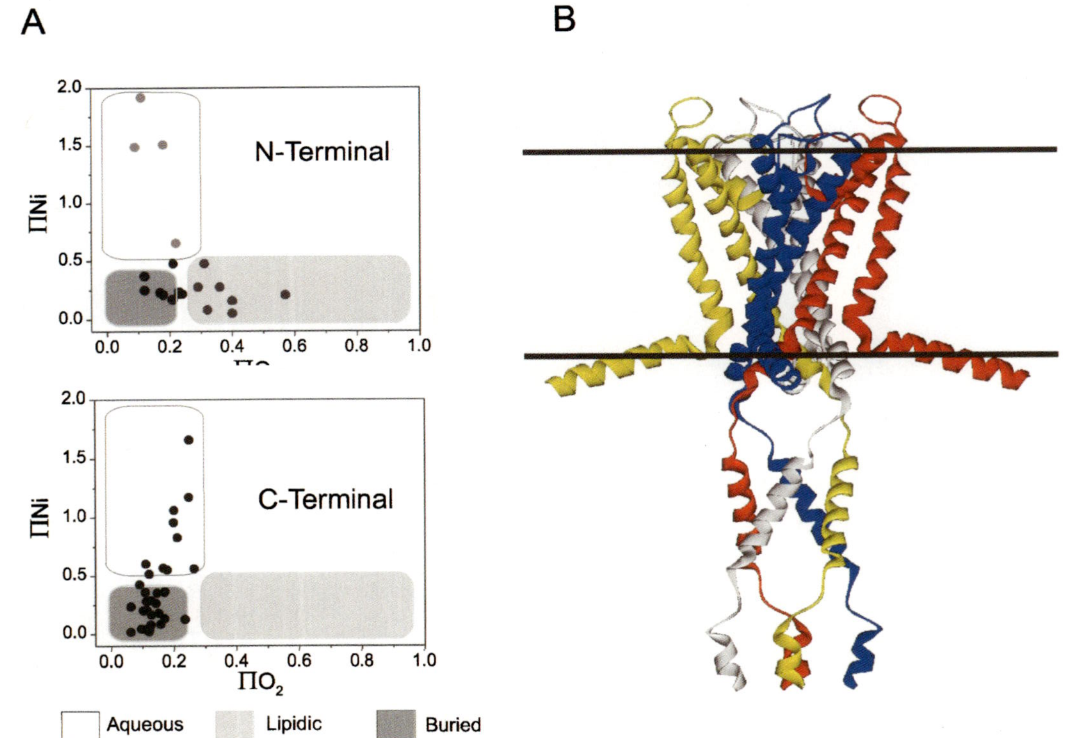

Bezanilla and Perozo, FIG. 6. Use of solvent accessibility in the structural analysis of KcsA cytoplasmic regions. (A) Solvent accessibility phase diagrams for the N terminus (top) and the C terminus (bottom). (B) Folding model of full-length KcsA derived from both X-ray crystallographic data and EPR spectroscopy.

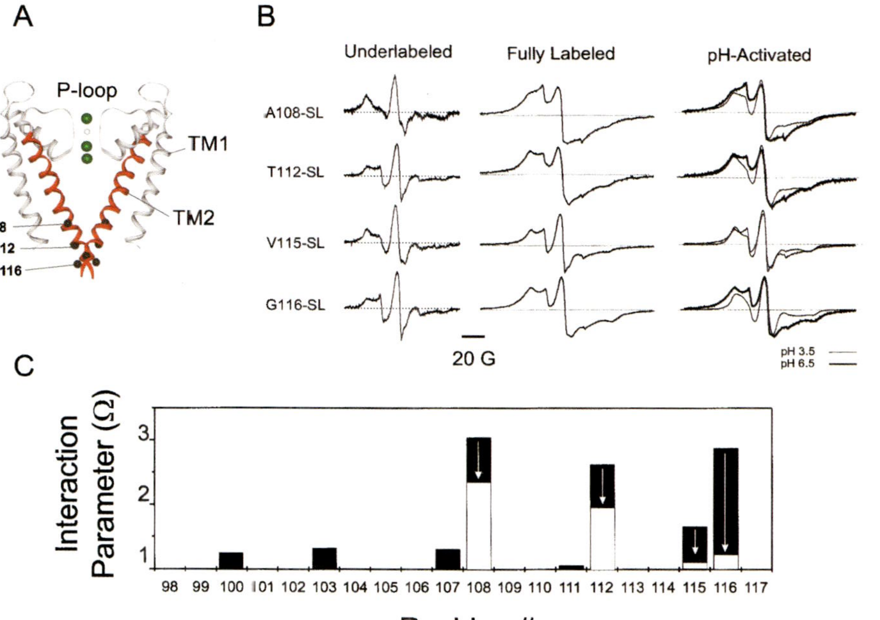

Bezanilla and Perozo, FIG. 7. Spin–spin interactions observed in TM2 and their pH-dependent changes. (A) Ribbon diagram of two diagonally related subunits in KcsA pointing the location of three residues with strong spin–spin coupling. (B) EPR spectra from key residues along the symmetry axis: Left, underlabeled channel; center, fully labeled channel at neutral pH; right, fully labeled channel at neutral (closed) and acidic (open) pHs. (C) Rough estimate of intersubunit proximities and their pH-dependent changes. Black bars represent spin coupling in the closed state (neutral pH). White bars show the extent of spin coupling after change to acidic pH. White arrows reveal the direction of the change.

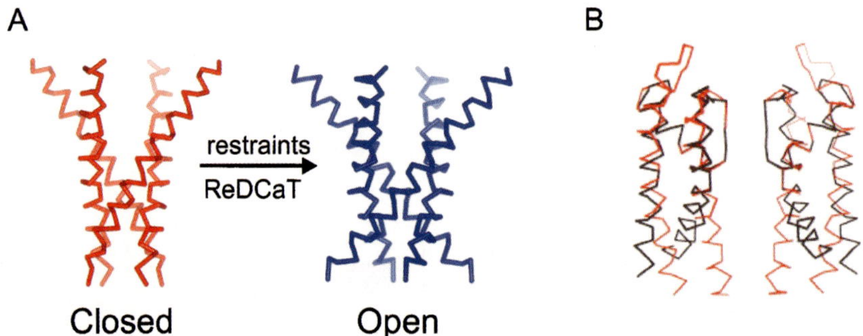

Bezanilla and Perozo, FIG. 9. Structural rearrangements associated with the opening of K channels. (A) Average structure of the inner TM2 helical bundle from the crystal structure, and the calculated open structure derived from intersubunit distances and the assumption of rigid-body conformational changes. ReDCaT refers to the computational approach used to calculate the open TM2 bundle (restraint-driven Cartesian transformation). From Liu et al. (2001). (B) Structure of the MthK channel (black trace) in its Ca^{2+}-bound form (open) overlapped to a closed structure represented by KcsA at neutral pH (red trace). From Jiang *et al.*, (2002).

measuring sensitized emission, and only a very small fraction contained more than one acceptor per donor, a situation that must be avoided because it complicates the computation of the distance. If the acceptor is receiving from three donors, as it was in these experiments, each donor is independent because the decay time of fluorescein is orders of magnitude faster than the decay of Tb. Under these conditions, there are two possible distances that could be measured: contiguous subunits or across-the-pore subunits. Therefore the sensitized emission should show two exponential components. In fact this was verified experimentally, and the two computed distances were found related to one another as predicted by the Pythagorean theorem (Fig. 4A). It is interesting to note that if the fluorophores are not centered in the corners of the triangle, but have a large rotational freedom with a radius *d*, then the two time constants do not convert to distances related by the Pythagorean theorem as shown in the diagram of Fig. 4B. As most of the distances measured were within 1 to 2 Å of the values predicted by the Pythagorean theorem (Cha *et al.*, 1999), the fluorophores did not have a large rotational freedom.

Several sites were measured and some showed changes in distance with voltage. The best characterized residue was position 346 in the S3–S4 loop that showed as much as 3.2 Å change in distance with voltage, and the voltage dependence of the change was the same as the voltage dependence of the charge movement. Another set of three consecutive sites in the same loop but closer to the N terminus of the S4 segment were found to change distances with voltage. Thus residue 351 increased distance with depolarization, 352 did not change, and 353 decreased. This pattern can be best explained by a rotation of an alpha helical segment where the residue 352 maintains the same distance to the pore (center of symmetry) regardless of the membrane potential. In addition, all the residues tested near S4 and S4 proper (363) indicate a change in the tilt of S4 with change in voltage.

The results of Glauner *et al.* (1999) and Cha *et al.* (1999) indicate a rotation of the S4 segment, but the experiments did not address directly a possible translation of the segment perpendicular to the plane of the membrane. The measurement of this possible movement has been addressed by Starace *et al.* (2002). The approach was to measure energy transfer efficiency between eGFP groups genetically encoded in the inside of Shaker to MTS-sulforhodamine attached to cysteine residues located on the outside loops. To calculate the energy transfer, the fluorescence of the donor (eGFP) was measured in the presence of the acceptor (F_{DA}) and then the acceptor was cleaved off by DTT allowing the measurement of the fluorescence of donor only (F_D). The distance

R was then calculated as

$$R = R_o \left(\frac{1}{F_D/F_{DA} - 1} \right)^{1/6}$$

The distances from the intracellular eGFP to the extracellular cysteines in the S1–S2 and S3–S4 loops were large but they showed very little voltage dependence. Some increased with depolarization and others decreased, but the changes were never larger than 2 Å. These results do not support a large translation of S4 perpendicular to the membrane.

4. *Role of the S3–S4 Linker*

If the S4 segment moves significantly on depolarization, the linker between S3 and S4 may become an important constraint in this movement if it is shortened. The case of Shaker that has a total of 31 amino acids has been investigated in detail by Gonzalez *et al.* (2000, 2001) by studying the effect of its shortening. The linker can be reduced to seven amino acids by splicing out the N-terminal part of the linker without noticeable effects in the ionic currents. A further shortening of the linker produces an effect on the speed of channel opening and stability of the closed state that is periodic with the number of amino acids left in the linker. This periodic behavior was studied in detail and it was found to be equal to 3.6 amino acids, indicating the presence of an alpha helix. It is important to note that the channel with only three amino acids in the linker behaves like the wild-type Shaker, indicating that the movement of S4 cannot be too large unless the S3 segment is also carried by the movement.

Gonzalez *et al.* (2001) developed a simple model that can explain the periodic behavior of the energy difference in opening the channel and the kinetics with the length of the linker. In this model, the maximum movement of the N-terminal part of S4 is only 3 Å based on the results of Cha *et al.* (1999), and the C-terminal part of S3 is not movable. Then, on depolarization, the movement of S4 will occur easily if the bonds of the residues in the linker can change their angles. However, certain linker lengths will require breaking one or two hydrogen bonds from the extracellular portion of S4 to allow for the movement, thus explaining the slower kinetics and stabilization of the closed state. In some cases, the closed or resting state has already required some bonds to be broken to unwind enough S4 to achieve the correct distance between S3 and S4. In some of the cases the unwinding is also enough to reach in the depolarizing case, making no difference in kinetics or closed-state stability. These cases will look like the wild type as it is when the linker is only three amino acids long. Interestingly, when the linker is eliminated completely the gating charge is reduced. This charge reduction is predicted by this model because the unwinding of the S4 is such that the first basic residue (R362) becomes part of the linker and does not participate in gating charge any longer.

D. *A Model of the Voltage Sensor*

We have seen that there are many types of experiments that show conformational changes during voltage sensing. The challenge is to build a plausible model of the operation of the voltage sensor that will account for all these observations. There are some other experiments that will impose constraints in the model such as the electrostatic interactions (Papazian *et al.*, 1995) and the amino acid scanning (Monks *et al.*, 1999; Li-Smerin *et al.*, 2000; Li-Smerin and Swartz, 2001).

Roux (2002) has used the experimental evidence available in the Shaker K channel to find the most probable three-dimensional arrangement of segments S1 through S4 around the KcsA pore structure. This procedure seems like a reasonable approach because it may be considered as an unbiased summary of all the available information that can be used as a starting point to propose the conformational changes during voltage sensing.

In the proposed model we will consider first that the gating charges of the S4 segment seem to be in contact with the aqueous environment connected to the intracellular or extracellular side, depending on the voltage (see Section II,C,1). A simple way to account for this is to hypothesize water crevices penetrating the core of the protein. These crevices may be stabilized by the presence of the charged groups of arginine and lysine and may disappear when those groups move away from the crevices. The second observation to consider is the small distance changes measured that do not seem to exceed about 3 Å. This small movement requires that the electric field be concentrated in the regions where the charges move (see Section II,C,2). The model presented in Cha *et al.* (1999) and expanded in Bezanilla (2000) fits within these requirements. However, when one considers the results obtained by Roux (2002) the model must be modified to account for the most probable distribution of membrane segments in the channel protein. This modified model is shown in Fig. 4. In this model many of the constraints have been considered including the electrostatic interactions among the S4, S3, and S2 segments (Papazian *et al.*, 1995; Tiwari-Woodruff *et al.*, 2000) as described at the beginning of Section II,C.

Roux used distance measurements from FRET and tethered blockers (Blaustein *et al.*, 2000) and found that the 45 Å measured with LRET (Section II,C,3) for position 363 (Cha *et al.*, 1999) imposes the restriction that the S4 segment must be located between subunits (Fig. 5, see color insert). It is interesting that a cavity connected to the intracellular solution is generated between S5 of one subunit and S6 of the contiguous subunit where one face of the S4 segment is located. Therefore if that face contains the first four charges they would be in an aqueous environment and connected to the intracellular side. This would be the closed state of the sensor. To move the charge across the field when depolarizing the membrane, the possibilities are a large translation of the segment toward the outside or a rotation with a change in tilt of the S4 segment to expose the charges to another crevice connected to the outside. Most likely the movement may involve a combination of both but with a rather restricted translation to account for the results of FRET (Section II,C,3) and short S3–S4 linker (Section II,C,4). In the model depicted in Fig. 5 only a simple rotation has been implemented just to illustrate that it is possible to translocate most of the charges with a small movement. A reliable model of voltage sensor movement is still in the future because the few measurements available are not consistent with one another and many more measurements are needed to precisely locate the charge positions in the closed and open configurations. The recently cloned sodium channel from *Bacillus halodurans* (Ren *et al.*, 2001) opens the possibility of precise measurements that may eventually give us a detailed picture of the conformational changes that occur during voltage sensing. Once these conformations are known it will be possible to finally link the kinetic models that have been developed in extreme detail with a physical model of charge movement.

III. The Channel Gate

A. *KcsA as a Model of Ionic Permeation and Gating*

The *Streptomyces lividans* K^+ channel (KcsA) is a 160-residue protein that forms homotetrameric channels closely related to the pore domain of larger voltage-dependent channels (Schrempf *et al.*, 1995). When purified and reconstituted in lipid bilayers, KcsA catalyzes single-channel activities with selectivity properties identical to those of other eukaryotic K^+ channels (Cuello *et al.*, 1998; Heginbotham *et al.*, 1999; Meuser *et al.*, 1999). The fact that KcsA is easily expressed in *Escherichia coli* at milligram levels made this protein an ideal target for structural analysis.

Crystallographic studies led to the high-resolution structure of KcsA by Doyle *et al.* (1998b), an achievement that lent a firm structural foundation to more than three decades of functional work on K^+ channels. The crystal structure revealed that KcsA is formed by the association of four subunits, contributing equally to form a water-filled pore. Each subunit has two transmembrane segments, TM1 in the periphery of the complex and TM2 lining the permeation path. Toward the extracellular face of the channel is the selectivity filter, where intimate contact with the permeant ions takes place.

B. *Site-Directed Spin Label Analysis of KcsA*

With the idea of obtaining structural dynamic information of the two transmembrane segments, an extensive site-directed spin label (SDSL) study of KcsA in its native environment was undertaken (Perozo *et al.*, 1998). Sixty-six single cysteine mutants were prepared and analyzed for residues 22–52 and residues 86–120 of KcsA, where the two putative transmembrane segments are thought to be located. Using discrete Fourier transform methods, angular frequency information was extracted from residue environmental profiles from each of the scanned protein segments and used in the assignment of secondary structure elements (Perozo *et al.*, 1998).

Collision of nitroxides with fast-relaxing radicals, such as oxygen and metal ion complexes, causes spin exchange that effectively shortens the spin–lattice relaxation time T_1 of the nitroxide (Hyde and Subczynski, 1989). This effect can be measured either by continuous wave (CW) power saturation techniques or by saturation recovery methods. Collision frequency is directly proportional to the accessibility of the paramagnetic reagent to the nitroxide radical and is defined as

$$\Pi = \frac{\Delta P_{1/2}(X)}{P_{1/2}(DPPH)} \frac{\Delta H_o(DPPH)}{\Delta H_o}$$

where $P_{1/2}$(DPPH) is the microwave power that saturates the signal relative to that without saturation for a 2,2-diphenyl-1-picrylhydrazyl crystal, $\Delta P_{1/2}(X) = [P_{1/2}(X) - P_{1/2}(N_2)]$, the difference in $P_{1/2}$ of a given sample exposed to a fast-relaxing paramagnetic species (such as O_2 or NiEdda) and N_2 (Altenbach *et al.*, 1990). ΔH_o is an estimate of probe mobility (Mchaourab *et al.*, 1996; Columbus *et al.*, 2001). This Π parameter can be used to deduce secondary and tertiary structure information, which is derived from the periodic behavior of a series of sequential mutants in the presence of relaxing agents. Solvent accessibility information derived from either O_2 or NiEdda collision frequencies was in agreement with the crystal structure, suggesting that TM1 has a large lipid-exposed surface, while TM2 is more protected from membrane lipid. An excellent example of the use of solvent accessibilities to describe unknown structures is found in the recent SDSL analysis of KcsA cytoplasmic domains (Cortes *et al.,* 2001). Because of the requirements specified by the successful crystallization conditions, no structural information exists on these regions of KcsA (Doyle *et al.,* 1998). Thus, regions in the N and C termini of KcsA predicted to be soluble and exposed to the cytoplasm were probed by site-directed spin labeling. The results can be viewed in the form of a residue environmental phase diagram by plotting Ni accessibility (ΠNiEdda) versus O_2 accessibility (ΠO_2, Fig. 6, see color insert). The structural phases can be classified as aqueous (high ΠNiEdda and low ΠO_2), lipid exposed (high ΠO_2 and low ΠNiEdda), and buried (both low ΠNiEdda and low ΠO_2). From these diagrams it is clear that the N-terminal region of KcsA resides at the membrane interface (straddling both membrane-exposed and water-exposed regions, Fig. 6A, top), while the C terminus is wholly exposed to the aqueous environment (Fig. 6A, bottom). Using this data set, together with the known coordinates from the crystal structure, it was possible to develop a backbone three-dimensional model of full-length KcsA (Fig. 6B) to complete the structural description of KcsA with the membrane as a reference point and to provide a structural basis for understanding its mechanism of activation gating.

Spin–spin interactions have played a key role in the structural analysis of KcsA and in the molecular description of its activation mechanism (see below). Due to the tetrameric stoichiometry of KcsA, spin–spin coupling can originate either from interactions between next neighbors or from diagonally related subunits (Fig. 7A, see color insert; only the diagonal subunits are shown). Thus, in an ideally labeled KcsA, four spin labels would be contributing to the spectral broadening (Fig. 7B). Under these conditions, a qualitative estimate of the magnitude of spin–spin interactions can be obtained from the ratio of amplitudes of the central resonance line ($M = 0$) between the underlabeled and fully labeled mutant, both normalized to the total number of spins in the sample (Perozo *et al.,* 1998). This interaction parameter, Ω, was used to survey patterns of spin coupling along the length of TM2 and deduce possible interhelical orientations. Strong broadening was in fact observed every three or four residues on one face of the C-terminal half of TM2 (in residues A108, T112, V115, and G116), as shown from the comparison between their spectra at underlabeling conditions (Fig. 7B, left) and those obtained from fully labeled channels (Fig. 7B, center). In agreement with the relative intersubunit angle between TM2 segments, significantly weaker spin–spin coupling (as measured from the Ω parameter) was detected on the N-terminal half of the helix (Fig. 7C). In a parallel SDSL analysis of KcsA, Gross *et al.* (1999) have shown essentially the same type of results in TM2, both for solvent accessibility and for the distribution of spin–spin coupled residues.

C. *Helical Movement during KcsA Gating*

The structure of KcsA can be used as a remarkable blueprint to explain the principles of ion permeation (Doyle *et al.,* 1998). The gating mechanism, however, has been harder to deduce from direct examination of the KcsA structure. The key breakthrough that gave a handle on the spectroscopic analysis of KcsA gating was the discovery that KcsA can be opened at low pH (Cuello *et al.,* 1998). This was done using, bulk $^{86}Rb^{+}$ flux assays and at the single-channel level in bilayer systems (Fig. 8), and thus established that the KcsA crystal structure most likely represented a closed conformation. Although initially the proton site was thought to be extracellular (Cuello *et al.,* 1998), more recent studies have clearly demonstrated that channel opening is induced by *intracellular* changes in pH (Heginbotham *et al.,* 1999; Meuser *et al.,* 1999).

Earlier results from voltage-dependent channels suggested that channel opening was associated with global conformational changes in the intracellular face of the channel (Armstrong, 1971; Liu *et al.,* 1997). The first direct demonstration of the types of movements that may be associated with channel opening came from Perozo *et al.* (1998) as large pH-dependent changes in intersubunit spin–spin coupling were observed at selected residues along the C-terminal end of TM2 (Figs. 7B, 7C). From the crystal structure, these residues were shown to be located at the narrowest section of the internal helix bundle, close to the fourfold axis of symmetry. In each case, spectra obtained at low pH showed a reduction in the magnitude of the intersubunit dipolar coupling (Fig. 7C), leading to the suggestion that channel opening was a consequence of a widening in the inner vestibule of the channel.

A more through analysis of these movements was subsequently carried out by studying changes in mobility and dipolar coupling along the main structural elements in the channel: the two transmembrane segments TM1 and TM2, as well as the regions flanking the selectivity filter (Perozo *et al.,* 1999). The changes in spectral line shape were analyzed using discrete Fourier transform methods to detect possible patterns that could indicate specific types of movements in each of the TM segments

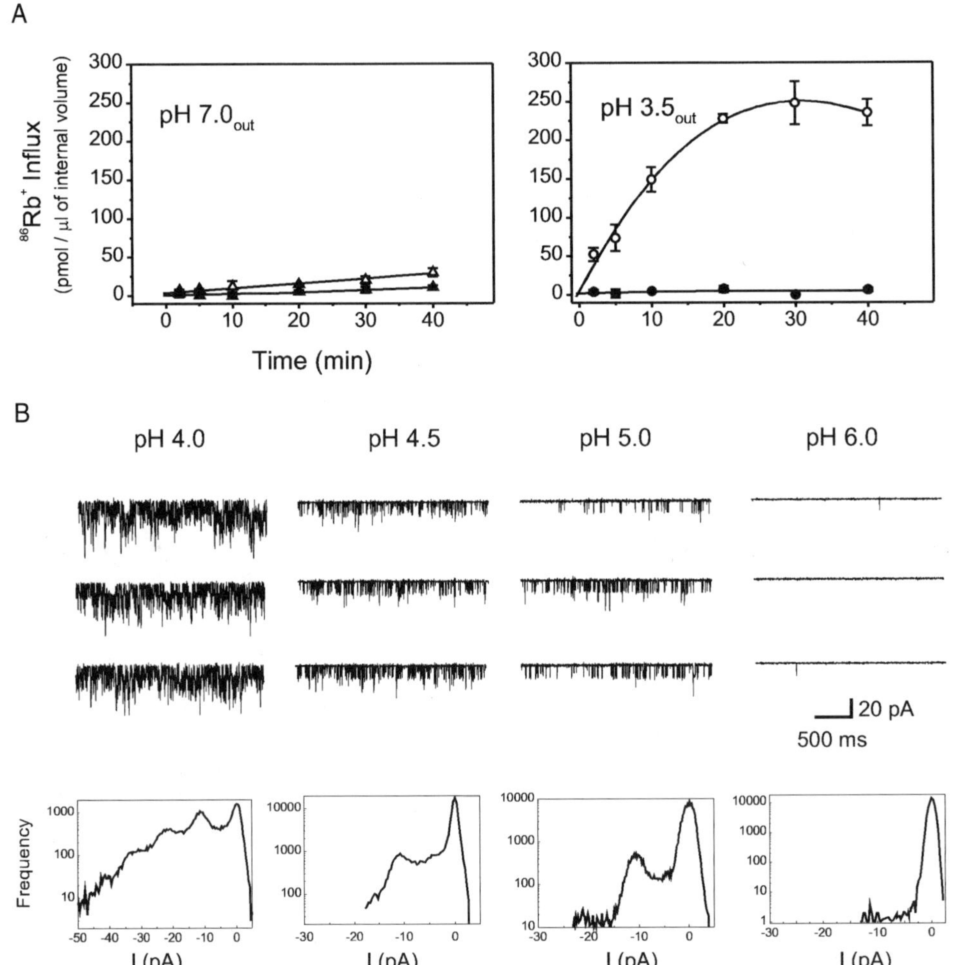

FIG. 8. Proton activation of KcsA. (A) Time course of ^{86}Rb$^+$ uptake depends on extravesicular pH. Open circles, uptake at pH 7.0 (left) or 3.5 (right). Filled circles, uptake in the presence of 10 mM $BaCl_2$ in both sides of the membrane. (B) pH effect on single-channel properties of KcsA in planar lipid bilayers. Top: Consecutive single-channel traces obtained at pH 4.0, 4.5, 5.0 and 6.0 for a multichannel asolectin bilayer. Bottom panels: All-point histogram corresponding to each set of single-channel records.

originating from pH acidification. The majority of the residues along TM2 showed a reduction in the extent of spin–spin coupling as measured by the Ω parameter, indicating a translational movement or helical tilt away form the symmetry axis that would indicate an increase in the diameter of the internal vestibule of the channel. Vector analysis of these changes in the context of a helical wheel projection indicated that the pattern of change in intersubunit proximity could be explained, in addition to the helical tilt, by a rotation of the individual TM2 helices in an anticlockwise direction (Perozo *et al.*, 1999).

Similar analysis of the spectral changes in the "external" transmembrane segment TM1 also revealed helix rotations and tilts compatible with a rigid-body movement of the entire helix. However, because TM1 is located further away from the axis of symmetry, changes in intersubunit residue proximity are more subtle and harder to detect than in TM2. For this reason, and given that less than half of TM1 residues are involved in any extensive tertiary contacts, changes in probe mobility ($\Delta\Delta H_o$) are a very sensitive indicator of helix tilting and rotation. As in TM2, these data suggested an anticlockwise rotation of TM1 on channel gating. Limited data on the types of conformational changes occurring in the vicinity of the selectivity filter showed subtle changes in both probe mobility and spin–spin interaction (Perozo *et al.*, 1999). Interestingly, spin labels located at the C-terminal end of the pore helix reported larger changes in spectral line shape than did those in the extracellular vestibule of the channel. Taken together, these spectral changes might be suggestive of possible conformational rearrangements that could affect the energetics of ion coordination at the selectivity filter.

As stated above, distance determinations in KcsA (and in other oligomeric proteins in general) are complicated by the geometric relationship between subunits, as a fully spin-labeled channel will exhibit dipolar interactions between both neighboring and diagonally related subunits. This problem has been solved by the use of tandem dimer constructs to determine changes in intersubunit distances at the inner helical bundle of KcsA (Liu *et al.*, 2001). Using tandem dimers a series of constructs were generated so that pairs of spin labels, located along the critical TM2 bundle crossing, were used to calculate distance changes in the channel gate between the closed and open conformations.

Two nitroxides, separated by the distance r, are coupled through space via electron–electron dipolar interactions arising from the unpaired electrons. Spin coupling induces line splitting dependent on their separation and their orientation with respect to the magnetic field according to

$$2B = (3/2)g_e\beta(3\cos^2\theta - 1)/r^3$$

where θ is the angle between the interspin vector and the magnetic field, g_e the electron g value, and β the Bohr magneton. In solution, θ is distributed isotropically, and thus line splitting is observed instead as an overall spectral broadening. The extent of the broadening is directly correlated to the interspin separation, which can be extracted by Fourier deconvolution of the broadened and unbroadened (noninteracting) EPR spectra with a ~2 Å accuracy within the 8–25 Å range (Rabenstein and Shin, 1995). These intersubunit distances were then used to derive a backbone structure for the KcsA inner helical bundle in the open state with the aid of a novel computational algorithm. This computational approach (restraint-driven Cartesian transformation, ReDCaT) is based on an exhaustive sampling of rigid-body movement in Cartesian space and was used to determine the type, direction, and magnitude of the conformational changes in TM2 using the limited distance information (Sompornpisut *et al.*, 2001). The overall picture of KcsA gating emerging from these studies is summarized in Figure 9 (see color insert). Assuming complete rigidity of the helical TM segments, the gating model proposes that on lowering the pH, TM1 and TM2 undergo an anticlockwise rotation and a simultaneous movement away from the symmetry axis of the tetramer, opening the intracellular side of the permeation path (Fig. 9A). Quite recently, however, the crystal structure of another prokaryotic K^+ channel, MthK from Methanobacterium thermoautotrophicum (a Ca^{2+}-activated channel), has been obtained in its open conformation (Jiang *et al.*, 2002b). With a >12 Å wide opening, this remarkable structure revealed larger overall movements than those modeled in KcsA (Fig. 9B), and this correlates well with the differences in maximal P_o between KcsA (low maximal P_o) and MthK (high $P_{o\,max}$). Nonetheless, the type and direction of the helical movements at the level of the gate seen in the MthK structure are in complete agreement with those derived from EPR measurements (Fig. 9A). A key finding in the new open structure is the fact that in MthK, gating may originate as a consequence of a glycine hinge in the middle of the helix that helps bend the C-terminal half of TM2 away from the permeation path.

D. *The Role of the Selectivity Filter in the Gating Mechanism*

Once significant pH-dependent movements in TM2 had been demonstrated, an important point that needed to be addressed was, how do these rearrangements fit within the overall gating mechanism of KcsA? That is, are the conformational changes in TM2 necessary *and* sufficient to open the channel? Given the structure of full-length KcsA, we thought of two additional regions that could participate in KcsA gating. On one hand, since the C terminus forms a helical bundle aligned to the fourfold symmetry axis, it could restrict ion permeation by reversibly plugging the entrance to the internal vestibule. This possibility was examined in a series of deletion experiments (Perozo *et al.*, 1999; Cortes *et al.*, 2001), which demonstrated that although complete truncation of the C terminus affects both pH dependence and the stability of the closed state, the channel was able to fully close by moving to a slightly more alkaline pH. Therefore, the C terminus cannot be the activation gate of the channel.

The second region under consideration was the selectivity filter. Coordination of K^+ ions at the selectivity filter establishes strict geometric constraints in order to optimally select against impermeable ions while allowing for fast K^+ ion translocation. Therefore, small perturbations in selectivity filter geometry can have significant consequences for its ability to conduct ions. This fact makes the selectivity filter an energetically economic gate. Indeed, a number of independent pieces of evidence seem to point to the selectivity filter as a region with a great deal of influence over the gating behavior of a channel. There are well-documented examples of permeant ion effects on gating (Swenson and Armstrong, 1981; Spruce *et al.*, 1989; Shapiro and Decoursey, 1991; Demo and Yellen, 1992; Clay, 1996; Mienville and Clay, 1996; Chen *et al.*, 1997; Pusch *et al.*, 2000). Ions with long occupancy times (Rb^+, Cs^+, $NH_4{}^+$) tend to stabilize the open state through a "foot in the door" effect of the gate, even though the only region of channel-ion interaction occurs at the selectivity filter. Additionally, channels seem to populate subconducting states on the way to the open state (Chapman *et al.*, 1997; Zheng and Sigworth, 1997), and these subconducting states show altered selectivities. More recently, unnatural amino acid mutagenesis targeted to the signature sequence of an inward rectifier K^+ channel had dramatic effects on the rapid gating transitions of an inward rectifier channel (Lu *et al.*, 2001), again pointing to the selectivity filter as a contributor to the gating process.

Small conformational changes were detected in a few residues near the selectivity filter when the gating mechanism was under examination (Perozo *et al.*, 1999), but the number of residues was too limited and the magnitude of the conformational changes too small to allow these changes to be interpreted in terms of specific protein motions. Recently, the overall conformational rearrangements associated with pH-dependent gating within the P-loop were analyzed in a more complete set of residues: residues 64–74 in the pore helix and residues 81–84 in the outer vestibule of the channel (Cuello and Perozo, 2002). EPR spectra obtained at neutral and acidic pH revealed changes in spin label mobility on each side of the signature sequence. Although subtle, these rearrangements were remarkably reproducible and were always observed under the same conditions that promoted major movements of the TM2 segments.

Although it was clear that there were reproducible changes around the selectivity filter on gating, additional evidence was needed to actually link the P-loop to the gating transitions in KcsA. It has been known for many years that some permeant ions can have profound effects on the gating behavior of a variety of K^+ channels. These effects seem to be correlated with the degree of occupancy of the ion in the selectivity filter (Demo and Yellen, 1992). Given that in K^+ channels intimate ion–protein interactions only take place at the selectivity filter (Doyle *et al.*, 1998b), the role of the P-loop in activation gating and its influence on other moving parts of KcsA were probed using different permeant ions (Cuello and Perozo, 2002). With K^+ as a charge carrier, the single-channel behavior of KcsA is characterized by fast gating transitions (mean open times 1–2 ms) clustered in bursts of activity separated by long silent periods. However, when Rb^+ is the charge carrier, open times were increased more than 10-fold (main $\tau_{open} = 165$ ms) with a sharp reduction in single-channel conductance (from ~100 to ~8 pS), yet the closed times remained broadly unchanged. Open probabilities increased from the standard $P_o = 0.05 - 0.1$ in symmetric K^+ to $P_o \sim 0.5$ in symmetric Rb^+. The next question was how these gating effects, induced by different permeant ions, are reflected in the structural rearrangements measured by EPR. Control experiments in high $[K^+]$ revealed only residual conformational changes in both the P loop and the intracellular gate (Fig. 10, see color insert). However, when the experiment was repeated in high $[Rb^+]$ (Fig. 10, right), significant changes in line shape were observed both in the P loop and in TM2. Thus, by promoting a small structural rearrangement in the selectivity filter with a higher affinity permeant ion (Rb^+), it was possible to detect significant conformational rearrangements in TM2, *at the opposite end of the molecule.* These results suggest that the internal gate and the P-loop are coupled both structurally and functionally, lending support to the notion that the selectivity filter *does* participate in gating during normal channel operation.

Additional evidence linking the selectivity filter to the gating machinery was obtained from single-channel analysis of site-directed mutants. Reasoning that mutations that stabilize channel–ion interactions will produce a gating behavior similar to Rb^+ ions, Cuello and Perozo (2002) searched for mutants that alter KcsA gating properties at the single-channel level. In the P-loop, they studied mutations targeted to the pore helix (T72C, A73C, T74C) and the external vestibule (Y82C). These mutations produced no major changes in the ability of KcsA to sense proton concentrations, as determined from macroscopic $^{86}Rb^+$ uptake experiments. However, cysteine substitutions in these pore helix

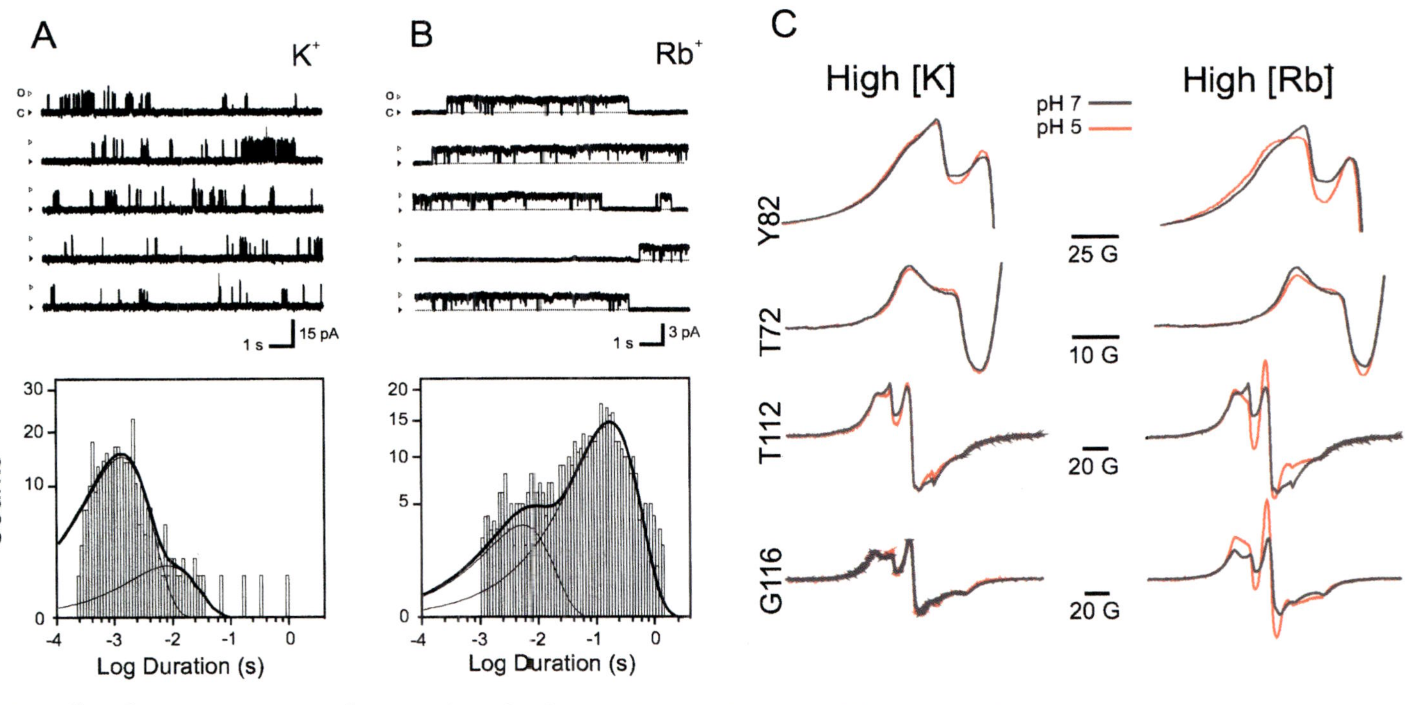

Bezanilla and Perozo, FIG. 10. Conformational coupling between the intracellular gate (TM2 bundle) and the selectivity filter. Single channel traces obtained in symmetric solutions and their respective open time distributions K^+ (A) and Rb^+ (B). (C) Structural changes near the selectivity filter (at T72 and Y82) and at the intracellular gate (T112 and G116) show direct correlation with the functional observations.

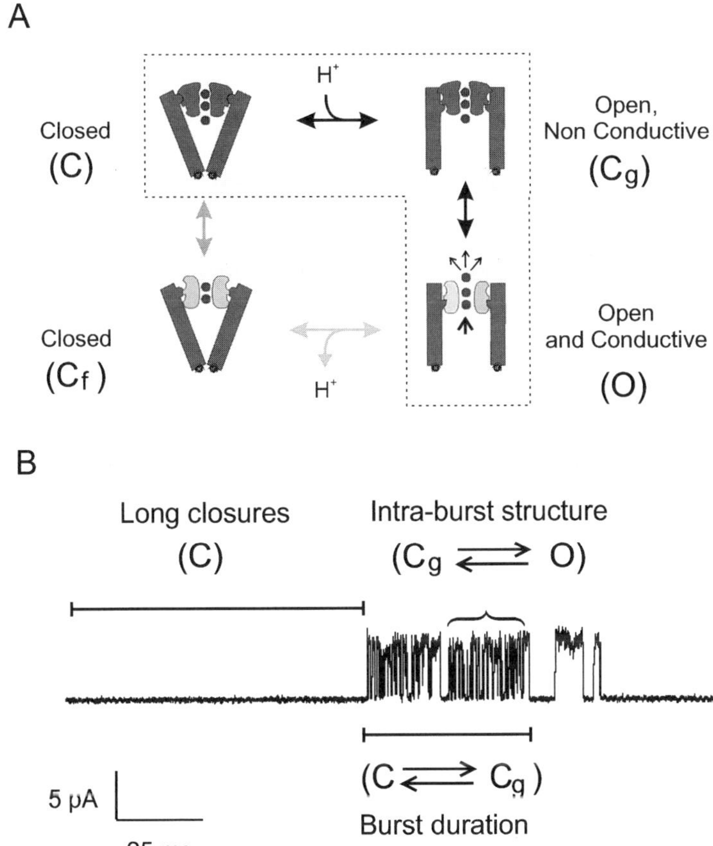

FIG. 11. Gating as a consequence of the structural interplay between two "gates". (A) The present data are interpreted according to a minimal four-state sequential model in which movement of the channel intracellular gates (TM2) allows the selectivity filter to fluctuate and enter a "conductive" conformation that corresponds to the open state. In transit from the closed state the channel must populate an intermediate open, non-conductive state. As these states are conformationally coupled, the model predicts that the intracellular gate is unlikely to close before the selectivity filter returns to its "non-conductive" state. (B). The model suggests that the duration of a burst is governed by the $C \leftrightarrow C_g$ equilibrium, occurring in the intracellular gate, while the intraburst activity (channel flicker) is determined by the $C_g \leftrightarrow O$ transition.

residues had substantial effects on the kinetics of KcsA. This particularly applies at position 72 where the mean open time increases from ~2 ms in wt-KcsA to ~55 ms in T72C. Open times were also lengthened for mutants A73C and T74C, while Y82C, the only mutation tested in the outer vestibule, showed no significant kinetic differences compared to wt-KcsA. It must be noted that these mutations increase mean open times even under symmetric K^+, producing an overall stabilization of the open state in a manner similar to that described above for Rb^+ above. High-affinity permeant ions such as Rb^+ and Cs^+ and specific mutations in the pore helix seem to be influencing gating through a "foot in the door" effect, as suggested a long time ago by Swenson and Armstrong (1981). This phenomenon was thought to be analogous to the actions of large quaternary ammonium ions in preventing the closure of the internal gate of many K^+ channels (Armstrong, 1971). However, although they are mechanistically similar, permeant ions and pore mutants can only exert their effect by affecting the conformation of the selectivity filter. Thus, all of the measured functional and structural effects observed in the internal gate might originate as a consequence of the conformational coupling between these two structures. In this interplay, it has been suggested (Cuello and Perozo, 2002) that opening of the intracellular gate (TM2) defines the length of a given single-channel burst and sterically allows conformational flexibility in the selectivity filter. The profound influence of permeant ions on KcsA open time distributions and its consequent effects on conformational changes indicate that the P-loop may be responsible for defining the fast gating properties of the channel (flicker) and thus, the mean open times (Fig. 11).

References

Aggarwal, S. K., and Mackinnon, R. (1996). Contribution of the S4 segment to gating charge in the *Shaker* K^+ channel. *Neuron* **16,** 1169–1177.

Almers, W. (1978). Gating currents and charge movements in excitable membranes. *Rev. Physiol. Biochem. Pharmacol.* **82,** 96–190.

Altenbach, C., Marti, T., Khorana, H. G., and Hubbell, W. L. (1990). Transmembrane protein structure: spin labeling of bacteriorhodopsin mutants. *Science* **248,** 1088–1092.

Armstrong, C. M. (1971). Interaction of tetraethylammonium ion derivatives with the potassium channels of giant axons. *J. Gen. Physiol.* **58,** 413–437.

Asamoah, O. K., Bezanilla, F., and Loew, L. (2002). Probing the electric field in the *Shaker* potassium channel. *Biophys. J.* **82,** 233a.

Baker, O. S., Larsson, H. P., Mannuzzu, L. M., and Isacoff, E. Y. (1998). Three transmembrane conformation and sequence-dependent displacement of the S4 domain in Shaker K^+ channel gating. *Neuron* **20,** 1283–1294.

Bezanilla, F. (2000). The voltage sensor in voltage-dependent ion channels. *Physiol. Rev.* **80,** 555–592.

Blaustein, R. O., Cole, P. A., Williams, C., and Miller, C. (2000). Tethered blockers as molecular "tape measures" for a voltage-gated K^+ channel. *Nat. Struct. Biol.* **7,** 309–311.

Cantor, C. R., and Schimmel, P. R. (1980). *Biophysical Chemistry,* Part II. *Techniques for the Study of Biological Structure and Function.* W. H. Freeman and Co., New York.

Cha, A., and Bezanilla, F. (1997). Characterizing voltage-dependent conformational changes in the *Shaker* K^+ channel with fluorescence. *Neuron* **19,** 1127–1140.

Cha, A., and Bezanilla, F. (1998). Structural implications of fluorescence quenching in the *Shaker* K^+ channel. *J. Gen. Physiol.* **112,** 391–408.

Cha, A., Snyder, G. E., Selvin, P. R., and Bezanilla, F. (1999). Atomic scale movement of the voltage-sensing region in a potassium channel measured via spectroscopy. *Nature* **402,** 809–813.

Chapman, M. L., VanDongen, H. M., and VanDongen, A. M. (1997). Activation-dependent subconductance levels in the drk1 K channel suggest a subunit basis for ion permeation and gating. *Biophys. J.* **72,** 708–719.

Chen, F. S. P., Steele, D., and Fedida, D. (1997). Allosteric effects of permeating cations on gating currents during K^+ channel deactivation. *J. Gen. Physiol.* **110,** 87–100.

Clay, J. R. (1996). Effects of permeant cations on K^+ channel gating in nerve axons revisited. *J. Membrane Biol.* **153,** 195–201.

Columbus, L., Kálai, T., Jekö, J., Hideg, K., and Hubbell, W. L. (2001). Molecular motion of spin labeled side chains in α-helices: Analysis by variation of side chain structure. *Biochemistry* **40,** 3828–3846.

Conti, F., and Stuhmer, W. (1989). Quantal charge redistribution accompanying the structural transitions of sodium channels. *Eur. Biophys. J.* **17,** 53–59.

Cortes, D., Cuello, L., and Perozo, E. (2001). Molecular architecture of full-length KcsA: role of cytoplasmic domains in ion permeation and activation gating. *J. Gen. Physiol.* **117,** 165–180.

Cuello, L., and Perozo, E. (2002). Conformational coupling between elements of the permeation pathway defines activation gating in K^+ channels. *J. Gen. Physiol.,* in press.

Cuello, L. G., Romero, J. G., Cortes, D. M., and Perozo, E. (1998). pH-dependent gating in the *Streptomyces lividans* K^+ channel. *Biochemistry,* in press.

Demo, S. D., and Yellen, G. (1992). Ion effects on gating of the Ca^{2+}-activated K^+ channel correlate with occupancy of the pore. *Biophys. J.* **61,** 639–648.

Doyle, D. A., Cabral, J. M., Pfuetzner, R. A., Kuo, A., Gulbis, J. M., Cohen, S. L., Chait, B. T., and MacKinnon, R. (1998). The structure of the potassium channel: molecular basis of K^+ conduction and selectivity. *Science* **280,** 69–77.

Glauner, K. S., Mannuzzu, L. M., Gandhi, C. S., and Isacoff, E. Y. (1999). Spectroscopic mapping of voltage sensor movement in the *Shaker* potassium channel. *Nature* **402,** 813–817.

Gonzalez, C., Rosenman, E., Bezanilla, F., Alvarez, O., and Latorre, R. (2000). Modulation of the Shaker K^+ channel kinetics by the S3–S4 linker. *J. Gen. Physiol.* **115,** 193–207.

Gonzalez, C., Rosenmann, E., Bezanilla, F., Alvarez, O., and Latorre, R. (2001). Periodic perturbations in *Shaker* K^+ channel gating by deletions in the S3–S4 linker. *Proc. Natl. Acad. Sci. USA* **98,** 9617–9623.

Gross, A., Columbus, L., Hideg, K., Altenbach, C., and Hubbell, W. L. (1999). Structure of the KcsA potassium channel from *Streptomyces lividans:* a site-directed spin labeling study of the second transmembrane segment. *Biochemistry* **38,** 10324–10335.

Heginbotham, L., LeMasurier, M., Kolmakova-Partensky, L., and Miller, C. (1999). Single *streptomyces lividans* K^+ channels: functional asymmetries and sidedness of proton activation. *J. Gen. Physiol.* **114,** 551–560.

Hirschberg, B., Rovner, A., Lieberman, M., and Patlak, J. (1996). Transfer of twelve charges is needed to open skeletal muscle Na^+ channels. *J. Gen. Physiol.* **106,** 1053–1068.

Hodgkin, A. L., and Huxley, A. F. (1952). A quantitative description of membrane current and its application to conduction and excitation in nerve. *J. Physiol.* **117,** 500–544.

Hyde, J. S., and Subczynski, W. K. (1989). Spin label oxymetry. In *Biological Magnetic Resonance,* Vol. 8 (L.J.a.R. Berliner, Ed.), pp. 399–425. Plenum, New York.

Jiang, Y., Lee, A., Chen, J., Cadene, M., Chait, B. T., and MacKinnon, R. (2002a). Crystal structure and mechanism of a calcium-gated potassium channel. *Nature* **417,** 515–522.

Jiang, Y., Lee, A., Chen, J., Cadene, M., Chait, B. T., and MacKinnon, R. (2002b). The open pore conformation of potassium channels. *Nature* **417,** 523–526.

Larsson, H. P., Baker, O. S., Dhillon, D. S., and Isacoff, E. Y. (1996). Transmembrane movement of the *Shaker* K^+ channel S4. *Neuron* **16,** 387–397.

Li-Smerin, Y., and Swartz, K. (2001). Helical structure of the COOH terminus of S3 and its contribution to the gating modifier toxin receptor in voltage-gated ion channels. *J. Gen. Physiol.* **113,** 415–423.

Li-Smerin, Y., Hackos, D. H., and Swartz, K. J. (2000). Alpha-helical structural elements within the voltage-sensing domains of a K^+ channel. *J. Gen. Physiol.* **115,** 33–50.

Liu, Y., Holmgren, M., Jurman, M. E., and Yellen, G. (1997). Gated access to the pore of a voltage-dependent K^+ channel. *Neuron* **19,** 175–184.

Liu, Y. S., Sompornpisut, P., and Perozo, E. (2001). Structure of the KcsA channel intracellular gate in the open state. *Nat. Struct. Biol.* **8,** 883–887.

Loots, E., and Isacoff, E. Y. (1998). Protein rearrangements underlying slow inactivation of the *Shaker* K^+ channel. *J. Gen. Physiol.* **112,** 377–389.

Lu, T., Ting, A. Y., Mainland, J., Jan, L. Y., Schultz, P. G., and Yang, J. (2001). Probing ion permeation and gating in a K^+ channel with backbone mutations in the selectivity filter. *Nat. Neurosci.* **4,** 239–246.

Mannuzzu, L. M., Moronne, M. M., and Isacoff, E. Y. (1996). Direct physical measure of conformational rearrangement underlying potassium channel gating. *Science* **271,** 213–216.

Mchaourab, H. S., Lietzow, M. A., Hideg, K., and Hubbell, W. L. (1996). Motion of spin-labeled side chains in T4 lysozyme. Correlation with protein structure and dynamics. *Biochemistry* **35,** 7692–7704.

Meuser, D., Splitt, H., Wagner, R., and Schrempf, H. (1999). Exploring the open pore of the potassium channel from *Streptomyces lividans. FEBS Lett.* **462,** 447–452.

Monks, S. A., Needleman, D. J., and Miller, C. (1999). Helical structure and packing orientation of the S2 segment in the Shaker K^+ channel. *J. Gen. Physiol.* **113,** 415–423.

Mienville, J. M., and Clay, J. R. (1996). Effects of intracellular K^+ and Rb^+ on gating of embryonic rat telencephalon Ca^{2+}-activated K^+ channels. *Biophys. J.* **70,** 778–785.

Papazian, D. M., Shao, X. M., Seoh, S.-A., Mock, A. F., Huang, Y., and Wainstock, D. H. (1995). Electrostatic interactions of S4 voltage sensor in *Shaker* K^+ channel. *Neuron* **14,** 1293–1301.

Perozo, E., MacKinnon, R., Bezanilla, F., and Stefani, E. (1993). Gating currents from a non-conducting mutant reveal open-closed conformations in Shaker K^+ channels. *Neuron* **11,** 353–358.

Perozo, E., Cuello, L., and Cortes, D. (1998). Three-dimensional architecture and gating mechanism of a K^+ channel studied by EPR spectroscopy. *Nat. Struct. Biol.* **5,** 459–469.

Perozo, E., Cortes, D. M., and Cuello, L. G. (1999). Structural rearrangements underlying K^+-channel activation gating. *Science* **285,** 73–78.

Pusch, M., Bertorello, L., and Conti, F. (2000). Gating and flickery block differentially affected by rubidium in homomeric KCNQ1 and heteromeric KCNQ1/KCNE1 potassium channels. *Biophys. J.* **78,** 211–226.

Rabenstein, M. D., and Shin, Y. K. (1995). Determination of the distance between two spin labels attached to a macromolecule. *Proc. Natl. Acad. Sci. USA* **92,** 8239–8243.

Ren, D., Navarro, B., Xu, H., Yue, L., Shi, Q., and Clapham, D. E. (2001). A prokaryotic voltage-gated sodium channel. *Science* **294,** 2372–2375.

Roux, B. (2002). What can be deduced about the structure of Shaker from available data? *Novartis Found. Symp.* **245,** 84–101.

Schoppa, N. E., McCormack, K., Tanouye, M. A., and Sigworth, F. J. (1992). The size of gating charge in wild-type and mutant Shaker potassium channels. *Science* **255,** 1712–1715.

Schrempf, H., Schmidt, O., Kummerlen, R., Hinnah, S., Muller, D., Betzler, M., Steinkamp, T., and Wagner, R. (1995). A prokaryotic potassium ion channel with two predicted transmembrane segments from *Streptomyces lividans. EMBO J.* **14,** 5170–5178.

Selvin, P. R. (1996). Lanthanide-based resonance energy transfer. *IEEE J. Selected Top. Quantum Electron.* **2,** 1077–1087.

Seoh, S.-A., Sigg, D., Papazian, D. M., and Bezanilla, F. (1996). Voltage-sensing residues in the S2 and S4 segments of the *Shaker* K^+ channel. *Neuron* **16,** 1159–1167.

Shapiro, M. S., and Decoursey, T. E. (1991). Permeant ion effects on the gating kinetics of the type-L potassium channel in mouse lymphocytes. *J. Gen. Physiol.* **97,** 1251–1278.

Sigg, D., Stefani, E., and Bezanilla, F. (1994). Gating current noise produced by elementary transition in Shaker potassium channels. *Science* **264,** 578–582.

Sigg, D., and Bezanilla, F. (1997). Total charge movement per channel. The relation between gating displacement and the voltage sensitivity of activation. *J. Gen. Physiol.* **109,** 27–39.

Sompornpisut, P., Liu, Y.-S., and Perozo, E. (2001). Calculation of rigid-body conformational changes using restraint-driven Cartesian transformations. *Biophys. J.* **81,** 2530–2546.

Spruce, A. E., Standen, N. B., and Stanfield, P. R. (1989). Rubidium ions and the gating of delayed rectifier potassium channels of frog skeletal-muscle. *J. Physiol.* (*London*) **411,** 597–610.

Starace, D., and Bezanilla, F. (1999). Histidine at position 362 causes inwardly rectifying H^+ conduction in Shaker K^+ channel. *Biophys. J.* **76,** A266.

Starace, D. M., and Bezanilla, F. (2001). Histidine scanning mutagenesis of basic residues of the S4 segment of the Shaker K channel. *J. Gen. Physiol.* **117,** 469–490.

Starace, D. M., Stefani, E., and Bezanilla, F. (1997). Voltage-dependent proton transport by the voltage sensor of the Shaker K^+ channel. *Neuron* **19,** 1319–1327.

Starace, D. M., Selvin, P. R., and Bezanilla, F. (2002). Resonance energy transfer measurements of transmembrane motion in the Shaker K^+ channel voltage sensing region. *Biophys. J.* **82,** 174a.

Swenson, R. P., Jr., and Armstrong, C. M. (1981). K^+ channels close more slowly in the presence of external K^+ and Rb^+. *Nature* **291,** 427–429.

Tiwari-Woodruff, S. K., Lin, M. A., Schulteis, C. T., and Papazian, D. M. (2000). Voltage-dependent structural interactions in the *Shaker* K^+ channel. *J. Gen. Physiol.* **115,** 123–138.

Yang, N., and Horn, R. (1995). Evidence for voltage-dependent S4 movement in sodium channels. *Neuron* **15,** 213–218.

Yusaf, S. P., Wray, D., and Sivaprasadarao, A. (1996). Measurement of the movement of the S4 segment during activation of a voltage-gated potassium channel. *Pflugers Arch. Eur. J. Physiol.* **433,** 91–97.

Zheng, J., and Sigworth, F. J. (1997). Selectivity changes during activation of mutant Shaker potassium channels. *J. Gen. Physiol.* **110,** 101–117.

RHODOPSIN STRUCTURE, DYNAMICS, AND ACTIVATION: A PERSPECTIVE FROM CRYSTALLOGRAPHY, SITE-DIRECTED SPIN LABELING, SULFHYDRYL REACTIVITY, AND DISULFIDE CROSS-LINKING

By WAYNE L. HUBBELL,* CHRISTIAN ALTENBACH,* CHERYL M. HUBBELL,* and H. GOBIND KHORANA†

*Jules Stein Eye Institute and Department of Chemistry and Biochemistry, University of California, Los Angeles, California 90095, and †Departments of Biology and Chemistry, Massachusetts Institute of Technology, Cambridge, Massachusetts 02139

I. Introduction to Rhodopsin and Visual Signal Transduction

G-protein coupled receptors (GPCRs) are a large superfamily of integral membrane proteins that serve as input stages to molecular signal transduction systems, the outputs of which modulate various cellular responses. The input signals, which may be molecular ligands or physical stimuli, are highly diverse, and particular GPCRs have evolved to interact with specific signals. Rhodopsin, the photoreceptor of the vertebrate retina, defines the largest subfamily of GPCRs (Gether, 2000).

ADVANCES IN PROTEIN CHEMISTRY, Vol. 63

0065-3233/03 $35.00

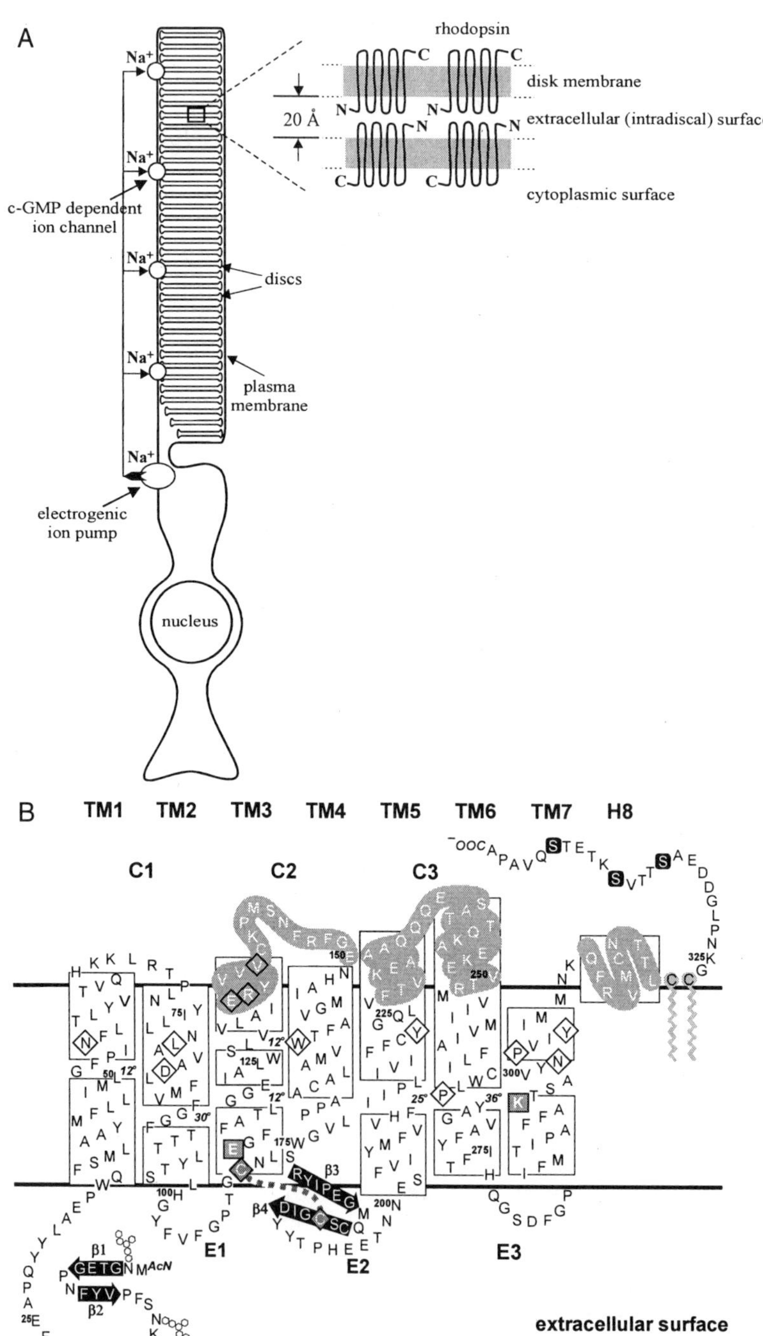
A
rhodopsin
disk membrane
extracellular (intradiscal) surface
cytoplasmic surface
20 Å
Na+
c-GMP dependent
ion channel
discs
plasma
membrane
electrogenic
ion pump
nucleus
B
TM1
TM2
TM3
TM4
TM5
TM6
TM7
H8
C1
C2
C3
E1
E2
E3
β1
β2
β3
β4
extracellular surface

The receptor is localized in highly organized intracellular organelles, the disk membranes, in the rod photoreceptor cell (Fig. 1A). Physiologically, the effects of photon absorption by rhodopsin are closure of ionic channels in the plasma membrane and a concomitant change in the voltage across the plasma membrane of the cell. The light-to-voltage transduction process is highly amplified and essentially thermally noiseless (Baylor, 1987; Burns and Baylor, 2001). For example, a single absorbed photon in a rod cell triggers an electrical response with a probability of at least 0.5, and during the response the flow of $\approx 10^7$ ions is blocked from passing through ion channels in the plasma membrane corresponding to a change of $\approx 3\%$ in the transmembrane potential. Despite an extremely high density of rhodopsin, $\approx 10^9$ per cell, the dark noise of the photoreceptor cell is the equivalent of only one photon per 100 s. The subject of this review is the molecular mechanism whereby rhodopsin initiates and shapes this remarkable process.

Rhodopsin holds a special place in the GPCR superfamily. It was the first GPCR to be sequenced (Ovchinnikov, 1982; Hargrave *et al.*, 1983), it has consistently led the way as a paradigm for other family members, and it was the first whose structure has been determined at the atomic level (Palczewski *et al.*, 2000). The general structural motif of rhodopsin, and all GPCRs, is a seven-transmembrane (TM) helix bundle (Fig. 1B) that was evident in early work from sequence and biochemical analysis

---

FIG. 1. (A) A schematic representation of the rod photoreceptor cell and the intracellular disk membrane organelles (for more detail, see Fein and Szuts, 1982). The inset represents a magnified view of a portion of a disk, showing the location of the rhodopsin molecule and the very small space separating the two membranes within a disk (the extracellular space). An electrogenic ATP-driven ion pump generates an extracellular Na^+ current (the "dark" current) and a transmembrane potential. The current returns to the cytoplasm via c-GMP-dependent ion channels in the plasma membrane. Light absorption by rhodopsin triggers closure of the channels, thereby hyperpolarizing the plasma membrane. (B) The secondary structure of rhodopsin. The transmembrane helices are TM1–TM7, and H8 is a short interfacial helix. The segments connecting the TMs at the cytoplasmic surface of the receptor are numbered C1–C3, and those at the extracellular surface E1–E3. The structure of C3 is based on site-directed spin labeling (see Section III,E). Kinks in the TM helices are indicated as breaks, and the kink angles indicated (Teller *et al.* 2001). The horizontal lines indicate the approximate location of the membrane/aqueous interface (see Section IV). Residue K296 and E113, identified with shaded square symbols, are the site of the protonated Schiff base attachment of the 11-*cis* retinal chromophore (not shown) and its counterion, respectively. Diamond symbols mark residues that are highly conserved throughout the rhodopsin subfamily of GPCRs. Sequences at the cytoplasmic surface directly involved in transducin interaction are shaded. These same sequences are also implicated in binding of the regulatory proteins arrestin and rhodopsin kinase. Zigzag lines at 322C and 323C represent palmitoyl residues, while phosphorylation sites in the C-terminal tail are marked by filled square symbols.

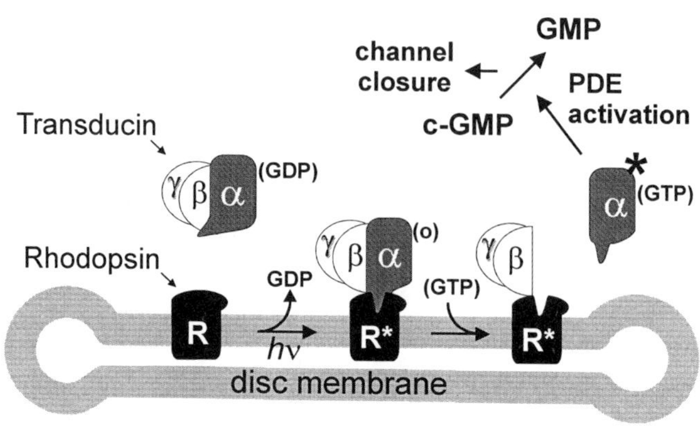

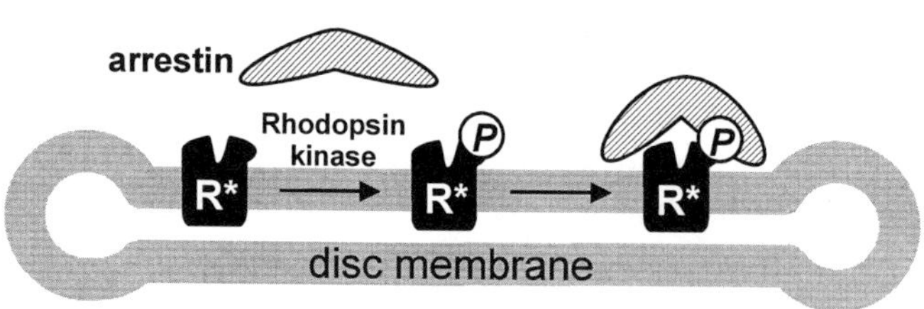

FIG. 2. The role of rhodopsin in visual signal transduction (see text). (A) Activation. (B) Inactivation.

(Hargrave, 2001), later strongly supported by cryoelectron microscopy (Unger *et al.*, 1997), and finally verified by the recent crystal structure.

The biochemistry that mediates photon absorption in the disk membranes and closing of the ionic channels in the plasma membrane is summarized in Fig. 2 and discussed in detail in reviews that provide access to the original literature (Helmreich and Hofmann, 1996; Menon *et al.*, 2001; Hamm 1998, 2001; Bunemann and Hosey, 1999; Krupnick and Benovic, 1998). In the dark (inactive) state, rhodopsin (R) contains a covalently bound 11-*cis* retinal chromophore buried within the protein core. In rhodopsin, the chromophore absorbs maximally around 500 nm, and absorption of a photon isomerizes the retinal to the all-*trans* form within about 200 fs (Peteanu *et al.*, 1993) with a quantum efficiency of 0.67 (Dartnall, 1972). The change in chromophore shape creates strain, and subsequent thermal relaxation of both chromophore and protein through a series of spectrally defined intermediates culminates in a conformational switch of rhodopsin to metarhodopsin II (MII), the active signaling species (R*). The signal is passed from R* to the heterotrimeric G protein, transducin [$G_{t\alpha\beta\gamma}$(GDP)], by a direct binding interaction. The R*-transducin interaction triggers a conformational switch in the α-subunit of transducin, causing dissociation of a bound GDP to form an empty complex [$G_{t\alpha\beta\gamma}$(o)]. Binding of GTP to the empty complex triggers yet another conformational switch in the α-subunit resulting in its dissociation from the complex in an active state [G_α*(GTP)]. In succeeding steps in the cascade, G_α*(GTP) activates a phosphodiesterase (PDE), thereby reducing the cytoplasmic level of c-GMP which closes a c-GMP regulated ion channel in the plasma membrane (Fig. 1A).

In this cascade, there are three stages of molecular amplification that account for the high gain of the rod cell response: (1) each R* activates multiple transducin molecules; (2) each activated PDE hydrolyzes many molecules of c-GMP, and (3) many channels respond to the drop in c-GMP. The high-gain amplifier is shut down at the rhodopsin level through phosphorylation of R* (in the C-terminal domain) by rhodopsin kinase, and subsequent binding of arrestin to the phosphorylated segments (Fig. 2B). Thus, the relatively small exposed cytoplasmic surface of rhodopsin ($\approx$1500 Å^2) must display recognition and interaction sites for transducin, arrestin and rhodopsin kinase each of which is of higher molecular weight than rhodopsin itself.

Each of the amplification stages is a potential source of the dark noise mentioned above (1 activation/100 s). The spontaneous activation of rhodopsin accounts for about 50% of the noise, and it can be estimated that a thermal activation event occurs on the average once every 2000 years (Burns and Baylor, 2001). This extremely low thermal activation rate is due to the covalently linked chromophore that provides extreme stability to the "off" state of the protein.

In a general sense, the transduction/amplification system and shutdown mechanisms outlined in Fig. 2 are relays of molecular switches, each triggered in succession. The input signal (light) triggers the rhodopsin switch, and the signal is passed along the relay by either protein–protein or small molecule–protein interactions that trigger the next element in the relay. To understand the operation of the system, it is then necessary to elucidate the molecular switch in each of the elements, and to map out the protein–protein interaction surfaces. At the present time, crystal structures are known for G_α(GDP) (Lambright, 1994), G_α(GTPγS) (Noel *et al.*, 1993), $G_{t\alpha\beta\gamma}$(GDP) (Lambright, 1996), rhodopsin (Palczewski *et al.*, 2000; Teller *et al.*, 2001), and arrestin

(Hirsch *et al.,* 1999). Because these are static images, they cannot by themselves reveal the mechanisms of molecular switching. However, they do form an essential foundation for the design of molecular genetic and spectroscopic experiments that will ultimately elucidate the dynamic events of molecular switching, as well as identify the protein–protein interactions. For example, site-directed mutagenesis and peptide competition studies have already identified key sequences in rhodopsin that are involved in transducin interaction and activation (Franke *et al.,* 1992; Konig *et al.,* 1989; Marin *et al.,* 2000; Yang *et al.,* 1996a). As shown in Figure 1B, these regions are confined to the cytoplasmic links C1–C3 and the H8 helix.

Over the past decade, cysteine scanning mutagenesis and associated techniques of site-directed spin labeling (SDSL), sulfhydryl reactivity, and disulfide cross-linking kinetics have been systematically employed to assign structural features to rhodopsin sequences throughout the entire cytoplasmic surface, and to monitor changes in the protein conformation in this critical region upon light activation. This comprehensive approach was made possible by the total synthesis of the rhodopsin gene (Ferretti *et al.,* 1988) and the development of expression systems and purification methods that provided mutant proteins in sufficient quantities and in a high state of purity (Oprian *et al.,* 1987; Reeves *et al.,* 1999). The strategies employed in this seminal work have been reviewed (Khorana, 2000).

The results of these studies revealed key features of the structure and the dynamic nature of the cytoplasmic surface of rhodopsin in solution. Most important, they provided the first direct structural evidence for helix movements underlying receptor activation in a GPCR (Farahbakhsh *et al.,* 1993; Altenbach *et al.,* 1996; Farrens *et al.,* 1996), and more recently a complete description of the spontaneous structural changes that occur at the cytoplasmic surface on formation of the activated state (Altenbach *et al.,* 1999a,b, 2001b,c; Cai *et al.,* 1999a,b, 2001; Klein-Seetharaman *et al.,* 1999, 2001; Langen *et al.,* 1999). This review will be focused on a collective interpretation of data from SDSL, sulfhydryl reactivity, disulfide cross-linking, and the recent crystal structure to provide a view of the resting and active state of rhodopsin at the cytoplasmic surface and to identify at least part of the rhodopsin switch.

In Sections II and III, the crystal structure of rhodopsin is briefly reviewed and compared with the dynamic structure in micellar solutions and membranes as inferred from the biophysical methods mentioned above. A structural model of the cytoplasmic surface derived from solution NMR of peptides has been presented (Yeagle *et al.,* 1997; Katagadda *et al.,* 2001), but this approach does not provide direct information on functional dynamics in the intact molecule or conformational changes leading to the activated state and will not be described in this article. Section IV examines the supramembrane organization of the disk membrane as interpreted from the recent rhodopsin crystal structure. In Section V, experimental evidence bearing on the rhodopsin conformational switch is reviewed. The final section, Section VI, reviews structure–function relationships deduced from a collective consideration of the crystal structure, the dynamics of the solution structure, and the conformational changes underlying activation.

II. The Rhodopsin Crystal Structure: The Inactive State

A structural model of rhodopsin in the dark (inactive) state derived from electron density maps refined at 2.8 Å is shown in Fig. 3 (see color insert) (Teller *et al.*, 2001). As is customary, the structure is discussed with respect to three topological (as opposed to folding) domains: transmembrane, cytoplasmic, and extracellular (intradiscal). The transmembrane helical segments (i.e., the segments of the helices *within* the bilayer interior) are referred to as "TM1–TM7," and the single helix lying parallel to the bilayer surface is designated "H8." The sequences corresponding to these segments are given in the legend to Fig. 3, and the basis of the assignments is discussed below in Section IV. The segments that link the TM helices on the cytoplasmic and extracellular surfaces will be referred to as C1–C3 and E1–E3, respectively. Finally, the sequence beyond the palmitoylation sites from 324 to 348 is designated the C-terminal tail.

In the crystals used to obtain the model of Fig. 3, there are two molecules in the asymmetric unit, molecules A and B. Molecules A and B differ in the degree of order within the crystal, as reflected in the plot of thermal factors in Fig. 4A. Gaps in the plot of Fig. 4A reflect portions of the chain that could not be modeled because of the extent of disorder. Molecule B has more extensive gaps and has generally more disorder throughout. The model of Fig. 3 corresponds to the most highly ordered Molecule A, and unless stated otherwise, all models of rhodopsin in this review will be based on this structure. The A structure has been analyzed in detail in several recent publications with respect to helix tilts, kinks, and interactions stabilizing the inactive state (Teller *et al.*, 2001; Menon *et al.*, 2001; Okada *et al.*, 2001), the role of specific residues in "tuning" the absorption wavelength of the chromophore (Teller *et al.*, 2001; Menon *et al.*, 2001), the relationship to inherited visual diseases (Menon *et al.*, 2001; Okada *et al.*, 2001), and comparison with bacteriorhodopsin (bR) (Teller *et al.*, 2001). Some key points related to rhodopsin activation will be summarized below.

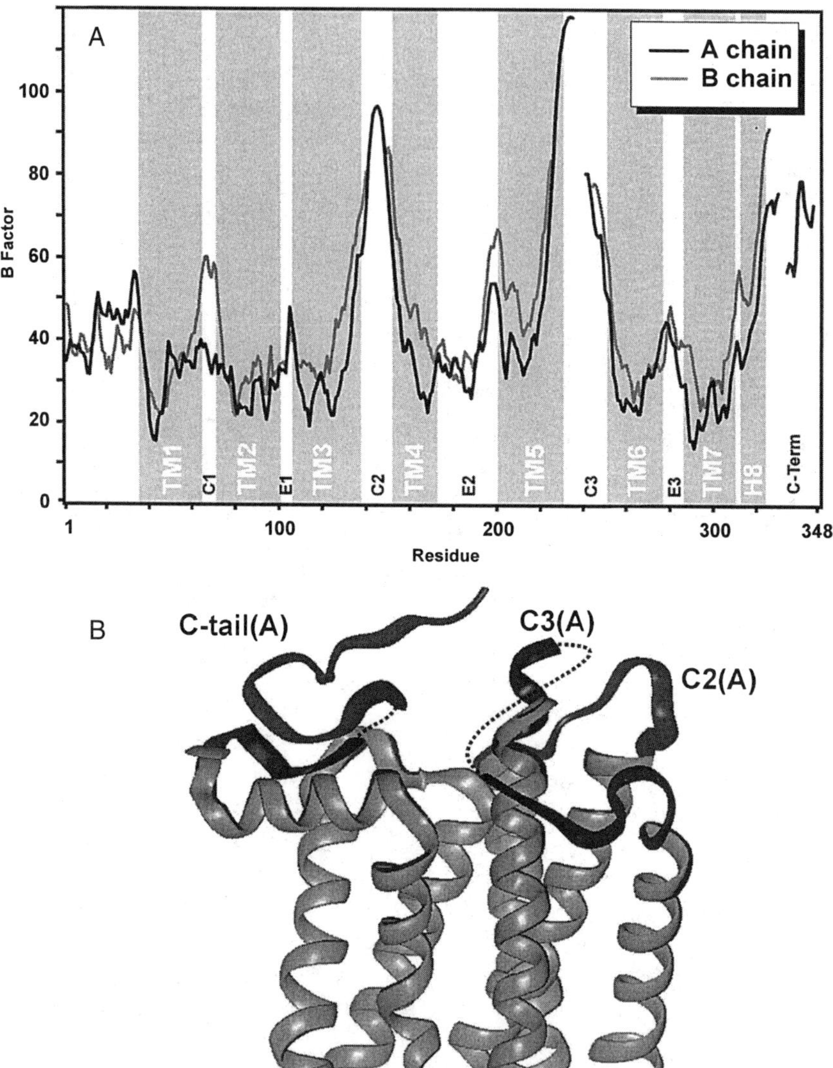

FIG. 4. (A) Plots of the thermal (*B*) factors as a function of sequence position for molecules A and B of the asymmetric unit (pdb entry 1HZX). The shaded and unshaded vertical bars mark the TM and interconnecting segments, respectively. (B) Aligned molecules A and B showing the "plastic" regions at the cytoplasmic surface where they differ. At the extracellular surface and in the TM sequences, they are well aligned. Dark and light ribbons mark the A and B chains, respectively.

The helices of the rhodopsin transmembrane domain are distinguished from those of bR by the large number of irregularities and kinks, due primarily to Pro and Gly residues (Figs. 1B and 3). These kinks provide potential points of flexibility in the otherwise rigid helical rods. One minor kink, at Ser-127 in TM3, may be due to H bonding of the side chain –OH with the backbone carbonyl of Ile-123, and as such would not constitute a point of flexibility. Of particular interest is Pro-267 in TM6. This residue is completely conserved throughout the family of rhodopsin-like receptors and is likely to be a fundamental element of the activation switch (Baldwin *et al.,* 1997; see Section V,C).

Many of the residues conserved throughout the rhodopsin receptor family are located in transmembrane helical segments near the cytoplasmic surface, or their extensions into the cytoplasmic space (Baldwin *et al.,* 1997; Gether, 2000). This is anticipated, because the cytoplasmic surface in the different receptors is designed for interaction with proteins of a common signal transduction cascade, whereas the extracellular surface is specialized for interaction with particular ligands. The highly conserved residues (>90%) are identified in Fig. 3 by spheres at the α-carbon atoms. One group is clustered at the interaction surfaces of TM1 (N55), TM2 (L79, D83), and TM7 (N302, P303, Y306, the NPXXY motif) and are involved in a network of interhelical interactions that stabilize this part of the helix bundle. Another group is at the cytoplasmic termination of TM3 (E134, R135, V139) (Palczewski *et al.,* 2000). The D134/R135 pair is salt-bridged or H-bonded (Honig and Hubbell, 1984) and buried within the hydrophobic environment of the helix bundle. While R135 may make interactions with TM6 and stabilize its position in the inactive receptor (Teller *et al.,* 2001), it is also clear that the E134/R135 pair is directly involved with functional coupling to the G protein (Franke *et al.,* 1990). Conserved Y223 in TM5 is located near the contact surface with TM6, but faces the lipid bilayer at the level of the polar head groups (see below). Tryptophan-161 in TM4 is hydrogen bonded to Asp-78 in TM2, a residue that is itself conserved as either Asp or Ser throughout the subfamily (88%). These residues face the hydrophobic interior of the lipid bilayer, an environment which favors strong H bonding, and this interaction may serve to fix the relative positions of TM2 and TM4 that sequester TM3 within the bundle interior at this level.

The thermal factors reveal that the extracellular domain, consisting of the N-terminal sequence and the extracellular links E1–E3, is the most highly ordered region of the structure in the crystal (Fig. 4). Together with the extracellular ends of the TM helices, the extracellular domain forms the binding pocket for the 11-*cis* retinal chromophore, shown as a space-filling model in Fig. 3. The chromophore is covalently bound to the protein through a protonated Schiff base linkage at K296 in TM7, with a counterion at E113 in TM3 (Sakmar *et al.*, 1989; Zhukovsky and Oprian, 1989; Nathans, 1990). The covalent bond between the chromophore and protein and the K296/E113 interhelical salt bridge, located in the hydrophobic interior of the protein, play crucial roles in determining both the wavelength of maximum absorbance and the extreme stability of the resting ("off") state of the receptor (Menon *et al.*, 2001). The latter point is appreciated from the astonishingly low rate of spontaneous activation of rhodopsin, one event every 2000 years, as estimated from the thermal noise of the photoreceptor (Burns and Baylor, 2001). Moreover, rupture of the K296/E113 salt bridge is apparently a requirement for reaching the active state of rhodopsin (Rao and Oprian, 1996).

One wall of the chromophore binding pocket is formed by beta strands $\beta3$ and $\beta4$ of extracellular loop E2. A highly conserved disulfide bond between Cys-110 and Cys-187 serves to firmly anchor $\beta4$ to the C-terminal end of TM3. If the chromophore were to enter and exit the binding pocket from the extracellular, surface, this rigidly held $\beta3/\beta4$ wall would be required to move. Other salient features of the extracellular surface include the carbohydrate residues linked to Asn-2 and Asn-15 (S1 and S2, Fig. 3) and the fingerlike loop projecting away from the bundle. The glycosylation at Asn-15 appears to have a structural role, because mutations at Asn-15 produce rhodopsins defective in G-protein activation (Kaushal *et al.*, 1994), and deglycosylation at this site produces structural changes at the cytoplasmic surface of rhodopsin as detected by SDSL (Zhang, C., Ridge, K, Khorana, H. G. and Hubbell, W. L., unpublished results). As will be discussed further below, the projecting loop, defined as the sequence between Pro-12 and Pro-27, may serve to stabilize the unique structure of the disk membrane.

The cytoplasmic surface of rhodopsin is of particular interest because C2, C3, and H8 form the recognition/interaction surface for the G protein (Fig. 2B). Thus, the activating conformational switch must be ultimately expressed as structural changes at the cytoplasmic surface. As can be seen in Fig. 4A, the interhelical links C2 and C3 are the most disordered and hence uncertain regions in the structure (aside from the C-terminal tail). The maximum B factor in each of the interconnecting links increases with length, and the B factors in the transmembrane helices rise sharply as the cytoplasmic links are approached. As noted above, a number of residues in C3 could not be modeled at all, and in the B molecule this is also the case in C2. This suggests plasticity in C2 and C3, and perhaps enhanced dynamics in solution. The axis of helix H8 lies approximately perpendicular to that of the TM helices at the cytoplasmic surface. The helix itself is amphipathic with the nonpolar surface exposed, suggesting that this surface is solvated by the lipid bilayer in the native environment. Arginine-314 is ideally positioned to interact with negatively charged lipids in this model.

The C-terminal tail of rhodopsin extends from residue 324 to residue 348 and is relatively poorly determined in the crystal structure (thermal factors >50 in the A chain, and not resolved in the B chain). From 324 to 330, the C-terminal tail curls back over H8. Residues 331–333, near the C-terminus of TM7, could not be modeled. From 334 to 338 the chain passes over the end of TM2, turns, and finally extends in the direction of TM4 but away from the body of the helix bundle.

In the following section, the average structure of the cytoplasmic domain in solution, as seen by SDSL, sulfhydryl reactivity, and disulfide cross-linking kinetics, is compared with the crystal structure. In addition, the dynamics of C1–C3 and H8 are evaluated with SDSL.

III. Structure and Dynamics of Rhodopsin in Solutions of Dodecyl Maltoside: The Cytoplasmic Surface in the Inactive State

Cysteine-scanning mutagenesis, involving more than 100 mutations, has been systematically carried out through C1–C3, the cytoplasmic terminations of TM1–TM7, H8, and the C-terminal tail. In addition, more than 40 pairs of cysteines have been introduced at the cytoplasmic face. With these mutants as a basis set, three classes of experiments have been carried out, namely SDSL, sulfhydryl reactivity, and disulfide cross-linking kinetics. A global comparison of the results provides a unique view of the solution state, its dynamics, and its correlation with the crystal structure. By solution state is meant, in all cases, rhodopsin solubilized in dodecyl maltoside (DM) micelles. The measured functional properties of rhodopsin, namely transducin activation (Resek *et al.*, 1993) and phosphorylation by the rhodopsin kinase (Thurmond *et al.*, 1997), are conserved in this detergent, and it is presumed to be a reasonable approximation to the bilayer environment.

Before discussing the results from these methods, a brief review of the measured quantities and interpretations for each, as applied to rhodopsin, is presented.

A. *Overview of Methods for Exploring the Solution Structure of Rhodopsin: Site-Directed Spin Labeling, Sulfhydryl Reactivity, and Disulfide Cross-Linking*

1. *Site-Directed Spin Labeling*

The developing methodology and expanding applications of SDSL have been chronicled in a series of reviews, and these provide access

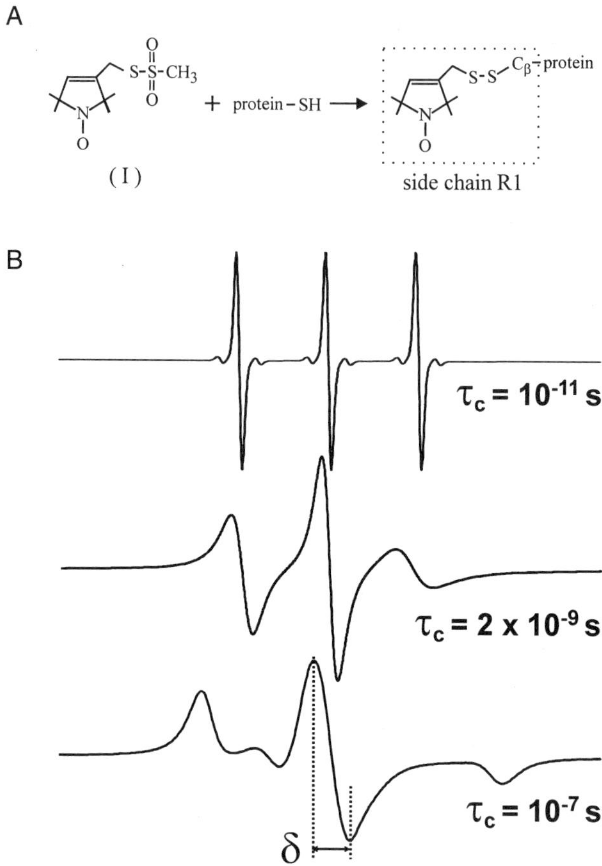

FIG. 5. (A) Reaction of (1-oxyl-2,2,5,5-tetramethylpyrroline-3-methyl)methanethiosulfonate to generate the nitroxide side chain R1. (B) Simulated first-derivative EPR spectra for a nitroxide showing the characteristic changes in line shape that accompany changes in correlation time for isotropic motion. These are conveniently categorized in terms of the central linewidth (δ).

to the original literature in the field (Hubbell and Altenbach, 1994; Hubbell *et al.*, 1996, 1998, 2000; Feix and Klug, 1998). In the method, a substituted cysteine residue is reacted with a selective nitroxide reagent to generate a nitroxide side chain, the most common of which is designated R1 (Fig. 5A). The electron paramagnetic resonance (EPR) spectrum of the R1 side chain can be analyzed in terms of three quantities relevant to examining the solution structure. These are (1) the mobility of the side chain; (2) the accessibility of the nitroxide to paramagnetic reagents free in solution; and (3) the distance between the nitroxide and another paramagnetic center in the protein, either another nitroxide or a bound metal ion.

The accessibility to paramagnetic reagents is measured by the experimental "accessibility parameter" Π, a quantity directly proportional to the collision frequency of the nitroxide with the reagent. As is intuitively reasonable, Π is directly proportional to the exposure of R1 on the surface of the protein. The commonly used paramagnetic reagents include O_2 and the metal ion complex NiEDDA. Because oxygen has sufficient solubility in both membrane interiors and water, $\Pi(O_2)$ is useful to explore the exposure of R1 on both hydrophobic and hydrophilic surfaces of a membrane protein.

It has been shown that $\Pi(O_2)$ for R1 is in fact proportional to the fractional solvent accessibility (f_{SA}) of the native side chain at the same site computed from the corresponding crystal structure (Isas *et al.,* 2002). The sequence dependence of the solvent accessibility, measured by either f_{SA} or $\Pi(O_2)$, is a "fingerprint" for a protein fold. For example, solvent accessibility is periodic through regular secondary structure, and the period and phase of the function identify the type of secondary structure and its orientation in the protein, respectively. In nonregular secondary structure encountered in loops, the solvent accessibility is not necessarily periodic, but the functional dependence on sequence remains characteristic of the fold. Thus, comparison of f_{SA} computed from a crystal structure and $\Pi(O_2)$ determined experimentally for the protein in solution is a convenient and efficient way of comparing the solution and crystal structures. This will be the method used below for rhodopsin.

Unlike the case for oxygen, NiEDDA has limited solubility in a bilayer interior and can be used together with $\Pi(O_2)$ to distinguish R1 residues facing lipid from those facing water. Such discrimination provides a convenient means of locating sequences that cross membrane/aqeuous interface. For example, the contrast parameter, defined as $\Phi = \ln[\Pi(O_2)/\Pi(\text{NiEDDA})]$, changes sharply for an R1 residue as it is moved along the outside surface of a helix in the region where it crosses the interface between lipid and water. For a relatively diffuse interfacial region, the simple ratio of accessibility parameters (e^{Φ}) is employed. In addition, Φ is a function of depth within the bilayer and can be used to estimate the immersion depth of R1 in a bilayer (Altenbach *et al.* 1994).

Two R1 side chains in a protein, or an R1 side chain and a paramagnetic metal ion in an engineered site, interact by a distance-dependent magnetic dipolar interaction. For the usual case where there is a distribution of interspin distances, the dipolar interaction leads to an EPR spectral broadening that can be analyzed in terms of the average distance and the distribution. The measurements can be made under physiological conditions (Altenbach *et al.,* 2001a; Voss *et al.,* 1995), and the effective distance range up to about 20 Å is ideal for mapping helix proximity relationships in membrane proteins. Recently developed time-domain EPR techniques extend this range beyond 50 Å, but the measurements must be carried out at low temperatures (Eaton *et al.,* 2001).

Rotational motion of a nitroxide modulates the anisotropic electron–nuclear magnetic dipolar interaction, giving rise to electron relaxation that affects the EPR spectral line shape. At X-band frequency, the spectra are sensitive to motions with correlation times in the range of $10^{-11} < \tau_c < 10^{-7}$, but also reflect the anisotropy of the motion. Figure 5B shows simulated EPR spectra for isotropic motion in the fast, intermediate, and slow motional regimes and illustrates the high sensitivity of the line shape to motional rate.

The motion of the R1 nitroxide in a protein has contributions from the overall tumbling of the protein, the internal motions of the side chain, and fluctuations in the backbone structure. For membrane proteins such as rhodopsin, the correlation time for molecular tumbling is slow on the EPR time scale defined above and can be ignored. The internal motion of the R1 side chain is due to torsional oscillations about the bonds that connect the nitroxide to the backbone, and the correlation times for these motions lie in the nanosecond regime where the EPR spectra are highly sensitive to changes in rate.

The motion of R1 in helices has been studied in detail, and this is the case relevant for analysis of rhodopsin. At helix surface sites where R1 has no interactions with other side chains or main-chain atoms, the motion is anisotropic and can be accurately modeled by a single order parameter (S) and effective correlation time (τ_c) (Columbus *et al.,* 2001). This simple anisotropic motion is expected to be the same at all helical surface sites unless modulated by direct interactions of R1 with other groups in the protein and/or by local backbone fluctuations. Interactions of R1 with the environment and local backbone fluctuations are qualitatively distinguishable by their opposite effects on motion: the former reduces and the latter increases the mobility relative to a noninteracting reference on the surface of a rigid helical segment.

The motion of a nitroxide side chain in terms of correlation times and order parameters can be deduced from spectral simulations (Budil *et al.,* 1996). On the other hand, a simple semiempirical quantity, the "scaled mobility" (M_s), suffices for many purposes (Hubbell *et al.,* 2000). The

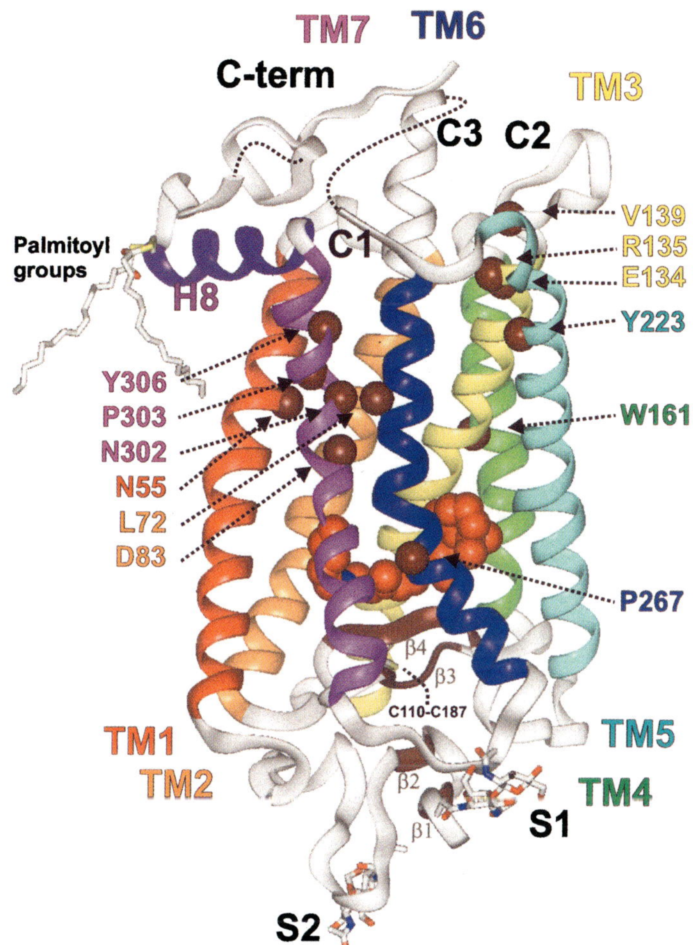

Hubbell *et al.*, FIG. 3. A structural model of rhodopsin (A chain of pdb entry 1HZX). The transmembrane helical segments TM1-TM7 and the helix H8 are shown in unique colors, and the interconnecting segments are shown in white (dotted lines indicate sequences not modeled). The TM segments are defined as those sequences included within the membrane interior (TM1: 35–64; TM2: 71–100; TM3: 106–136; TM4: 152–172; TM5: 200–230; TM6: 251–276; TM7: 286–310). Brown spheres on the alpha-carbons identify highly conserved residues in the rhodopsin family (sequence numbers are color-coded according to the helix in which they lie). Also shown are the thioester-linked palmitoyl groups at C322 and C323, the carbohydrate moieties S1 and S2, and the 11-*cis* retinal chromophore (space-filling model).

A

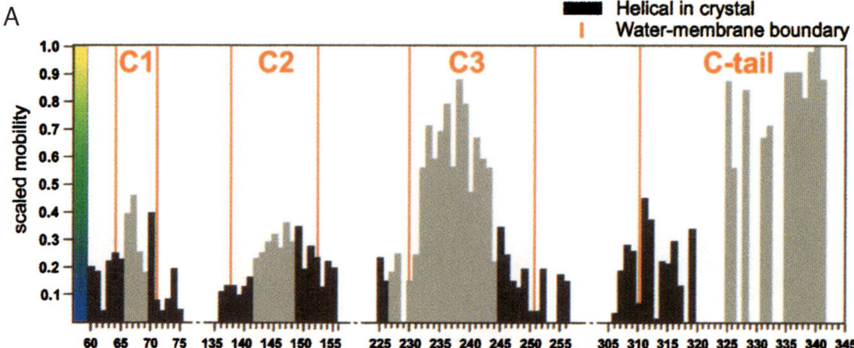

B

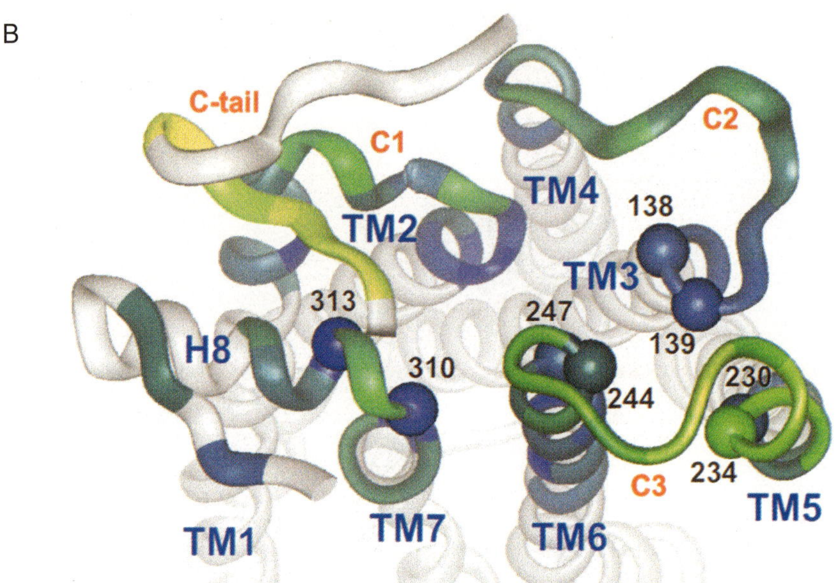

Hubbell *et al.*, FIG. 15. (A) The scaled mobility of R1 throughout the rhodopsin cytoplasmic surface. Values were computed from published linewidth data (Altenbach *et al.*, 1996, 1999a,b; Farahbakhsh *et al.*, 1995; Langen *et al.*, 1999) using $\delta_i^{-1} = 0.12$ and $\delta_m^{-1} = 0.65$ (see Section III,A,l). The shaded vertical bars identify residues in the TM helices (Fig. 3 legend). The black data bars indicate helical residues assigned on the basis of the crystal structure of rhodopsin (1HZX, A chain). The color fountain codes the scaled mobility axis. (B) A ribbon model of rhodopsin with the C3 loop modified from the crystal structure according to the SDSL data (Fig. 11). The ribbon is color-coded according to the graded scale for M_s shown in (A). Spheres at the C_α carbon atoms identify residues that make up the core of the helical triad formed from extensions of TM3, TM5, and TM6 (138, 139, 230, 234, 244, 247). The relatively high mobility of R1 at these sites is consistent with the dynamic state of the TM6 extension in C3. Also shown are spheres at the C_α carbon atoms of 310 and 313, sites where R1 is highly immobilized.

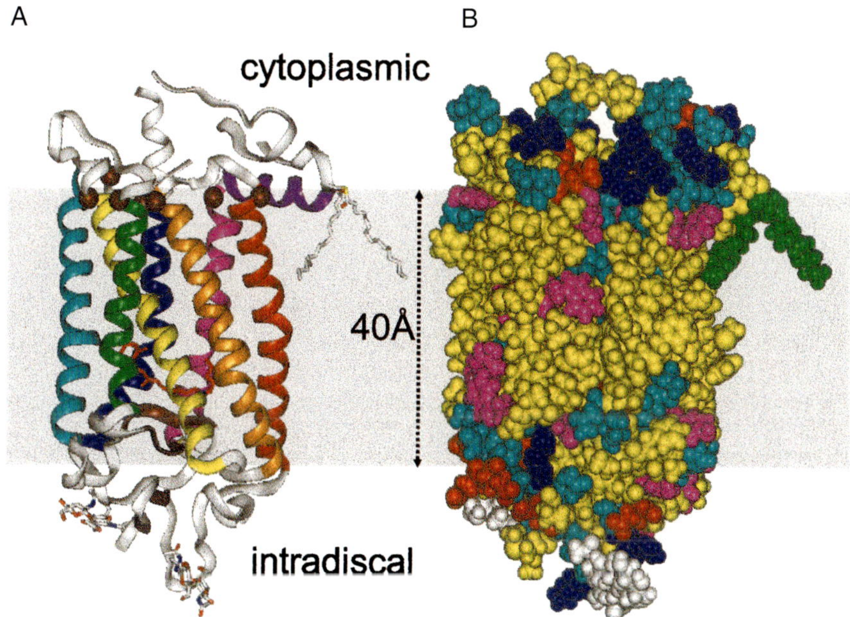

Hubbell *et al.*, Fig. 17. (A) Ribbon model of rhodopsin showing spheres at the alpha carbon atoms of residues in the membrane/aqueous interface as identified by SDSL; these residues define a unique plane of intersection of rhodopsin with the bilayer surface. The phospholipid bilayer is modeled with a thickness of 40 Å (phosphate-to-phosphate) and placed such that the plane of the polar head groups coincides with the plane defined by the alpha carbons. (B) A space-filling model of rhodopsin. The color code identifies nonpolar (yellow), basic (blue), acidic (red), polar (cyan), and tyrosine and tryptophan residues (magenta).

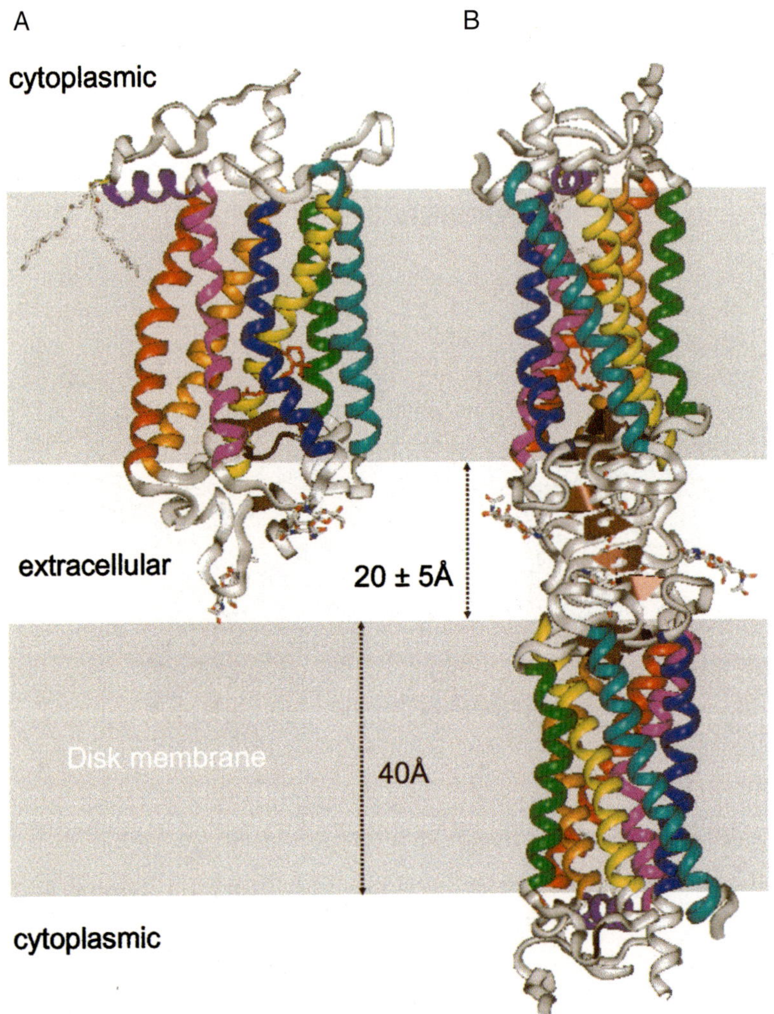

Hubbell *et al.*, FIG. 18. Models of the disk membrane based on the crystal structure. (A) The 12–23 loop of the N terminus projects to the opposing membrane and determines the extracellular spacing. (B) Putative tail-to-tail dimer of rhodopsin (see text).

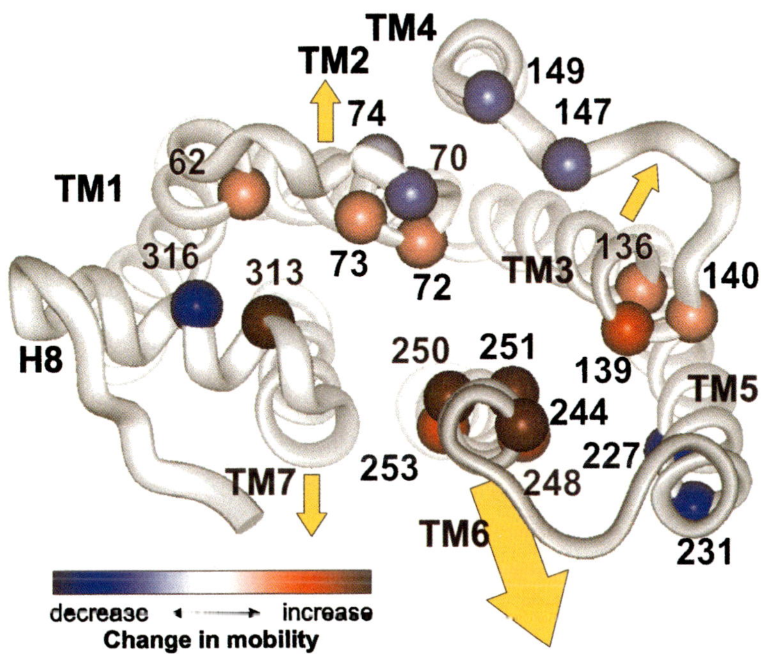

Hubbell *et al.*, FIG. 20. Summary of changes in mobility of R1 throughout the cytoplasmic face of rhodopsin due to photoactivation. The magnitudes of the mobility changes are color-coded as shown, with blue and red representing decreases and increases in mobility, respectively. The arrows indicate the direction and relative magnitude of helix movements inferred from the changes.

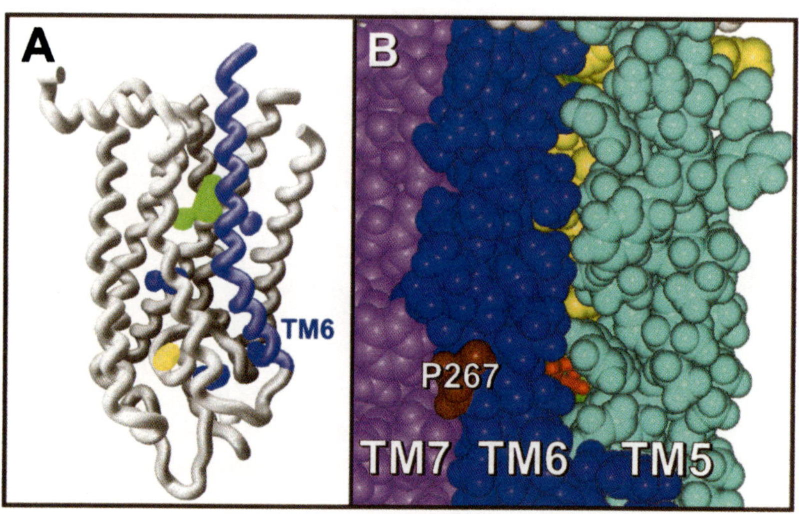

Hubbell *et al.*, FIG. 23. (A) Wire-frame model of rhodopsin showing the location of the cavities within the structure located by GRASP (reference) with a probe radius of 1.4 Å. The largest cavity is 145 $Å^3$. (B) Space-filling model of rhodopsin, looking at the TM5, TM6 interface. Helices are color coded as in Fig. 3.

scaled mobility is a function of both rate and order and is defined as

$$M_s = \frac{\delta^{-1} - \delta_i^{-1}}{\delta_m^{-1} - \delta_i^{-1}}$$

where δ is the central resonance linewidth indicated in Figure 5B, and the subscripts m and i refer to the most mobile and immobile residues encountered in proteins. The M_s value for R1 in a random coil is $M_s \approx 1$, at a noninteracting helix surface site on a rigid helical backbone is $M_s \approx 0.39$, and for R1 at a buried site on a rigid helical segment is $M_s \approx 0$. Thus, in an R1 scan along a helical sequence, M_s is periodic in position because of periodic modulation of the tertiary interaction strength.

Contributions from backbone fluctuations will increase M_s values for both surface or buried sites, relative to a rigid reference helix. The backbone contribution is thus reflected in the average value of M_s along one turn of the helix, the mean scaled residue mobility ($\langle M_s \rangle$). Alternatively, if noninteracting helix surface sites can be identified, deviations of M_s from the rigid helix reference value may be taken as a direct measure of backbone contributions. Noninteracting helix surface sites are readily recognized by their diagnostic EPR spectral line shape (Mchaourab *et al.*, 1996; Columbus *et al.*, 2001). In the following sections, M_s and $\langle M_s \rangle$ will be employed to map backbone dynamics for rhodopsin in solution.

2. *Sulfhydryl Reactivity and Disulfide Cross-Linking*

The solvent accessibility of particular sites may be assessed by the reaction rate of substituted cysteines with the sulfhydryl reagent 4,4′-dithiopyridine (4-PDS) (Grassetti and Murray, 1967; Chen and Hubbell, 1978) (Fig. 6A). The products of the reaction are a mixed disulfide

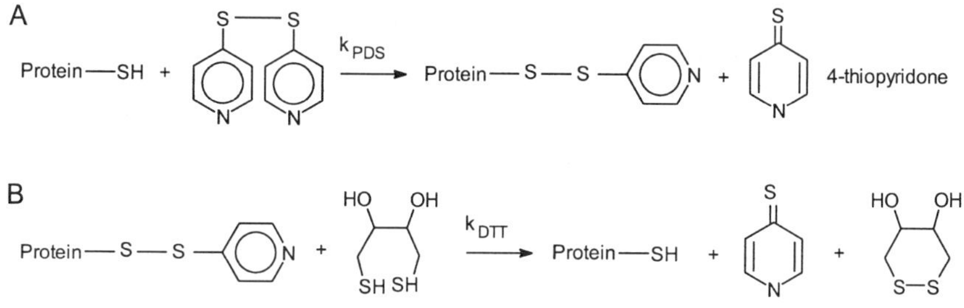

FIG. 6. (A) Reaction of 4,4′-dithiopyridine with cysteine to produce a mixed disulfide and 4-thiopyridone; (B) the subsequent reduction of the mixed disulfide with dithiothreitol (DTT). Rate constants k_{PDS} and k_{DTT} characterize these second-order reactions.

between cysteine and 4-thiopyridine, and 4-thiopyridone (4-TP). The reaction rate may be conveniently followed by appearance of the 4-TP absorbance at 324 nm, and the rate constant for the reaction, k_{PDS}, taken as a measure of relative accessibility. The mixed disulfide may be subsequently cleaved with dithiothreitol (DTT) to yield the original cysteine residue and 4-TP as shown in Fig. 6B. The rate of this reaction may again be monitored by the appearance of 4-TP absorbance, and the rate constant, k_{DTT}, provides a second measure of solvent accessibility. Both k_{SH} and k_{DTT} have been used as measures of solvent accessibility in rhodopsin to complement those of $\Pi(O_2)$ and Π(NiEDDA).

The k and Π values give different views of residue "accessibility" in a protein. Π values, because of the nature of the measurement, reflect a state in proportion to its equilibrium population. Thus, rare fluctuations in the protein structure that transiently expose a buried residue will not be seen. On the other hand, the covalent reaction of 4-PDS with a buried cysteine during a fluctuation will trap the state, and the event will be counted and added to previous such events. Thus, the sulfhydryl reactivity method is an "integrating" method that can reveal low-frequency fluctuations much like hydrogen exchange. A comparison of relative Π and k values thus may provide information on the existence of low-frequency structural fluctuations (Altenbach *et al.*, 1999a,b), and this point will be discussed further below.

In SDSL, proximity relationships are inferred from direct distance measurements between R1 residues. A complementary proximity measurement using the same double cysteine mutants as for SDSL is the spontaneous rate of disulfide formation, measured by the rate constant k_{ss} (Falke and Koshland, 1987). The spontaneous rate of oxidation depends on many features of the structure, including (1) the distance and relative orientation of the residues, (2) the flexibility of the sequences containing the cysteines, and (3) the effective pK_a of the residues as determined by local electrostatic potentials. A comparison of distances from SDSL with k_{ss} can provide information on the structure as well as dynamics. For example, relatively large values of both k_{ss} and the corresponding interspin distance suggest that the cysteines in question are not in proximity on the average, but that one or both are in flexible sequences and come into proximity during rare fluctuations in structure.

B. *Structure of the C-Terminal Tail in Solution*

As noted above, the C-terminal tail is poorly determined in the crystal structure. In order to examine the structure in solution, cysteine residues were substituted, one at a time, at sites in the C-terminal tail indicated in Fig. 7A (325, 326, 328, 331, 332, 335–340) (Cai *et al.*, 1997).

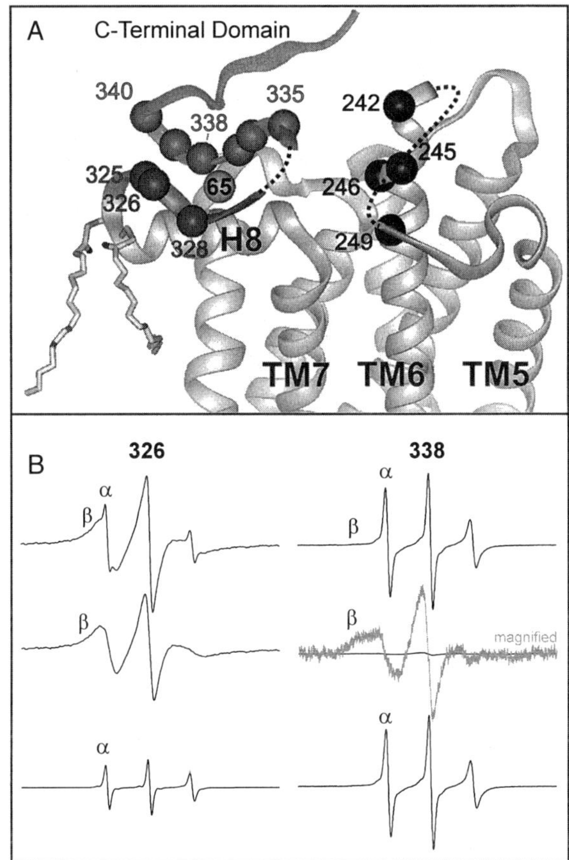

FIG. 7. (A) Structural model of rhodopsin showing the location of cysteine substitution mutants in the C-terminal tail (325, 326, 328, 331, 332, 335–340). Mutants at 331 and 332 are not represented because these sites were not modeled in the crystal structure. Also shown are sites in C3 (dark spheres at 242, 245, 246, 249) and 65 in C1 discussed in the text. Dotted lines indicate sequences not modeled in the crystal structure. (B) Top row: EPR spectra of 326R1 and 338R1, representing sites in the proximal and distal portions of the C-terminal tail, respectively; center and bottom rows: the two components of α and β, respectively, resolved by spectral subtraction. In 338R1, the immobilized component β is minor and is nearly invisible in the experimental spectrum.

The cysteine residues were derivatized to introduce the R1 side chain (Fig. 5A), and the EPR spectra analyzed in terms of R1 mobility (Langen *et al.*, 1999). The EPR spectra each revealed two components reflecting R1 populations of different mobility. Using simulation and subtraction techniques, the two spectral components (α and β) were resolved and analyzed separately (see Fig. 7B for examples). In each case, the most mobile component (α) has effectively isotropic motion with correlation time in the range 0.3–0.5 ns, similar to R1 in a unfolded protein. The more immobilized component (β) varies in mobility from site to site, reflecting an ordered secondary and tertiary structure. Thus, the C-terminal tail is in equilibrium between an ordered and disordered conformation. The equilibrium constant, however, is strongly position dependent. For 325R1, 326R1, and 328R1, the ordered state strongly dominates ($\approx$5% disordered), and it may be concluded that this sequence has a stable configuration.

In the crystal structure, the 325–328 sequence folds back and packs against H8 in a conformation where the side chains of 325R1, 326R1, and 328R1 are solvent exposed (Fig. 7A). The R1 spectra are consistent with this assignment. Additional support for an interaction of the 325–328 sequence with H8 will be discussed below.

In the sequence beyond 331, there is a dramatic increase in the population of the highly mobile, disordered component. For 338, for example, the disordered state accounts for $\approx$70% of the population (Fig. 7B), increasing to essentially 100% at 340. In the crystal structure, the sequence from 331 to 340 is modeled to curl back over the top of TM2 in contact with the cytoplasmic surface of the helix bundle. Data from disulfide cross-linking and R1 magnetic dipolar interactions in double mutants argue against this conformation in solution (Cai *et al.,* 1997). One such mutant investigated was 338C/65C. The crystal structure model would place the sulfur atoms of the two cysteines in at an optimal distance and orientation for disulfide formation (see Fig. 7A), but in fact essentially no cross-linking was observed. Given the dynamic properties of this domain observed in solution, it is extremely unlikely that these residues are in proximity. On the other hand, when one cysteine was placed at 338, and the second scanned through 240–250 in C3, local maxima in cross-linking rates were found between 338C and residues 242C, 245C, and 248C. In either the crystal structure or the proposed helical extension of TM6 based on SDSL (see Section III,E below), these residues lie on the exterior surface of the molecule, facing the aqueous solution (Fig. 7A). In addition, some of the double cysteine mutants in the latter experiment were derivatized with spin labels, and it was found that the strongest spin–spin interaction occurred between 338C and 245C and 246C. These residues lie on the same face of TM6 defined by the residues marking local maxima in cross-linking with 338C. Thus, it is most likely that in solution a dynamic C-terminal sequence from 331 on extends toward TM6 rather than curling around over the top of TM2. Interestingly, removal of the palmitoylation sites has little effect on the structure or dynamics of the C-terminal tail, including the structure from 325 to 328 that lies close to these sites (Langen *et al.,* 1999).

C. *Structure of C1 and Adjacent Sequences in TM1 and TM2*

C1 (65–71) is the shortest of the cytoplasmic TM links and does not contain residues essential for transducin activation, although mutations K67C and L68C do reduce transducin activation to ≈30% of that for the wild-type protein (Klein-Seetharaman *et al.,* 1999). Figure 8A identifies the sequence through which single cysteine substitution mutants were prepared, relative to the crystal structure (Klein-Seetharaman *et al.,* 1999). Each of these mutants was derivatized to introduce R1, and the EPR spectra analyzed in terms of $\Pi(O_2)$, Π(NiEDDA), and R1 mobility (Altenbach *et al.,* 1999b).

As discussed in Section III,A,1, comparing the sequence dependence of f_{SA} and $\Pi(O_2)$ is an efficient means of comparing a crystal structure with features of the solution structure derived from SDSL data. Figure 8B shows f_{SA} for both the A and B molecules of the crystal (upper panel) and $\Pi(O_2)$ (center panel) from 60 to 75, which includes parts of TM1 and TM2 as well as C1. The A and B molecules have similar structures in C1 (Fig. 4B), and this is reflected in the similarity of f_{SA} throughout the sequence, except for small differences around 63 and 64 due to the absence of the C-terminal domain in the B chain.

The pattern of experimental solvent accessibilities in solution measured by $\Pi(O_2)$ clearly mirrors that of f_{SA}, indicating the similarity of the crystal and solution structures for the C1 linker. The correspondence between solvent accessibility and the crystal structure is illustrated in Fig. 8A by the C_α–C_β bonds, which are shaded according to accessibility (see figure legend). As expected, residues that project into regions of extensive tertiary contact lie at local minima in $\Pi(O_2)$, whereas residues that are fully solvent exposed lie at local maxima. The periodicity of the TM1 (60–64) and TM2 (72–75) sequences is extended through C1, as observed in other helical hairpin motifs with short loops (Isas *et al.* 2002). The accessibility to collision with the highly polar reagent Ni(EDDA) clearly shows that loop residues 64–66 and 70 are solvent exposed (Fig. 8B, center panel). Residues 68 and 69 in the loop point into the fold of the protein and are hence sequestered from solvent to a large extent, as readily identified by both $\Pi(O_2)$ and Π(NiEDDA).

Both k_{SH} and k_{DTT} have been employed to investigate the structure of C1, but quantitative data at more sites were obtained with k_{DTT} (Klein-Seetharaman *et al.,* 1999). Figure 8B (lower panel) shows values of k_{DTT} for sites in C1 and adjacent helices. Clearly, k_{DTT} and Π(NiEDDA) identify the same sites as being highly solvent exposed, although the relative values differ.

The crystal and solution structures in the vicinity of C1 may be compared in another way, namely the determination of proximity

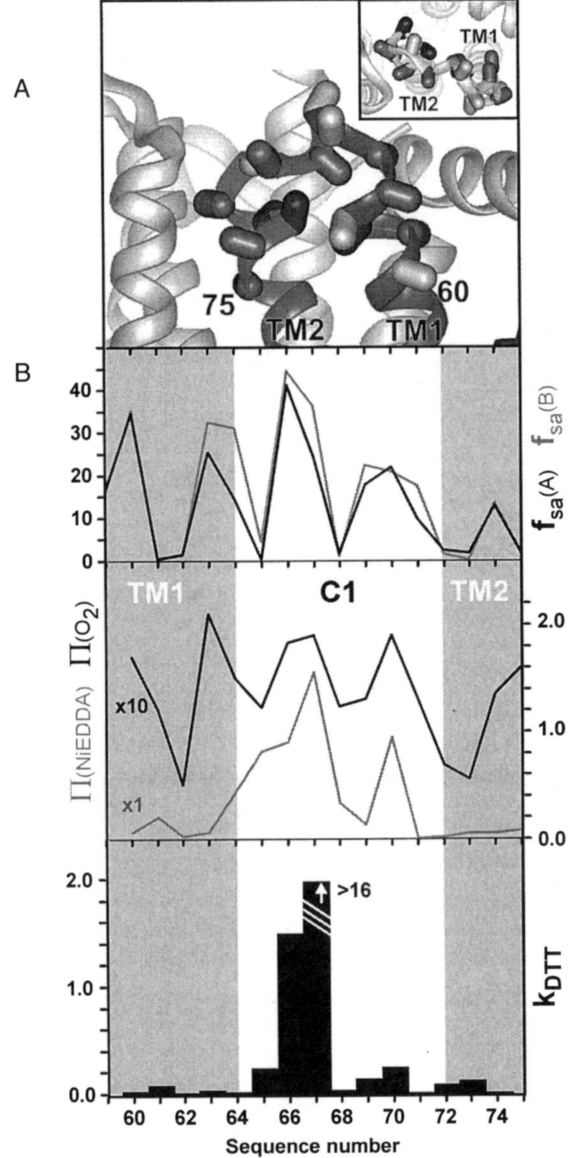

FIG. 8. (A) Structural model of rhodopsin showing the location of cysteine substitution mutants in C1 and adjacent sequences in TM1 and TM2 (60–75). Individual residues are represented by the C_α–C_β bonds in order to indicate that direction in which the side chains project. To illustrate the correspondence between experimental values of solvent accessibility and structure, the bonds are coded according to local maxima (white), minima (black), or intermediate (gray) values of $\Pi(O_2)$. (B) f_{SA} values computed from the crystal structure for both A (dark trace) and B (light trace) molecules (upper panel), $\Pi(O_2)$ (dark trace) and Π(NiEDDA) (light trace) for R1 at each site (center panel), and k_{DTT} for each site (lower panel). The shaded vertical bars mark residues in TM1 and TM2, as indicated.

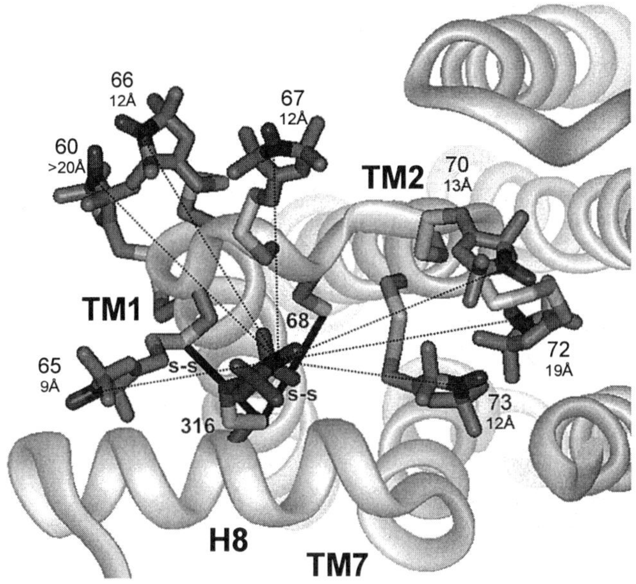

FIG. 9. Proximity relationships for residues in C1 relative to 316 in H8. Example of R1 distance mapping in rhodopsin. For each distance measurement, only two R1 side chains were in the protein, one fixed at the reference site 316, and the other at a site in the sequence 55–75. The R1 side chains were modeled based on crystal structure data with energy minimization subject to the experimentally determined distance constraint (shown). In each case, the measured distances in solution were in good agreement with those expected from the rhodopsin crystal structure. Substituted cysteine residues 65 and 68 most rapidly formed disulfide cross-links with the reference cysteine at 316 in H8. This is indicated by the dark bars connecting the potential disulfide partners.

relationships based on direct distance measurements between pairs of R1 residues, or on determination of k_{SS}, the rate constant for spontaneous disulfide formation. Figure 9 shows an example of results obtained from direct distance measurements with SDSL (Altenbach *et al.*, 2001c). In this experiment, a reference nitroxide was placed at residue 316 in H8, and a second nitroxide scanned through the sequence 60–75. The distances determined between each pair in solution were consistent with the topology of the crystal structure and known conformations of the R1 side chain (Langen *et al.*, 2000). Using the same pairs of cysteines, proximity was estimated by the rate of disulfide formation, and this was again consistent with the crystal structure (Klein-Seetharaman *et al.*, 2001) (Fig. 9).

Collectively, results from sequence-correlated solvent accessibility, direct distance measurement, and disulfide cross-linking rates all suggest that the crystal and solution structures are similar in the C1 region.

D. Structure of C2 and Adjacent Sequences in TM3 and TM4

C2 constitutes part of the transducin interaction domain (Fig. 1B). Figure 10A shows the sequence in C2, TM3, and TM4 (136–155) that has been investigated by cysteine scanning mutagenesis and SDSL (Ridge *et al.*, 1995; Farahbakhsh *et al.*, 1995). Figure 10B (upper panel) shows the fractional solvent accessibilities for residues in the 136–155 sequence

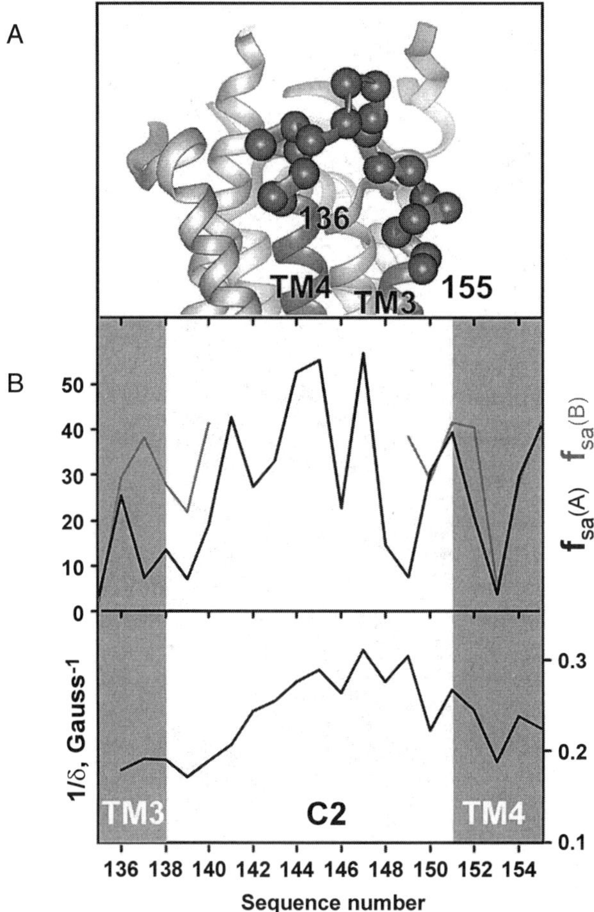

FIG. 10. (A) Structural model of rhodopsin showing the location of cysteine substitution mutants (136–155) in C2 and adjacent sequences in TM3 and TM4. (B) f_{SA} values computed from the crystal structure for both A and B molecules (upper panel), and δ^{-1} (the inverse central linewidth) (bottom panel) for R1 at each site. The shaded vertical bars mark residues in TM3 and TM4.

for both the A and B chain. In the B chain, a structure was not assigned for 140–149, presumably because of disorder. In the A chain, the C2 structure is complete, but different from that in the B chain in segments where they can be compared. These properties suggest plasticity in C2.

Experimental solvent accessibilities [$\Pi(O_2)$] are not available for C2, so a direct comparison between the crystal and solution structures cannot be made on this basis. However, R1 side-chain accessibility and mobility are closely correlated (Isas *et al.,* 2002), and Fig. 10B (lower panel) shows a plot of R1 mobility, measured by the inverse linewidth, δ^{-1}, for comparison with f_{SA}. In the TM4 helix from 153 to 155, the patterns of f_{SA} and δ^{-1} are similar and reflect the helical structure observed in the crystal. However, from 144 to 150, δ^{-1} has a period of 2, characteristic of a β-like structure that only partially agrees with the f_{SA} pattern, and this section appears to have a different structure in solution than it does in the crystal. Interestingly, this section overlaps with that of unassigned structure in the B molecule. Along the sequence 139–143, a gradient of increasing f_{SA} is also reflected in δ^{-1}. A regular helical structure was inferred for residues 130–142 in solution from patterns of disulfide bond formation between engineered cysteines (Yu and Oprian, 1999).

E. Structure of C3 and Adjacent Sequences in TM5 and TM6

The cytoplasmic interhelical linker C3 is a key recognition/binding site for transducin (Fig. 1B) as well as for the rhodopsin kinase and arrestin. In the crystal, the structure of this important segment is poorly determined because of disorder in the lattice (Fig. 4), presumably corresponding to flexibility in the structure. Figure 11A shows ribbon models for the C3 region based on the crystal structure (right panel) and a tentative solution structure based on cysteine scanning mutagenesis and SDSL data (left panel) (Yang *et al.,* 1996b; Altenbach *et al.,* 1996). Figure 11B shows data for f_{SA} (upper panel) and Π values (lower panel) for the sequence 225–256 which encompasses parts of TM5, TM6, and the whole of C3 (231–251). In the B molecule of the crystal, a long stretch of amino acids, residues 227–243, could not be modeled. The A molecule is more highly ordered, but is still missing 236–240.

In the sequences corresponding to TM5 (225–230) and TM6 (252–256), f_{SA} for the A molecule reflects the expected helical periodicity, as does that for the residues of the B chain that are modeled. For residues in these regions that can be compared, the f_{SA} and $\Pi(O_2)$ data are in good agreement, suggesting a similar structure in the crystal and in

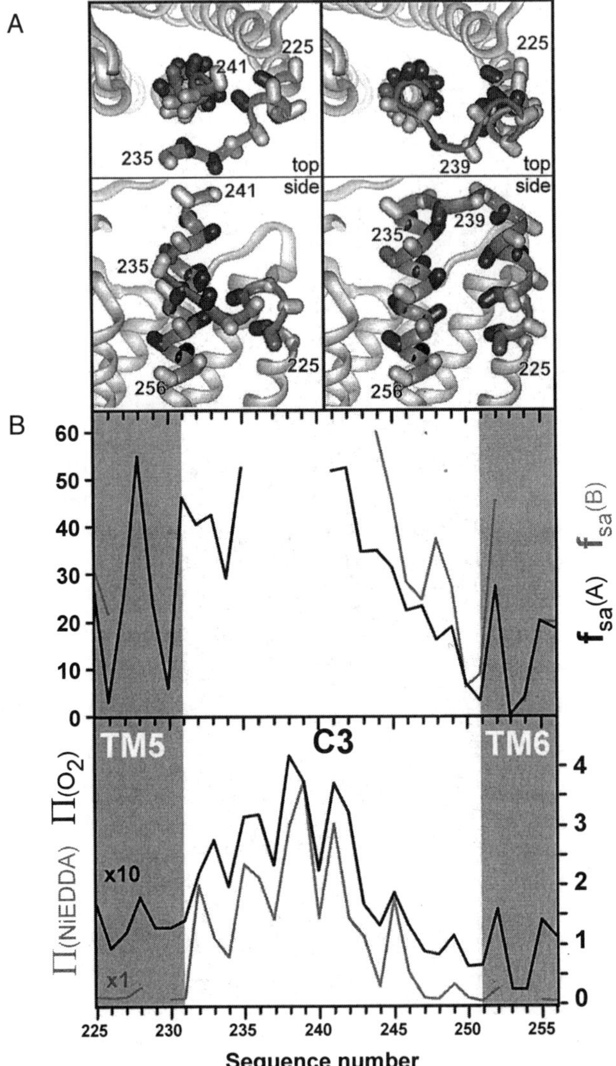

FIG. 11. (A) Structural models of rhodopsin showing the location of cysteine substitution mutants in C3 and adjacent sequences in TM5 and TM6 (residues 225–256). The left panel is based on the crystal structure (1HZX, A chain), and the right panel on SDSL data for the solution structure. In each panel, both top (upper) and side (lower) views of the structure are shown. Individual residues in the sequence are represented by the C_α–C_β bonds in order to indicate the direction in which the side chains project. The bonds are coded according to local maxima (white), minima (black), or in-between (gray) values of $\Pi(O_2)$. (B) f_{SA} values computed from the crystal structure for both A and B molecules (upper panel), and $\Pi(O_2)$ and Π(NiEDDA) for R1 at each site (lower panel). The shaded vertical bars mark residues in TM5 and TM6, as indicated.

solution [the $\Pi(O_2)$ points for 229 and 254 are missing because of lack of reactivity of the corresponding cysteine residues with the spin label]. In addition, the pattern of cross-linking observed in cysteine and disulfide scans is consistent with helical secondary structure from 218 to 225 (Yu and Oprian, 1999). Direct distance measurements between pairs of R1 side chains, one in TM3 and a second in the sequence 247–252 in TM6, are also compatible with a regular helical structure for TM6 near the cytoplasmic surface (Farrens *et al.,* 1996). These latter measurements will be discussed in Section V,A in relation to photoinduced conformational changes.

In the solvent-accessible C3 sequence (231–251), the crystal structure is incomplete and different between the A and B molecules (Fig. 4B). On the other hand, the Π data in solution suggest a simple helix–turn–helix motif for C3, formed from extensions of the TM helices, as shown in Fig. 11A (right panel). The helical periodicity of both $\Pi(O_2)$ and Π(NiEDDA) from 231 to 239 (TM5 extension) and from 241 to 250 (TM 6 extension) is clearly seen in the data (Fig. 11B, lower panel). The interruption of the regular periodicity (a phase shift) between these two sequences at 240 is due to a short turn, probably involving three residues. The C_α–C_β bonds that mark the location and direction of spin-labeled side chains in Fig. 11A are shaded according to $\Pi(O_2)$ values (see figure legend). The excellent correspondence between $\Pi(O_2)$ values and the proposed helix–turn–helix motif is obvious (right panel, top view). That is, residues that project into the protein interior are at local minima in $\Pi(O_2)$ and residues fully exposed to solvent are at local maxima in $\Pi(O_2)$. On the other hand, the crystal structure does not correspond as well to the experimental solvent accessibilities in the C3 sequence 231–235 (for example, 233 is fully exposed in the structure, but $\Pi(O_2)$ for this site is near a local minimum), and 236–240 are missing and cannot be compared. In the crystal structure model, TM5 terminates within the bilayer interior.

The structure assignment in C3 based on SDSL data must be considered tentative, because it is based on periodicity alone. In addition, it is not possible to predict the topography of the motif. That is, the putative helical hairpin could project straight upward as regular extensions of the TMs, as shown in Fig. 11A (right panel), or the helical extensions could bend toward the membrane. However, the fact that the oscillatory $\Pi(O_2)$ and Π(NiEDDA) functions are in phase demonstrates that the C3 structure is homogeneously solvated by an aqueous environment, and none of the residues enter the membrane phase (Hubbell *et al.,* 1998; Oh *et al.,* 1996). Future distance mapping experiments will allow a more definitive structure and topographical description of C3.

F. Structure of H8 and Adjacent Sequence in TM7

Figure 12A shows the structural model of rhodopsin marking the sequence 306–321 through which cysteine scanning mutagenesis was carried out for investigation of sulfhydryl reactivity (Cai *et al.,* 1999a) and SDSL analysis (Altenbach *et al.,* 1999a). This sequence spans the C-terminal end of TM7 (306–309) and helix H8 (312–321). The H8 helix is of particular interest because it is apparently involved in direct interaction with transducin (Fig. 1B) as well as with the lipid bilayer.

The structure of the H8 helix and the adjacent sequence in 306–311 is essentially identical in the A and B chains of the crystal lattice (Fig. 4B). The differences in the computed values of f_{SA} between the A and B molecules along this sequence (Fig. 12B, upper panel) are simply due to the absence of the C-terminal domain in the B chain model, which in the A chain folds back along H8 and partially occludes side chains from about N311 to T320.

Data for experimental solvent accessibilities [$\Pi(O_2)$] for R1 side chains at 308–316 are reproduced in Fig. 12B (center panel), where they can be compared with calculated f_{SA} values. As can be seen, the oscillatory behavior of f_{SA} from 308 to 315 in the A chain is reproduced by $\Pi(O_2)$ in both period and phase. The difference between relative f_{SA} and Π values at 316 could be due to a real structure difference, or to differences in the molecular size of the native (cysteine) and R1 side chains (Isas *et al.,* 2002).

Evidence was presented in Section III,B that in solution the sequence 325–328 in the C-terminal tail is ordered and compatible with the crystal structure in that region, but beyond 328 the tail is largely disordered in solution. This model is consistent with a number of observations regarding the H8 sequence. For example, the extremely low chemical reactivity of a cysteine residue substituted at position 318, and the modest reactivity of a cysteine at 319, both on the solvent-exposed surface of H8, can be accounted for by the interaction with 325–328 (Fig. 12A). On the other hand, site 315, also on the solvent-exposed surface of H8, shows high reactivity as well as high Π values. Interestingly, residue 315 lies just under residue 328 of the C-terminal tail, the position marking the end of the ordered state in the solution structure. The EPR spectral line shape for 315R1 is consistent with this environment. In the absence of the C-terminal domain, 315 should be a helix surface site with a single dominant component (Columbus *et al.,* 2001). However, the spectrum of R1 (Fig. 13A, black trace) shows two distinct spectral components, one of which corresponds to a strongly immobilized population of R1 (α), and the other to a dynamic mode similar to that expected for

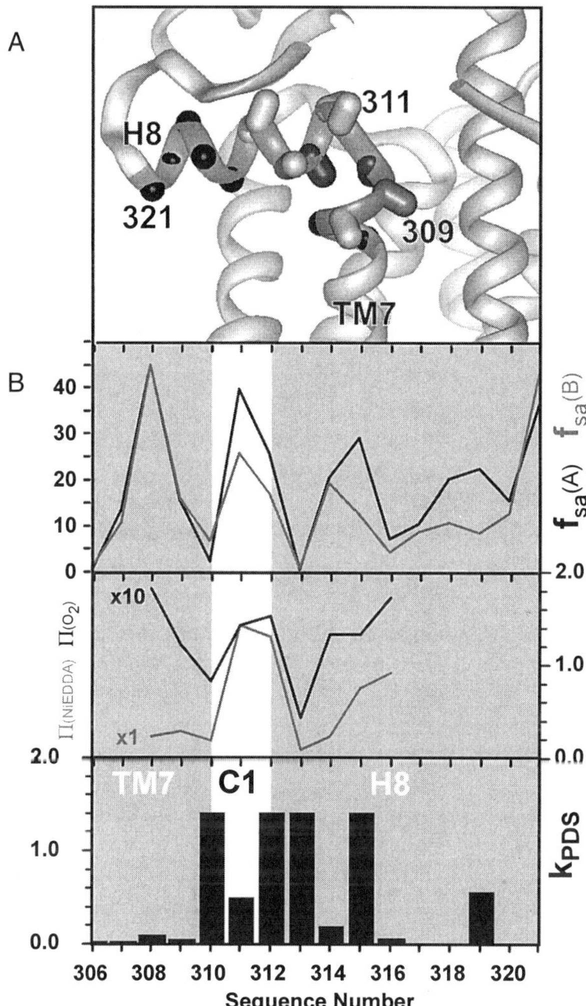

FIG. 12. (A) Structural model of rhodopsin showing the location of cysteine substitution mutants in TM7 (306–309) and H8 (310–321). C_α–C_β bonds indicate the direction in which the side chain projects. The bonds are shaded according to local maxima (white), local minima (black), and intermediate (gray) in $\Pi(O_2)$. Cysteine residues that were unreactive to the spin label reagent and 4-PDS are marked by black spheres at the corresponding C_α. (B) f_{SA} for the A and B molecules (upper panel), $\Pi(O_2)$ and Π(NiEDDA) (center panel), and k_{PDS} (lower panel) for the sequence 306–321.

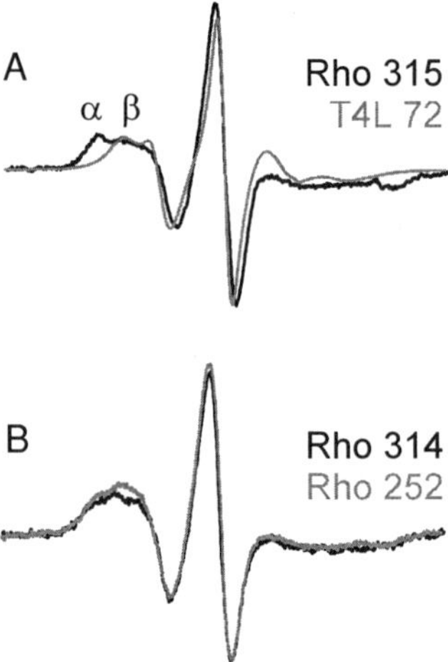

FIG. 13. (A) EPR spectra for 315R1 in rhodopsin (dark trace) and 72R1 in T4 lysozyme (light trace). Component α corresponds to an immobilized population of R1 side chains, while β corresponds to a population with a dynamic mode similar to a solvent-exposed helix surface site, such as 72R1 in T4 lysozyme. (B) EPR spectra for 252R1 and 314R1 in rhodopsin.

a noninteracting, helix surface site (β). For comparison, the spectrum of a helix surface site in T4 lysozyme is also shown. A likely origin of the immobilized state is interaction of 315R1 with the end of the ordered sequence in the C-terminal domain. The fact that there are two states suggests an equilibrium mixture of either R1 rotamers, or possibly two conformations of the C-terminal domain.

The H8 helix is amphipathic, and given its location in the crystal structure, it is logical to assume that the hydrophobic surface is solvated by the lipid bilayer in the native membrane (or by the micelle in micellar solutions). There is direct evidence that this is the case for rhodopsin in micelles of dodecyl maltoside. For example, cysteine residues on the exposed hydrophobic face (M317, T320, L321) are highly unreactive to either 4-PDS (Fig. 12B, bottom panel) or to the spin label reagent (Altenbach *et al.*, 1999a). This would be expected for cysteine facing the hydrophobic interior of the micelle because of both the suppressed ionization of SH groups and the low solubility of the reagents employed in a hydrophobic medium. Residue 314 lies on an exposed surface of H8 at the boundary between the polar and nonpolar faces. The high and low values for $\Pi(O_2)$ and Π(NiEDDA) clearly place 314R1 in the hydrophobic micelle interior, accounting for the low value of k_{PDS} at this site (Fig. 12B, lower panel).

Additional direct evidence for the location of 314 on a lipid-solvated helical surface is provided by the spectrum of 314R1, which has a line shape characteristic of other lipid-facing helix surface sites in rhodopsin such as 252R1 (Fig. 13B). Although, the native side chain at 314 is arginine, the α-carbon is located on the hydrophobic face of H8, and the side chain bends upward so that the positively charged atoms would be located in the polar head-group region of the bilayer, while the nonpolar carbons of the side chain remain lipid-solvated. The uncharged and relatively hydrophobic R1 side chain would project directly into the hydrophobic interior of the DM micelle, as indicated by all of the evidence outlined above.

As for C1, the global topology of the TM7–H8 sequence was mapped in detail by direct distance measurements between pairs of spin labels (Altenbach *et al.,* 2001c) and by disulfide cross-linking (Klein-Seetharaman *et al.,* 2001). For direct distance measurement using spin–spin interactions, a reference R1 side chain was placed at site 65 in C1, and a second R1 placed at each site in the sequence 306–316. Figure 14 shows the pairs and indicates the interspin distances

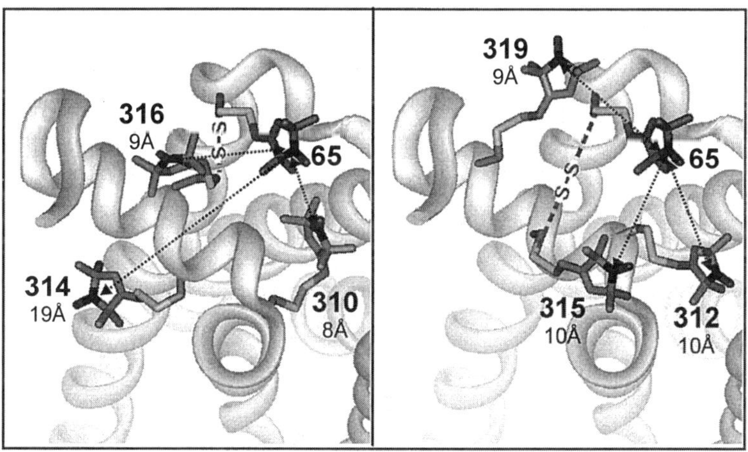

FIG. 14. Proximity relationships in H8 and TM7 relative to 65 in TM1. Examples of interspin distances measured in solution. In each distance measurement, only two R1 side chains were in the protein, one on the reference site 65 and the other at a site in the sequence 306–319. The R1 side chains were modeled based on crystal structure data with minimization subject to the experimentally determined distance constraints (indicated in Å). In each case, the measured distance in solution was in good agreement with that expected from the rhodopsin crystal structure. Substituted cysteine residues 315, 316, and 319 most rapidly formed disulfide cross-links with a cysteine at 65 in TM1. The potential disulfide bonds are indicated as gray dashed lines.

measured. In each case, the distances were in good agreement with predictions based on the rhodopsin crystal structure and the structure of the R1 side chain. With the same double-cysteine mutants used for spin labeling, the spontaneous rate of disulfide formation (k_{ss}) was determined. The cysteine pairs with most rapid cross-linking, 65C/316C and 65C/315C, are in fact those expected from the rhodopsin structure. Surprisingly, essentially no cross-linking was found between 65C and 319C, although the interspin distance in 65R1/319R1 was short (≈9 Å) and the same as for 65R1/316R1 where cross-linking was rapid. This interesting difference will be discussed further below.

Collectively, all of the data obtained on the solution structure with SDSL, sulfhydryl reactivity, and disulfide cross-linking kinetics strongly support the conclusion that the structure of the TM7–H8 sequence investigated is very similar in the crystal and micellar solution state. In solution, the H8 helix is sandwiched between the hydrophobic/aqueous interface on one side and residues 325–328 of the C-terminal tail on the other.

G. Dynamics of C1–C3, H8, and the C-Terminal Tail

The preceding sections highlighted the similarities and differences between the static crystal structure and the average solution structure of rhodopsin in the cytoplasmic domain. This section considers fluctuations in the structure of the cytoplasmic domain relative to the average. This is of interest, because dynamics sequences are often correlated with function, and that appears to be the case in rhodopsin.

Backbone fluctuations with correlation times in the nanosecond regime are revealed in variations of the R1 residue scaled mobility, M_s, along the sequence (see Section III,A,1 for definition). Figure 15A (see color insert) shows a plot of M_s versus sequence for C1–C3, H8, and adjacent sequences in the TM helices. Figure 15B shows the cytoplasmic surface of rhodopsin color-coded according to M_s values. In this figure, C3 was modeled from the SDSL data, as in Fig. 9.

There are several important features in Fig. 15. First, the sequence dependence of M_s through C1, C2, C3, and H8 follows a pattern essentially identical to that of $\Pi(O_2)$ (compare with Figs. 8, 11, and 12). This is particularly obvious in the extended helical periodicity of both M_s and $\Pi(O_2)$ throughout the long C3 segment. This correlation is expected, because the most immobile (buried) and mobile (surface) residues are also the most inaccessible and accessible, respectively. Second, the highest values of M_s are reached in the C terminus, beyond 328, and in the center of C3. As discussed above, the C-terminal sequence is dynamically disordered beyond 328, similar to an unfolded protein. Residue 340R1 defines the highest mobility in the molecule and $M_s \equiv 1$. Surprisingly, the central part of C3 has M_s values in the range of 0.5–0.9, approaching those of the disordered C terminus. Thus, although the periodic variation of M_s, $\Pi(O_2)$, and Π(NiEDDA) all confirm a regular structure with period $\approx$3.6, the helix is a dynamic structure, and clearly not tightly packed with the remainder of the structure. For comparative purposes, the mean scaled residue mobility through one turn of a C3 helix segment (232–235) is $\langle M_s \rangle = 0.63$ compared to $\langle M_s \rangle \approx 0.15$ for a turn through a well-packed helix in annexin XII or T4 lysozyme (Isas *et al.*, 2002; Mchaourab *et al.*, 1996). The gradients in M_s along the C3 sequence suggest that the dynamic mode is fluctuation of backbone dihedral angles rather than rigid-body helical motion, because the latter would simply add a constant contribution to all positions.

It is interesting that the two most dynamic sequences identified by nitroxide scanning, C3 and the C-terminal domain, are also the most disordered in the crystal structure. Moreover, they are two of the most important functional domains of the molecule, both being involved in protein–protein interactions along the signal transduction pathway. This is in keeping with a growing body of data on other proteins that implicate dynamic sequences with protein–protein or ligand–protein interactions (Crump *et al.*, 1999; Duggan *et al.*, 1999; Bracken *et al.*, 1999).

Insufficient data are available to compute $\langle M_s \rangle$ through complete helical turns of each TM segment, but a useful estimate of $\langle M_s \rangle$ can be obtained as the pairwise average of buried and surface sites. On this basis, TM1, TM2, TM6, TM7, and surface helix H8 are all relatively rigid, having $\langle M_s \rangle \approx 0.1$. The corresponding segments in TM4 and TM5 are nearly twice as large, suggesting greater flexibility. Interestingly, TM4 and TM5 at the cytoplasmic face of rhodopsin have the largest surface area exposed to lipid, and thus have small areas of contact with protein, perhaps accounting for the flexibility. Helices TM3, TM5, and TM6 all have extensions into the aqueous phase of 1 or more turns. These extensions form a triad with a common core, and the mobility of R1 residues facing this core (138, 139, 230, 234, 244, 247; see Fig. 15B) is significantly greater than expected for R1 at buried sites in a well-packed protein. Presumably, this is due in large part to the backbone motion of the TM6 extension in C3.

For backbone motions to modulate M_s, they must have correlation times shorter than about 30 ns (for X-band EPR spectroscopy); low-frequency modes with microsecond or longer correlation times cannot be characterized by M_s. Correlation times in the microsecond regime can be investigated by saturation transfer EPR spectroscopy (Thomas *et al.*, 1976; Marsh and Horvath, 1989), but the competing rotational motion of the protein complicates interpretation. However, low-frequency fluctuations that transiently expose an otherwise buried cysteine can be detected by a finite reaction rate of the cysteine with a sulfhydryl reagent in solution. Likewise, the formation of a disulfide bond between pairs of cysteines otherwise too far apart to react in a static structure can serve to identify transient states (Careaga and Falke, 1992). Because these reactions are essentially irreversible, each event is summed over time in the formation of product. In this way, even rare states of the protein can be detected, and their probability of occurrence quantified by the reaction rate constant.

It is expected that sites buried within the protein should generally have low reactivity due to lack of accessibility. Sites 310 and 313 in the TM7–H8 loop region, identified by spheres at the α-carbons in Fig. 8B, are interesting exceptions. The very low accessibility to collision with either O_2 or NiEDDA (Fig. 12), and the low mobility (Fig. 15A) clearly demonstrate that residues 310R1 and 313R1 are buried in the protein interior, as predicted by the crystal structure ($f_{SA} < 0.05$ for both). Nevertheless, k_{PDS} is high, equal to that of some solvent-exposed sites. This suggests low-frequency fluctuations of the structure in the turn between TM7 and H8. This is further supported by a finite cross-linking rate between a cysteine at 65 or 246 and a second cysteine in the turn sequence 310–312 (Cai *et al.*, 1999a). The distances between these residues (C_α–C_α>10 Å) are greater than the average distance for disulfides in known proteins (5–6 Å) (Kosen, 1992). On the other hand, the flexibility is apparently limited to the vicinity of the turn, because cysteines 65 and 319, which are in close proximity as judged by spin–spin interaction between 65R1 and 319R1, show no spontaneous cross-linking (Cai *et al.*, 1999a). The C1 segment appears to be rigid on a low-frequency time scale as well, based on both k_{DTT} data (Fig. 8) and the paucity of cross-links formed between a reference cysteine at 316 and a second cysteine throughout C1 (Klein-Seetharaman *et al.*, 2001).

IV. Location of the Membrane–Aqueous Interface and the Structure of the Disk Membrane

The crystal structure provides no direct information on the location of the membrane–aqueous interface relative to the protein structure, although its location can be approximately inferred from sequence information (see below). A direct experimental identification of the residues located in the interface is possible with SDSL using the e^{Φ} function described in Section III,A,1. Figure 16 shows plots of e^{Φ} for the R1 side

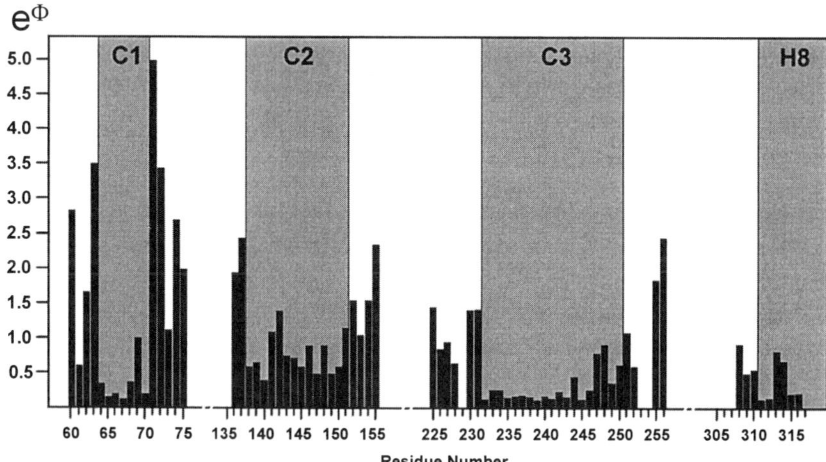

FIG. 16. Plots of e^{Φ} versus sequence for the solvent-exposed sites in each of the loops investigated. The approximate position of the membrane–aqueous interface relative to the rhodopsin sequence is located at the transitions between white and grey background.

chain at multiple positions throughout (only solvent-exposed sites are included in the analysis) (Altenbach *et al.*, 1994). The function e^{Φ} changes sharply when crossing a polarity boundary, and based on this criterion, the approximate location of the membrane–aqueous boundary (within one or two residues) is indicated in Fig. 16 for each sequence.

Fig. 17A (see color insert) shows a ribbon model of the rhodopsin structure indicating the residues assigned to the interface in each helix by a sphere centered on the corresponding α-carbon. Also shown is a sphere on the α-carbon of residue 314, which is located in the interface (see Section III,F). Clearly, these residues define a unique plane of intersection of the molecule with the membrane–aqueous interface. The shaded band in Fig. 17 represents a phospholipid bilayer with a phosphate–phosphate distance of $\approx$40 Å, the expected thickness of the bilayer in the disk membrane (Saiz and Klein, 2001). The outer interface of the bilayer is positioned so that the polar head groups coincide with the intersection plane defined by the data in Fig. 16. This procedure then fixes the intersection plane of the molecule on the extracellular surface as well.

The position of rhodopsin in the membrane deduced from SDSL data is compatible with the topography of surface residues in the molecule. Figure 17B shows a space-filling model of the rhodopsin structure with residues color coded according to charge, polarity, and identity of tyrosines and tryptophans. It is clear that the demarcation between the charged and hydrophobic residues on the cytoplasmic surface defines the same membrane–aqueous boundary shown in Fig. 16. With a fixed bilayer thickness of 40 Å, the intradiscal membrane–aqueous interface is also coincident with a plane that separates the charged, primarily anionic, from the hydrophobic residues.

In general, the aromatic residues tryptophan and tyrosine on the exposed surfaces of membrane proteins have a high propensity for the membrane–aqueous interface (Reithmeier, 1995; Yau *et al.*, 1998). Of the 10 solvent-accessible ($f_{SA} \geq 0.15$) tryptophan and tyrosine residues in rhodopsin, 8 are located within the membrane–aqueous interfaces predicted from SDSL and bilayer thickness data. Finally, the arrangement of rhodopsin and the bilayer shown in Fig. 17 places the palmitoyl groups at C322 and C323 (shown in green in Fig. 17B) in an ideal position to be solvated by the bilayer.

An unusual feature of the disk membrane system of the rod outer segment is the very small and uniform spacing between the surfaces of the membrane in the intradiscal space (Fig. 1A). This is of interest because the surface potential of the intradiscal surface is ≈ -40 mV at physiological ionic strengths (Tsui *et al.*, 1990), and the equilibrium spacing between the membrane surfaces is only ≈ 20 Å (Chabre and Cavaggioni, 1975), on the order of twice the double layer thickness. What attractive forces overcome the significant electrostatic repulsion to determine this close spacing? The structure of rhodopsin provides a suggestion. Figure 18 (see color insert) shows a model of the disk membrane based on the crystal structure of rhodopsin. A prominent feature of the intradiscal surface of rhodopsin is a finger-like loop projecting toward the opposite membrane surface. Prolines 12 and 23 flank the loop, which contains two positively charged residues (K16, R21) that point toward the opposing membrane surface, and a number of hydrophobic residues (F13, V19, V20). This sequence has among the lowest *B* factors in the molecule, and is thus presumed to be a relatively rigid feature. The projection of the loop from the membrane–aqueous interface is ≈ 20 Å, the same as the disk membrane spacing. Thus, the spacing of the disk may be determined by interaction of the loop with the opposing membrane, as shown in Fig. 18A, possibly involving the interaction of the positive charges with phosphatidylserine.

An alternative and more interesting model is the formation of rhodopsin dimers across the disk membranes, as shown in Fig. 18B. The putative tail-to-tail dimer interface is formed by side-by-side stacking of the $\beta 1$ and $\beta 2$ strands from each molecule to make a continuous antiparallel β-sheet. The Pro-16–Pro-23 loops from each monomer interdigitate and lay over one surface of the sheet. Intermolecular salt bridges and Ca^{2+} binding sites formed from pairs of carboxylate residues, one from each monomer, could aid in stabilizing the structure (not shown). It is interesting to note that the mutation of Pro-23, a key residue in forming the loop, gives rise to one type of autosomal dominant retinitis pigmentosa in which disk formation is abnormal (Liu *et al.,* 1997).

V. Photoactivated Conformational Changes: The Rhodopsin Activation Switch

A recent review has summarized a large body of experimental data, its relationship to the crystal structure, and what can be concluded about the mechanism of activation (Meng and Bourne, 2001). To date, assignment of specific structural changes in rhodopsin on photoactivation is primarily based on SDSL analysis using the set of spin-labeled mutants described above. Structural changes can be detected either by changes in mobility of R1 residues at tertiary contact sites, or by direct changes in distances between pairs of R1 residues, and both have been extensively employed in mapping structural changes in rhodopsin. As discussed in Section V,A, photoactivation results in the movement of specific helical segments. The functional relevance of these movements has been assessed by interhelical cross-linking experiments, and these are summarized in Section V,B. Finally, Section V,C explores the structure–function relationships evident from a comparison of the crystal structure and the conformational changes believed to underlie activation.

A. *Analysis of Structural Changes from R1 Mobility and Interhelical Distances Measured by Spin–Spin Interactions*

For each of the spin-labeled sites throughout the cytoplasmic domain of rhodopsin, EPR spectra were recorded in the dark and after photoactivation to the metarhodopsin II (MII) state, the state of rhodopsin competent for activation of transducin (Hofmann, 2000). Changes in the structure of the protein are reflected by changes in the mobility of the R1 side chains, which provide a map of the tertiary contact surfaces that rearrange during the transition to the activated form.

The most dramatic changes, illustrated, for example, in Fig. 19, are detected in and around TM6, and have been interpreted in terms of a rigid-body tilt of TM6 outward from the center of the protein (Altenbach *et al.,* 1996; Hubbell *et al.,* 2000). For example, residues 250R1 and 251R1 on the buried surface of TM6 dramatically *increase* in mobility, reflecting a decrease in tertiary contact interaction, while residues 227R1 and 231R1 on TM5 experience a *decrease* in mobility as they are engaged by the outward movement of TM6. On the other hand, residue 252R1 on

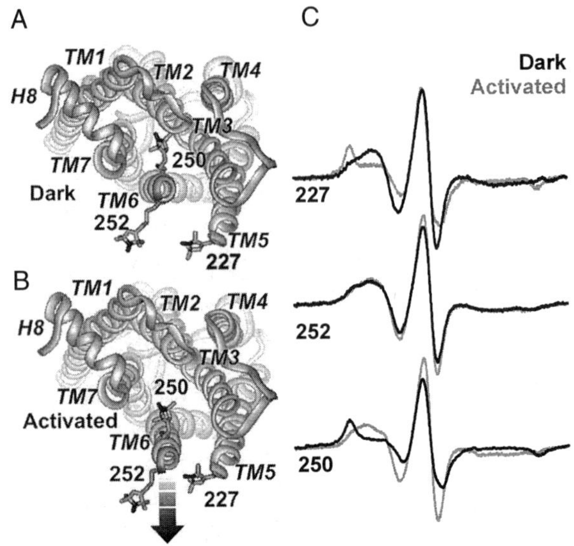

FIG. 19. Changes in R1 mobility in and around TM6 due to photoactivation. (A) A ribbon model of rhodopsin showing the location of 227R1, 250R1, and 252R1 in the dark state. (B) After photoactivation, showing the movement of TM6. (C) The corresponding EPR spectra in the dark (heavy trace) and after photoactivation (light trace).

the solvent-exposed surface of TM6 remains unchanged, demonstrating that the secondary structure of TM6 remains intact and that the motion is a rigid body motion of the helix.

Figure 20 (see color insert) shows a summary of the mobility changes detected by R1 sensors throughout the cytoplasmic surface of rhodopsin and their interpretation in terms of helical motions based on arguments similar to those given above for TM6 (Resek *et al.*, 1993; Farahbakhsh *et al.*, 1993, 1995; Yang *et al.*, 1996b; Altenbach *et al.*, 1999b). No significant changes were observed for residues not shown. The increases in mobility for R1 at the buried surfaces of TM1, TM2, TM3, TM6, and TM7 indicate reduced packing in the core of the protein on photoisomerization of retinal. In the case of residues that face TM6 (72R1 in TM2 and 136R1, 139R1 in TM3) the mobility increases could be coupled to the dominant outward movement of TM6. However, the concomitant decrease in mobility of residues on the outer surface of TM2 (70R1, 74R1), and residues in C2 that point toward TM2 (147R1, 148R1), suggests a displacement of TM2 toward TM4. The subtle mobility increases in 140R1 further suggest a rearrangement in structure at the cytoplasmic surface of TM3, near the highly conserved sequence 134E/135R/136Y. Finally, decreases in mobility at 313R1 in H8 have been linked to a possible movement of the segment of TM7 above Pro-303 (see below). The origin of the decrease in mobility of 316R1 is not clear, but could be due to a movement a H8 about a fulcrum located near 316 (such a motion of H8 would be coupled to the movement of TM7).

Collectively, the data suggest movement of the TM segments as suggested in Fig. 20 by arrows. All the data leading to this model were collected for rhodopsin in micellar solutions of DM. The mobility changes for R1 residues in TM6 and TM3 have been confirmed for rhodopsin reconstituted in lipid bilayers, but the changes at 316R1 are greatly reduced in magnitude (N.-J. Yu, K. Cai, H. G. Khorana, and W. Hubbell, unpublished). Thus, the dominant movement of TM6 and smaller movements in and around TM3 on photoactivation are preserved in the native membrane, but the environment may modulate the magnitude of the changes at 316, and possibly other sites.

The mobility changes shown in Fig. 20 clearly identify the type (rigid body helix motion) and direction of conformational changes in rhodopsin during activation, but cannot as yet provide quantitative measures of the magnitude of the helix movements. For this purpose, direct distance measurements between multiple pairs of R1 residues in helical segments were made before and after photoactivation. Helix pairs TM3/TM6 (Farrens *et al.*, 1996), H8/TM1 (Altenbach *et al.*, 2001b,c), and H8/TM2 and TM1/TM7 (Altenbach *et al.*, 2001c) have been investigated in detail. For example, Fig. 21 shows results of experiments to map distance changes between residue 139R1 in TM3 and a second residue in the sequence 247–251 in TM6. In the dark, the interspin distance between 139R1 and each of the other residues was consistent with the rhodopsin crystal structure (Fig. 21A). On light activation, there was a striking pattern of distance changes reflecting an outward motion of TM6 relative to TM3 of ≈8 Å to account for the decrease in distance between 139R1 and 250R1 (Fig. 21B). This represents the largest distance change observed and corresponds to the motion of TM6 inferred from changes in R1 mobility (Fig. 19). Similar studies were carried out with the double mutants shown in Fig. 9 and 14 to map relative movements among helices TM1, TM2, H8, and TM7 due to photoactivation, and the results support the movements summarized in Fig. 20, and provided the displacement amplitudes shown in the figure. Additional evidence for the movement of the C-terminal segment of TM7 away from the core of the protein is provided by the fact that a peptide segement near the TM7–H8 turn is only accessible to an antibody after photoactivation (Abdulaev and Ridge, 1998). The low-amplitude structural fluctuations in the TM7-H8 turn inferred from the reactivity of 310C and 313C also

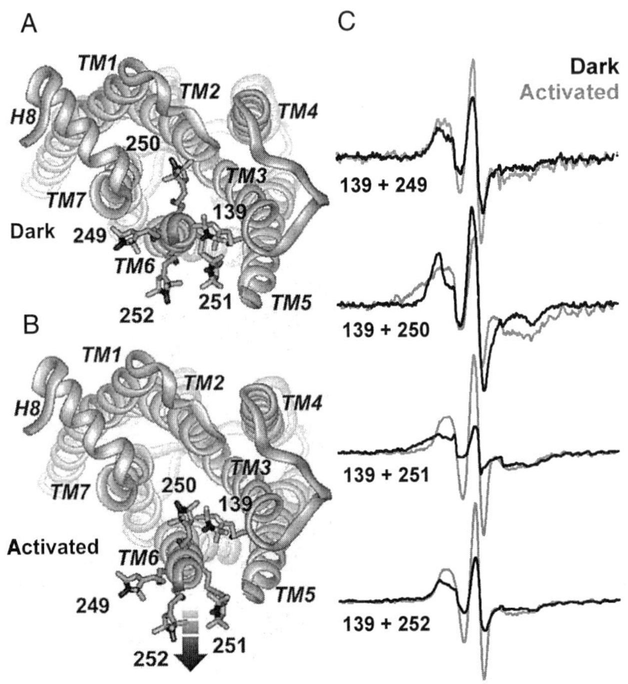

FIG. 21. Changes in distance between R1 residues in TM3 and TM6. (A) Ribbon model of rhodopsin in the dark state showing the location of R1 side chains. For each distance measurement, only two R1 side chains were in the protein, one fixed at the reference site 139 in TM3, and the other at a site in the sequence 248–252. (B) After photoactivation, showing the movement of TM6 and the new positions of the pairs. (C) The corresponding EPR spectra of the double mutants in the dark (solid traces) and after photoactivation (light traces). A decrease in spectral amplitude represents a decrease in interspin distance.

point to a flexible structure whose role may be to support the proposed movement of TM7.

This activated MII state is transient, and following the spontaneous decay of MII to opsin (the apoprotein), the receptor is once again in an inactive state, similar to rhodopsin itself. Interestingly, light-induced conformational changes detected by spin labels in the native membrane also reverse on the time scale of MII decay (Farahbakhsh *et al.*, 1993). The opsin molecule produced by the decay of MII is thermodynamically less stable than rhodopsin, as expected from the loss of the chromophore binding energy. This is reflected in a dramatically increased flexibility in the molecule. For example, only in the opsin state are cross-links formed between native residues C140 and C316, whose α-carbon atoms are nearly 30 Å apart (Yu and Oprian, 1999). Moreover, the structure of opsin, but not rhodopsin, is strongly coupled to the thickness of the lipid bilayer environment (Farahbakhsh *et al.*, 1992).

B. Interhelical Cross-Linking Experiments and the Functional Relevance of the Helical Movements

Are the helical movements detected by the spin labels functionally important? If so, a molecular cross-link between the helix in question and an adjacent helix that does not move or moves in a different direction should block activation of transducin. To implement this strategy to identify functionally relevant helix motions, two different kinds of interhelical cross-links have been employed. In the first, helices are cross-linked by spontaneous or catalyzed disulfide formation between two cysteines, one engineered in each helix. In the second, cross-links are introduced by binding of Zn^{2+} to engineered histidine ligands, one in each helix (Elling *et al.*, 1995).

Twelve cross-links designed to explore the role of helix movement at the cytoplasmic surface of rhodopsin are shown in Fig. 22. An important conclusion from these studies is that any interhelical cross-link involving TM6 blocks activation. These include the five disulfide cross-links between 139C in TM3 and a second cysteine in the sequence 247C–251C in TM6 (Farrens *et al.*, 1996; Cai *et al.*, 1999b), the cross-link between

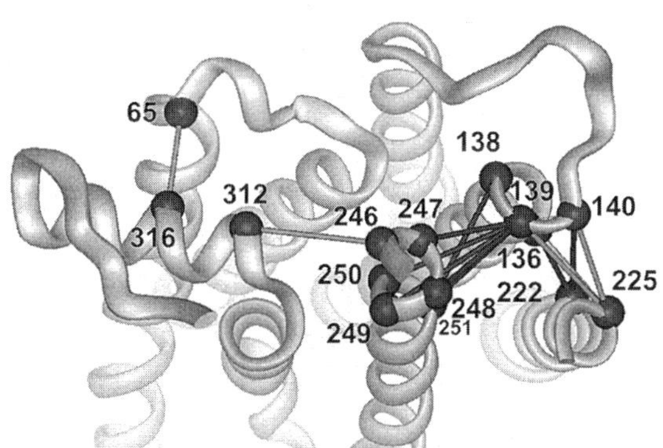

FIG. 22. Cross-links introduced into rhodopsin to test role of helix movement in activation. Of the 12 cross-links shown, 11 are disulfide bonds produced by oxidation of engineered cysteine pairs. The pairs are 139/247(i), 139/248(i), 139/249(i), 139/250(i), 139/251(i), 65/316(p), 136/222(i), 136/225(p), 140-222(i), 140/225(p). One cross-link, 138/251(i) was produced by Zn^{2+} binding (see text). The letter in parentheses following each pair identifies the cross-link as inhibitory (i) or permissive (p) with respect to transducin activation. Inhibitory bonds are shown in black, and permissive in gray.

312C in H8 and 246C in TM6 (Cai *et al.*, 2001), and the Zn^{2+} cross-link between 138H in TM3 and 251H in TM6 (Sheikh *et al.*, 1996). These results strongly support the notion that movement of TM6 away from both TM3 and H8, as suggested in Fig. 20, is required for activation. In contrast, disulfide cross-links between residues 242 or 245 in TM6 and 338 in the C-terminal tail have little effect on transducin activation (Cai *et al.*, 2001). As discussed above, the formation of these disulfide bonds is incompatible with the crystal structure, but is consistent with the highly dynamic nature of the C-terminal domain found in solution. Because of the high mobility of the C terminus, the cross-link with residue 242 or 245 on the outer surface of TM6 would not inhibit its movement during activation or interaction of transducin with residues on the internal surface.

Interestingly, disulfide bonds formed between residue 276C in TM6 and residue 198, 200, or 204 in TM5 on the extracellular surface were permissive with respect to transducin activation, indicating a lack of motion of TM6 relative to TM5 on this surface (Struthers *et al.*, 1999, 2000). As will be discussed below, this important result is consistent with the proposal that TM6 moves about a hinge point at Pro-267, so cross-links on one side of the hinge have no effect on motion.

Of five disulfide cross-links formed between residues in the cytoplasmic ends of TM3 and TM5 (136C/222C, 136C/225C, 139C/225C, 140C/222C, 140C/225C), the two involving 222C were inhibitory and the others permissive with respect to transducin activation (Yu and Oprian, 1999). The disulfides that would produce the largest distortions in the structure on formation are in fact permissive and the smallest distortions are inhibitory. Thus, the mixed results cannot be explained in terms of protein distortion, and all that can be concluded is that large-scale relative movements of TM3 and TM5 are not involved in rhodopsin activation. Finally, a cross-link between 316C in H8 and 65C in TM1 is permissive, and it is concluded that a relative movement between these residues is unimportant for activation of the receptor. However, 316C is at the contact point between H8 and TM1 and could be at a fulcrum about which H8 rocks relative to TM1.

Other evidence that the helical motions illustrated in Fig. 20 are related to activation comes from both kinetic studies and structural changes produced by constitutively active mutations. An important feature of SDSL is that conformational changes sensed either by side-chain mobility or changes in interspin distance can be monitored in real time in the millisecond domain. An early SDSL kinetic investigation reported that the light-activated structural change in TM3 sensed by R1 mobility changes had the same rate constant and activation energy as the appearance of MII (Farahbakhsh *et al.*, 1993). Moreover, the conformational change reversed with a decay time that matched the decay of MII. Similar experiments have not yet been carried out for the important movement of TM6.

The apoprotein opsin activates transducin with an efficiency 10^6 times less than that of MII (Melia *et al.,* 1997). However, a number of mutations have been found that constitutively activate the opsin to various degrees (Rao and Oprian, 1996; Han *et al.,* 1998). The structural origin of constitutive activation has been explored with the use of "sensor" R1 side chains placed either at site 140, 227, 250 or 316. As indicated in Fig. 20, R1 residues at these sites serve to monitor the status of the helix bundle and report movement of TM3 (140), TM6 (227, 250) or H8 (316). These sensor mutants were employed to monitor the response of the rhodopsin helix bundle to various mutations that constitutively activate the protein. The remarkable conclusion of these studies was that constitutively active mutations trigger the movement of specific helical segments in the dark state that closely resemble the corresponding movements due to photoactivation, supporting the functional importance of the helix movements (Kim *et al.,* 1997).

C. *Structure–Function Relationships*

As described in the previous sections, cysteine scanning mutagenesis and the associated techniques of SDSL, sulfhydryl reactivity, and disulfide cross-linking rates have provided a rather detailed view of rhodopsin dynamics in solution and conformational changes leading to the activated state. In this section, the structural origins of these functional properties in solution are examined from the point of view of the crystal structure.

As shown in Fig. 4, the crystallographic B factors for C3 are the highest in the structure. This disorder in the crystal is clearly mirrored by the dynamics in solution, as shown in the scaled mobility map of Fig. 15A. In both the crystal structure and the solution structure inferred from SDSL (Fig. 11A), residues in C3 make few tertiary contacts, a fact that accounts for the disorder/dynamics. What is the functional advantage for the dynamic flexibility of C3? One answer is that the large-scale movement of TM6 on photoactivation (Fig. 20) requires that C3 be flexible. C3 contains numerous sites that are apparently involved in interaction/activation of transducin (Fig. 1B). The interaction of the transducin α subunit with activated rhodopsin must be specific but not too strong, because, following GDP/GTP exchange, the subunit must dissociate from the receptor (Fig. 2). A high specificity can be achieved by requiring multiple attractive interactions, but this would lead to an enthalpic contribution favoring a high binding constant. However, the binding free energy, and hence the equilibrium constant, could be limited by a large entropic loss that would occur upon transducin binding to the flexible backbone of C3.

In the crystal structure of rhodopsin (1HZX, A chain), the C-terminal tail is modeled, but the *B* factors are high. Arguments against this structure in the solution state were given above, and all evidence favors a dynamically disordered state in solution from 328 to 348, with little change produced by photoactivation (Langen *et al.,* 1999). This sequence of the C-terminal tail is of functional importance because it contains residues S334, S338, and S343, the major sites of phosphorylation of rhodopsin by rhodopsin kinase during the deactivation phase of transduction (Ohguro *et al.,* 1994) (Fig. 1B). The high mobility of the C-terminal phosphorylation domain is similar to other substrates for protein kinases where the high mobility of the region surrounding the site of phosphorylation is likely to be important in recognition and affinity (Johnson *et al.,* 1998).

The extracellular domain of rhodopsin that defines the chromophore binding site has the lowest *B* factors in the structure, is tightly packed, and is characterized by numerous stabilizing interactions, many of which are between the chromophore, the transmembrane helices, and beta strands forming the floor of the binding cavity (Palczewski *et al.,* 2000; Teller *et al.,* 2001). On the other hand, the helix bundle at the cytoplasmic domain is stabilized by weak van der Waals and hydrophobic interactions, the interhelical links are dynamic in solution and disordered in the crystal structure, and the domain undergoes structural rearrangements upon photoactivation. This general picture is in keeping with the proposal of Khorana that the extracellular surface serves as a template for the folding of the protein, and mutations in this region of the molecule often lead to misfolded protein (Hwa *et al.,* 1997; Garriga *et al.,* 1996; Liu *et al.,* 1996).

The most dramatic change resolved by SDSL in the cytoplasmic domain due to photoactivation is the movement of TM6, with relatively smaller movements in the rest of the structure. It seems surprising that one helix of a bundle can move without destabilizing the entire structure. The structure of rhodopsin provides clues to how this may come about. First, TM6 makes fewer polar interactions with its neighbors than other helices. As discussed previously, there are networks of hydrogen-bonding interactions that connect TM1–TM5 and TM7, but TM6 is connected only to TM7 by only two side chain/backbone hydrogen bonds (Teller *et al.,* 2001). Second, TM6 is unique in its poor packing within the structure. Figure 23A (see color insert) shows cavities within the rhodopsin molecule identified with a probe of 1.4 Å (Teller *et al.,* 2001).

The largest cavity (145 Å^3) lies between TM3 and TM6 at the level of residue 258 in TM6. In addition, there is a cavity at the interface between TM6 and TM5, reflecting loose packing between TM6 and this nearest neighbor. This latter point is emphasized in Fig. 23B, which shows a view of rhodopsin facing the interface between TM5 and TM6. In this figure, the helices behind TM5 and TM6 (TM3, TM4) were removed to show the free volume at the interface. Between P267 and the cytoplasmic termination of TM6 the helix packing is less than optimal. However, between P267 and the extracellular surface, TM6 is tightly packed, and the highly conserved P267 is the likely hinge site for the rigid body tilting motion of TM6 observed in SDSL studies reviewed above (see Figs. 3, 23A, and 23B). Indeed, suitable rotations about the P267 backbone dihedral angles moves TM6 along a trajectory that accounts for the mobility changes at 250R1 and 227R1 (this is how the tilted helix in Fig. 21 was generated). This model is supported by the fact that disulfide cross-links involving TM6 sites between P267 and the cytoplasmic surface are inhibitory with respect to transducin activation, whereas those between P267 and the extracellular surface are permissive (Struthers *et al.*, 1999, 2000; Yu *et al.*, 1995). Other smaller helix motions observed in TM1–TM3 and TM7 could simply be the response of the helix bundle to the movement of TM6, or they may be associated with more fundamental events responsible for triggering the movement of TM6 itself.

VI. Summary: The Mechanism of Rhodopsin Activation and Future Directions

The focus of this review has been on rhodopsin structure and dynamics at the cytoplasmic face in solution, the comparison between the solution structure and the crystal structure, and the structural changes underlying receptor activation. The data from SDSL and disulfide cross-linking together indicate that the crystal and solution structures are very similar at the level of the backbone fold for C1, H8, and the cytoplasmic termination of TM1–TM7. However, substantial differences are seen in C3 and the C-terminal tail, wherein the backbones are flexible on the nanosecond time scale in solution. Not surprisingly, these are the most disordered regions in the crystal lattice and correspond to sequences important for function.

Two independent strategies, modulation of tertiary contact interaction and direct distance mapping, provide compelling evidence for a dominant rigid-body movement of TM6 near the cytoplasmic surface, with smaller but significant movements of TM1, TM2, TM3, and TM7.

The large-amplitude motion of TM6 is enabled by the highly flexible C3 link that connects TM5 and TM6 at the cytoplasmic surface. There is now abundant evidence for a similar movement of TM6 in other GPCRs (Gether, 2000), and it may be a "central dogma" of GPCR receptor activation as proposed by Khorana (2000). However, it is possible that additional induced structural changes may occur on binding of transducin.

The pattern of helix motions observed experimentally suggests a simple model for the activated state in which helices on opposite sides of the chromophore binding pocket move outward to open a cleft in the molecule (Fig. 20). In the prose of Henry Bourne, the molecule has "blossomed" (Meng and Bourne, 2001) to expose new surfaces for coupling of rhodopsin with transducin.

How are these helix movements coupled to chromophore isomerization? A detailed mechanistic answer is unknown, but a general view is that geometric isomerization results in unavoidable repulsive (steric) interactions with the protein, the relaxation of which removes a set of constraints on TM6 (and others), allowing for the thermal motions necessary to form the active state (for review, see Hofmann, 2000; Arnis and Hofmann, 1993). The challenge ahead is to identify the specific steric interactions between groups on the chromophore and protein, and how they guide the protein along the activation pathway to remove the necessary constraints, which also remain to be identified. It is already clear that the ionone ring and the 9-methyl group of the chromophore participate in critical steric interactions with the protein (Jager *et al.,* 1994; Borhan *et al.,* 2000; Meyer *et al.,* 2000; Vogel *et al.,* 2000), and that the salt bridge between the protonated Schiff base of the retinal at K296 (in TM7) and the counterion E113 (in TM3) is a critical constraint that must be removed to form the activated state of the protein (Rao and Oprian, 1996). Judging from the results reviewed above, systematic application of time-resolved SDSL to monitor helix movements, together with the use of chemically modified chromophore structures, is a promising approach to exploring the relationship between chromophore structure, the critical salt bridge, and helix movements leading to activation.

References

Abdulaev, N. G., and Ridge, K. D. (1998). *Proc. Natl. Acad. Sci. USA* **95,** 12854–12859.

Altenbach, C., Greenhalgh, D., Khorana, H. G., and Hubbell, W. L. (1994). *Proc. Natl. Acad. Sci. USA* **91,** 1667–1671.

Altenbach, C., Yang, K., Farrens, D. L., Farahbakhsh, Z., Khorana, H. G., and Hubbell, W. L. (1996). *Biochemistry* **35,** 12470–12478.

Altenbach, C., Cai, K., Khorana, H. G., and Hubbell, W. L. (1999a). *Biochemistry* **38,** 7931–7937.

Altenbach, C., Klein-Seetharaman, J., Hwa, J., Khorana, H. G., and Hubbell, W. L. (1999b). *Biochemistry* **38,** 7945–7949.

Altenbach, C., Oh, K.-J., Trabanino, R. J., Hideg, K., and Hubbell, W. L. (2001a). *Biochemistry* **40,** 15471–15482.

Altenbach, C., Cai, K., Klein-Seetharaman, J., Khorana, H. G., and Hubbell, W. L. (2001b). *Biochemistry* **40,** 15483–15492.

Altenbach, C., Klein-Seetharaman, J., Cai, K., Khorana, H. G., and Hubbell, W. L. (2001c). *Biochemistry* **40,** 15493–15500.

Arnis, S., and Hofmann, K. P. (1993). *Proc. Natl. Acad. Sci. USA* **90,** 7849–7853.

Baldwin, J. M., Schertler, G. F. X., and Unger, V. M. (1997). *J. Mol. Biol.* **272,** 144–164.

Baylor, D. A. (1987). *Invest. Ophthalmol. Vis. Sci.* **28,** 34–49.

Borhan, B., Souto, M. L., Imai, H., Shichida, Y., and Nakanishi, K. (2000). *Science* **288,** 2209–2212.

Bracken, C., Carr, P. A., Cavanagh, J., and Palmer, A. G. (1999). *J. Mol. Biol.* **285,** 2133–2146.

Budil, D. E., Lee, S., Saxena, S., and Freed, J. H. (1996). *J. Mag. Res. (A)* **120,** 155–189.

Bunemann, M., and Hosey, M. M. (1999). *J. Physiol.* **517,** 5–23.

Burns, M. E., and Baylor, D. A. (2001). *Annu. Rev. Neurosci.* **24,** 779–805.

Cai, K., Langen, R., Hubbell, W. L., and Khorana, H. G. (1997). *Proc. Natl. Acad. Sci. USA* **94,** 14267–14272.

Cai, K., Klein-Seetharaman, J., Farrens, D., Zhang, C., Altenbach, C., Hubbell, W. L., and Khorana, H. G. (1999a). *Biochemistry* **38,** 7925–30.

Cai, K., Klein-Seetharaman, J., Hwa, J., Hubbell, W. L., and Khorana, H. G. (1999b). *Biochemistry* **38,** 12893–12898.

Cai, K., Klein-Seetharaman, J., Altenbach, C., Hubbell, W. L., and Khorana, H. G. (2001). *Biochemistry* **40,** 12479–12485.

Careaga, C. L., and Falke, J. J. (1992). *J. Mol. Biol.* **226,** 1219–1235.

Chabre, M., and Cavaggioni, A. (1975). *Biochim. Biophys. Acta* **382,** 336–343.

Chen, Y., and Hubbell, W. L. (1978). *Membrane Biochem.* **1,** 107–129.

Columbus, L., Kalai, T., Jeko, J., Hideg, K., and Hubbell, W. L. (2001). *Biochemistry* **40,** 3828–3846.

Crump, M. P., Spyracopoulos, L., Lavigne, P., Kim, K. S., Clark-Lewis, I., and Sykes, B. D. (1999). *Protein Sci.* **8,** 2041–2054.

Dartnall, H. J. A. (1972). In: *Photochemistry of Vision* (H. J. A. Dartnall, Ed.), p. 122. Springer Verlag, New York.

Duggan, B. M., Dyson, H. J., and Wright, P. E. (1999). *Eur. J. Biochem.* **265,** 539–548.

Eaton, G. R., Eaton, S. S., and Berliner, L. J. Eds. (2001). *Biological Magnetic Resonance 19: Distance Measurements in Biological Systems by EPR.* Kluwer Academic, New York.

Elling, C. E., Nielsen, S. M., and Schwartz, T. W. (1995). *Nature* **374,** 74–77.

Falke, J. J., and Koshland, D. E., Jr. (1987). *Science* **237,** 1596–1600.

Farahbakhsh, Z., Altenbach, C., and Hubbell, W. L. (1992). *Photochem. Photobiol* **56,** 1019–1033.

Farahbakhsh, Z., Hideg, K., and Hubbell, W. L. (1993). *Science* **262,** 1416–1419.

Farahbakhash, Z., Ridge, K., Khorana, H. G., and Hubbell, W. L. (1995). *Biochemistry* **34,** 8812–8819.

Farrens, D. L., Altenbach, C., Yang, K., Hubbell, W. L., and Khorana, H. G. (1996). *Science* **274,** 768–770.

Fein, A., and Szuts, E. Z. (1982). *Photoreceptors: Their Role in Vision.* Cambridge University Press, Cambridge, UK.

Feix, J., and Klug, C. S. (1998). In: *Biological Magnetic Resonance 14: Spin Labeling: The Next Millenium* (L. J. Berliner, Ed.), pp. 251–281. Kluwer, New York.

Ferretti, K., Karnik, S. S., Khorana, H. G., Nassal, M. M., and Oprian, D. D. (1988). *Proc. Natl. Acad. Sci. USA* **83,** 599–603.

Franke, R. R., Konig, B., Sakmar, T. P., Khorana, H. G., and Hofmann, K. P. (1990). *Science* **250,** 123–125.

Franke, R. R., Sakmar, T. P., Graham, R. M., and Khorana, H. G. (1992). *J. Biol. Chem.* **267,** 14767–14774.

Garriga, P., Liu, X., and Khorana, H. G. (1996). *Proc. Natl. Acad. Sci. USA* **93,** 4560–4564.

Gether, U. (2000). *Endocr. Rev.* **21,** 90–113.

Grassetti, D. R., and Murray, J. F., Jr. (1967). *Arch. Biochem. Biophys.* **119,** 41–49.

Hamm, H. E. (1998). *J. Biol. Chem.* **273,** 669–672.

Hamm, H. E. (2001). *Proc. Natl. Acad. Sci. USA* **98,** 4819–4821.

Han, M., Smith, S. O., and Sakmar, T. P. (1998). *Biochemistry* **37,** 8253–8261.

Hargrave, P. A. (2001). *Invest. Ophthalmol. Vis. Sci.* **42,** 3–9.

Hargrave, P. A., McDowell, J. H., Curtis, D. R., Wang, J. K., Juszczak, E., Fong, S. L., Rao, J. K., and Argos, P. (1983). *Biophys. Struct. Mech.* **9,** 235–244.

Helmreich, E. J., and Hofmann, K. P. (1996). *Biochim. Biophys. Acta* **1286,** 285–322.

Hirsch, J. A., Schubert, C., Gurevich, V. V., and Sigler, P. B. (1999). *Cell* **97,** 257–269.

Hofmann, K. P. (2000). In: *Handbook of Biological Physics 3: Molecular Mechanisms in Visual Signal Transduction* (A. J. Hoff, D. G. Stavenga, W. J. de Grip, and E. N. Pugh, Eds.), pp. 91–142. Elsevier, Amsterdam.

Honig, B., and Hubbell, W. L. (1984). *Proc. Natl. Acad. Sci. USA* **81,** 5412–5416.

Hubbell, W. L., and Altenbach, C. (1994). *Curr. Opin. Struct. Biol.* **4,** 566–573.

Hubbell, W. L., Mchaourab, H. S., Altenbach, C., and Lietzow, M. (1996). *Structure* **4,** 779–782.

Hubbell, W. L., Gross, A., Langen, R., and Lietzow, M. (1998). *Curr. Opin. Struct. Biol.* **8,** 649–656.

Hubbell, W. L., Cafiso, D. S., and Altenbach, C. (2000). *Nat. Struct. Biol.* **7,** 735–739.

Hwa, J., Garriga, P., Liu, X., and Khorana, H. G. (1997). *Proc. Natl. Acad. Sci. USA* **94,** 10571–10576.

Isas, J. M., Langen, R., Haigler, H. T., and Hubbell, W. L. (2002). *Biochemistry* **41,** 1464–1473.

Jager, F., Jager, S., Krautle, O., Friedman, N., Sheves, M., Hofmann, K. P., and Siebert, F. (1994). *Biochemistry* **33,** 7389–7397.

Johnson, L. N., Lowe, E. D., Noble, M. E. M., and Owen, D. J. (1998). *FEBS Lett.* **430,** 1–11.

Katagadda, M., Chopra, A., Bennett, M., Alderfer, J. L., Yeagle, P. L., and Albert, A. D. (2001). *J. Peptide Res.* **58,** 79–89.

Kaushal, S., Ridge, K. D., and Khorana, H. G. (1994). *Proc. Natl. Acad. Sci. USA* **91,** 4024–4028.

Khorana, H. G. (2000). *J. Biomol. Struct. Dynamics* **11,** 1–16.

Kim, J. M., Altenbach, C., Thurmond, R. L., Khorana, H. G., and Hubbell, W. L. (1997). *Proc. Natl. Acad. Sci. USA* **94,** 14273–14278.

Klein-Seetharaman, J., Hwa, J., Cai, K., Altenbach, C., Hubbell, W. L., and Khorana, H. G. (1999). *Biochemistry* **38,** 7938–7944.

Klein-Seetharaman, J., Hwa, J., Cai, K., Altenbach, C., Hubbell, W. L., and Khorana, H. G. (2001). *Biochemistry* **40,** 12472–12478.

Koing, B., Arendt, A., Mcdowell, J. H., Kahlert, M., Hargrave, P. A., and Hofmann, K. P. (1989). *Proc. Natl. Acad. Sci. USA* **86,** 6878–6882.

Kosen, P. A. (1992). In: *Stability of Protein Pharmaceuticals, Part A* (T. J. Ahern and M. C. Manning, Eds.), Plenum Press, New York.

Krupnick, J. G., and Benovic, J. L. (1998). *Annu. Rev. Pharmacol. Toxicol.* **38,** 289–319.

Lambright, D. G., Noel, J. P., Hamm, H. E., and Sigler, P. B. (1994). *Nature* **369,** 621–628.

Lambright, D. G., Sondek, J., Bohm, A., Skiba, N. P., Hamm, H. E., and Sigler, P. B. (1996). *Nature* **379,** 311–319.

Langen, R., Cai, K., Altenbach, C., Khorana, H. G., and Hubbell, W. L. (1999). *Biochemistry* **38,** 7918–7924.

Langen, R., Oh, K. J., Cascio, D., and Hubbell, W. L. (2000). *Biochemistry* **39,** 8396–8405.

Liu, X., Garriga, P., and Khorana, H. G. (1996). *Proc. Natl. Acad. Sci. USA* **93,** 4554–4559.

Liu, X., Wu, T. H., Stowe, S., Matsushita, A., Arikawa, K., Naash, M. I., and Williams, D. S. (1997). *J. Cell Sci.* **110,** 2589–2597.

Marin, E. P., Khrishna, A. G., Zvyaga, T. A., Isele, J., Siebert, F., and Sakmar, T. P. (2000). *J. Biol. Chem.* **275,** 1930–1936.

Marsh, D., and Horvath, L. I. (1989). In: *Advanced EPR Applications in Biology and Biochemistry* (A. J. Hoff, Ed.), pp. 702–752. Elsevier, Amsterdam.

Mchaourab, H. S., Lietzow, M. A., Hideg, K., and Hubbell, W. L. (1996). *Biochemistry* **35,** 7692–7704.

Melia, T. J. Jr., Cowan, C. W., Angleson, J. K., and Wensel, T. G. (1997). *Biophys. J.* **73,** 3182–3191.

Meng, E. C., and Bourne, H. R. (2001). *Trends Pharmacol. Sci.* **22,** 587–593.

Menon, S. T., Han, M., and Sakmar, T. P. (2001). *Physiol. Rev.* **81,** 1659–1688.

Meyer, C. K., Bohme, M., Ockenfels, A., Gartner, W., Hofmann, K. P., and Ernst, O. P. (2000). *J. Biol. Chem.* **275,** 19713–19718.

Nathans, J. (1990). *Biochemistry* **29,** 9746–9752.

Noel, J. P., Hamm, H. E., and Sigler, P. B. (1993). *Nature* **366,** 654–663.

Oh, K. J., Zhan, H., Cui, C., Hideg, K., Collier, R. J., and Hubbell, W. L. (1996). *Science* **273,** 810–812.

Ohguro, H., Johnson, R. S., Ericsson, L. H., Walsh, K. A., and Palczewski, K. (1994). *Biochemistry* **33,** 1023–1028.

Okada, T., Ernst, O. P., Palczewski, K., and Hofman, K. P. (2001). *Trends Biochem. Sci.* **26,** 318–324.

Oprian, D. D., Molday, R. S., Kaufman, R. J., and Khorana, H. G. (1987). *Proc. Natl. Acad. Sci. USA* **84,** 8874–8878.

Ovchinnikov, Y. A. (1982). *FEBS Lett.* **148,** 179–191.

Palczewski, K., Kumasaka, T., Hori, T., Behnke, C. A., Motoshima, H., Fox, B. A., Le Trong, I., Teller, D. C., Okada, T., Stenkamp, R. E., Yamamoto, M., and Miyano, M. (2000). *Science* **289,** 739–745.

Peteanu, L. A., Schoelein, R. W., Wang, Q., Mathies, R. A., and Shank, C. V. (1993). *Proc. Natl. Acad. Sci. USA* **24,** 11762–11766.

Rao, V. R., and Oprian, D. D. (1996). *Annu. Rev. Biophys. Biomol. Struct.* **25,** 287–314.

Reeves, P. J., Klein-Seetharaman, J., Getmanova, E. V., Eilers, M., Loewen, M. C., Smith, S. O., and Khorana, H. G. (1999). *Biochem. Soc. Trans.* **27,** 950–955.

Reithmeier, R. A. (1995). *Curr. Opin. Struct. Biol.* **5,** 491–500.

Resek, J. F., Farahbakhsh, Z., Hubbell, W. L., and Khorana, H. G. (1993). *Biochemistry* **32,** 12025–12032.

Ridge, K., Zhang, C., and Khorana, H. G. (1995). *Biochemistry* **34,** 8804–8811.

Saiz, L., and Klein, M. L. (2001). *Biophys. J* **81,** 204–216.

Sakmar, T. P., Franke, R. R., and Khorana, H. G. (1989). *Proc. Natl. Acad. Sci. USA* **86,** 8309–8313.

Sheikh, S. P., Zvyaga, T. A., Lichtarge, O., Sakmar, T. P., and Bourne, H. R. (1996). *Nature* **383,** 347–350.

Struthers, M., Yu, H., Kono, M., and Oprian, D. D. (1999). *Biochemistry* **38,** 6597–6603.

Struthers, M., Yu, H., and Oprian, D. D. (2000). *Biochemistry* **39,** 7938–7942.

Teller, D. C., Okada, T., Behnke, C. A., Palczewski, K., and Stenkamp, R. E. (2001). *Biochemistry* **40,** 7761–7772.

Thomas, D. D., Dalton, L. R., and Hyde, J. S. (1976). *J. Chem. Phys.* **65,** 3006–3024.

Thurmond, R. L., Creuzenet, C., Reeves, P. J., and Khorana, H. G. (1997). *Proc. Natl. Acad. Sci. USA* **94,** 1715–1720.

Tsui, F. C., Sundberg, S. A., and Hubbell, W. L. (1990). *Biophys. J.* **57,** 85–97.

Unger, V. M., Hargrave, P. A., Baldwin, J. M., and Schertler, G. F. X. (1997). *Nature* **389,** 203–206.

Vogel, R., Gan, G.-B., Sheves, M., and Siebert, F. (2000). *Biochemistry* **39,** 8895–8908.

Voss, J., Salwinski, L., Kaback, H. R., and Hubbell, W. L. (1995). *Proc. Natl. Acad. Sci.* **92,** 12295–12299.

Yang, K., Farrens, D. L., Hubbell, W. L., and Khorana, H. G. (1996a). *Biochemistry* **35,** 12464–12469.

Yang, K., Farrens, D. L., Altenbach, C., Farahbakhsh, Z., Hubbell, W. L., and Khorana, H. G. (1996b). *Biochemistry* **35,** 14040–14046.

Yau, W. M., Wimley, W. C., Gawrisch, K., and White, S. H. (1998). *Biochemistry* **37,** 14713–14718.

Yeagle, P. L., Alderfer, J. L., and Albert, A. D. (1997). *Biochemistry* **36,** 9649–9654.

Yu, H., and Oprian, D. D. (1999). *Biochemistry* **38,** 12033–12040.

Yu, H., Kono, M., McKee, T. D., and Oprian, D. D. (1995). *Biochemistry* **34,** 14963–14969.

Zhukovsky, E. A., and Oprian, D. D. (1989). *Science* **246,** 928–930.

THE GLYCEROL FACILITATOR GlpF, ITS AQUAPORIN FAMILY OF CHANNELS, AND THEIR SELECTIVITY

By ROBERT M. STROUD, PETER NOLLERT,[1] and LARRY MIERCKE

Department of Biochemistry and Biophysics, School of Medicine, University of California, San Francisco, California 94143

I. An Ancient and Long Recognized Channel

A century ago Alfred Fischer lysed pathogenic bacteria by placing them in hyperosmotic solutions, and found that they did not undergo lysis when placed in hyperosmotic glycerol solutions (Fischer, 1903). They did not shrink as expected from the osmotic imbalance. He concluded correctly that they had to be very permeable to glycerol. The "glycerol facilitator" was identified in a brilliant series of studies pioneered by E. C. C. Lin and colleagues that began in the mid-1960s (Sanno *et al.*, 1968) and led to the conducting specificity of GlpF (Heller *et al.*, 1980), to cloning and sequencing of the gene (Sweet *et al.*, 1990), and to characterization of the 281 amino acid 29,780 Da GlpF protein (Weissenborn *et al.*, 1992). The seemingly incongruous genetic relatedness between GlpF, a soybean nodulin-26, and the major intrinsic protein (MIP) of the eye lens was identified (Baker and Saier, 1990) before the first functional

[1]Current affiliation: Emerald BioStructures, Bainbridge Island, WA 98110

ADVANCES IN PROTEIN CHEMISTRY, Vol. 63

0065-3233/03 $35.00

characterization of a water channel. Today they are all in the aquaporin channel family.

The "glycerol facilitator" GlpF, a highly selective transmembrane channel that conducts glycerol and certain other small uncharged organic molecules, was the first member of the large aquaporin family (AQP family) to be identified at the genetic level and functionally characterized (Eze and McElhaney, 1978). Once glycerol is passaged inside the cell, it is rapidly phosphorylated by glycerol kinase to produce glycerol 3-phosphate that, in turn, proceeds by dehydrogenation to dihydroxyacetone phosphate (DHAP), or to phospholipid synthesis where glycerol provides the basis for attachment of fatty acid chains and phophatidyl head groups in ~2/3 of cellular phospholipids. The glp regulon is inducible by glycerol 3-phosphate which provides an advantage for growth of *Escherichia coli* on glycerol. The glpF gene is the first gene on the glp operon that also contains the gene for glycerol kinase.

The lack of an energy-dependent process for glycerol accumulation was inferred by the lack of any ability to concentrate glycerol in kinase-negative mutants (Lin, 1976). The rate of GlpF-mediated glycerol influx is 100- to 1000-fold greater than expected for a transporter and is non-saturable to a glycerol concentration of >200 m*M* (Heller *et al.*, 1980), providing early evidence that it is a highly selective channel. In addition there is no counterflow of radioactive substrate against a concentration gradient, again indicating that GlpF acts as a selective channel (Maurel *et al.*, 1993).

GlpF not only conducts glycerol but also can conduct urea, glycine, and DL-glyceraldehyde, and is both stereo- and enantioselective in conductance of linear carbohydrates called alditols (Heller *et al.*, 1980). These results have been extended using highly purified glpF in reconstituted vesicles (Fu *et al.*, 2000). The cyclized alditols, then termed aldoses or sugars, are not conducted through the GlpF channel, and the structure of the GlpF channel shows that it is too small to conduct the cyclic molecules and beautifully explains aspects of the stereo- and enantioselectivity (Fu *et al.*, 2000; Nollert *et al.*, 2001). GlpF also conducts water at about 1/6 of the rate that the water-conducting homologue AQPZ does.

Though the similarity between GlpF, nodulin-26, and MIP had been recognized in 1990, it was not until the first water-conducting channel had been identified based on function (Preston and Agre, 1991) that the larger family had a functional definition as a superfamily of membrane channels that conduct water and other small uncharged molecules. The first cDNA of a water-conducting channel to be identified as a protein channel, a protein abundant in erythrocytes, and previously termed CHIP28 (abbreviated from channel-like integral membrane protein, a 28 kDa protein), was discovered by Agre and colleagues (Preston and

Agre, 1991). The family was subsequently given the functional name "aquaporins" (Agre *et al.,* 1993). Since then AQPs have been found in all species from bacteria to yeast, plants, and human. The sequences of now more than 200 AQP aquaporins have been determined. Thus, because GlpF is in this family, most of what can be said about the structure of GlpF will carry over directly to other AQPs and vice versa and can be seen in the context of its selectivity versus other family members.

The conservation of amino acid sequence between any two AQPs is typically in the range 28–32%; thus they are all constructed around a highly conserved structural backbone fold. The family arose by tandem intragenic duplication (Pao *et al.,* 1991) to give proteins that have an internal repeat where the N-terminal segment displays ~20% conservation with the C-terminal segment (Wistow *et al.,* 1991). This duplication event would seem to have appeared early in evolution since bacteria such as *Escherichia coli* contain both a water plus glycerol conducting channel (GlpF) and a separate water channel (AQPZ). Each segment contains a conserved -Asn-Pro-Ala- signature sequence (-NPA-) near its center, and several other conserved residues including a Glu near the beginning of each segment. While conservation between the two repeats suggests similarity of roles in a bidirectional conducting channel, the C-terminal segment plays a more dominant role in solute specificity and regulation (Reizer *et al.,* 1993).

The aquaporins (AQPs) are a large family of genetically divergent channels that conduct their substrates at close to the diffusion-limited rate. There are 11 members with key roles in human physiology numbered from AQP-0 to AQP-10. AQPs are composed of functionally distinct subgroups that include transmembrane water-conducting channels (aquaporins) and channels that, like GlpF, conduct glycerol as perhaps the most relevant physiological substrate in humans, but also conduct urea, DL-glyceraldehyde, linear polyalcohols, (called alditols), and certain other small organic molecules (Preston *et al.,* 1992; Maurel *et al.,* 1994; Park and Saier, 1996). In humans AQP-3, AQP-7, AQP-9, and AQP-10 fall into this subclass (Verkman *et al.,* 1995). The divergence between the subfamilies appears to have occurred ~2 billion years ago, and representative ancestors of both subclasses are found in *E. coli* (Heymann and Engel, 2000).

The major intrinsic protein (MIP) of the eye lens, once thought to play a primary role as a gap junction between cells of the eye lens, a biophysically characterized aqueous channel that formed arrays in the eye lens (Gorin *et al.,* 1984), was brought back to the fold as AQP-0. Many of the AQPs in eukaryotes are regulated by phosphorylation, pH, osmolarity, or binding of other proteins or ligands (Anthony *et al.,* 2000; Engel *et al.,* 2000).

AQPs are highly insulating to the electrochemical potential across the cell membrane by sharp selectivity that absolutely excludes conduction of all charged molecules, all ions including hydroxide and hydronium ions (Finkelstein, 1987), and protons. An important fundamental question is how this is accomplished, and structural data on GlpF and AQP-1 have given rise to suggested mechanisms for this (Mitsuoka *et al.*, 1999; Fu *et al.*, 2000). Molecular mechanics has been harnessed to identify, and then evaluate the contributions of each factor to this property (Tajkhorshid *et al.*, 2002).

There is only one recognized exception to the absolute insulation to ions found in human AQPs, where AQP-6 conducts anions. This adaptation is particularly interesting as a means of defining the basis for this selectivity. AQP-6 is activated by low pH, most probably to control its action in membranes of intracellular organelles (Yasui *et al.*, 1999).

GlpF, AQP-0, AQP-1, and (it is expected) all AQPs form tetramers of four independent water channels in membranes (Smith and Agre, 1991). GlpF remains a tetramer throughout purification and crystallization, and the structure of GlpF at 2.2 Å resolution shows that it crystallizes as a tetramer (Fu *et al.*, 2000). In reconstituted membranes these tetramers come together alternately cytoplasmic and extracellular side up. Electron crystallography of these reconstituted arrays (Cheng *et al.*, 1997; Li *et al.*, 1997; Walz *et al.*, 1997; Hasler *et al.*, 1998b; Ringler *et al.*, 1999) at increasing resolutions down to 4.5 Å in-plane ($4.5 \times \sim 9$ Å in 3D) resolution (Heymann and Engel, 2000) reveals a right-hand twisted arrangement of six, and two half-spanning α-helices that surround a central channel (Heymann and Engel, 2000). These pioneering structural investigations defined the relationship of the protein fold to the lipid bilayer and provided the first views of how the helices are folded around the central "hourglass"-like channel.

The structures of AQPs at increasing resolutions from the electron microscopic structures reconstituted into phospholipid-based membranes for GlpF (Braun *et al.*, 2000; Stahlberg *et al.*, 2000) and culminating in atomic structures from X-ray crystallography at a uniform 2.2 Å for detergent solubilized GlpF (Fu *et al.*, 2000) and for deglycosylated AQP-1 (Sui *et al.*, 2001) beautifully illustrate the basis for observations of their selectivity and rates of conductance. The atomic structure of GlpF also served as a template for molecular mechanics treatments of the theoretical basis for the rates of glycerol conductance in quantitative terms (Jensen *et al.*, 2002) and as a basis for assembling models of all other AQPs including AQP-1 (de Groot *et al.*, 2001). Though now superseded by the crystal structure of AQP-1 (Sui *et al.*, 2001), these models provide a quantitative diagnostic as to how good such models of other AQPs are likely to be when modeled on GlpF or AQP-1.

Throughout simulations based on the high-resolution GlpF structure it is clear that the GlpF channel is a remarkably well behaved template that remains stable to molecular mechanics using current force fields (de Groot and Grubmuller, 2001; Jensen *et al.*, 2002). One reason is the degree of correctness of the final refined structure, but another is perhaps that a channel that conducts molecules of neutral charge, maintained by an appropriate approximation to a lipid bilayer, is in fact very stably restrained by the balance of hydrophobic forces in the membrane.

II. Three-Dimensional Structure of GlpF with Glycerol in Transit

Structures of channels are extremely difficult to obtain for numerous reasons. However the first structures in each class have been critical to understanding mechanisms(Rees *et al.*, 2000). The structure of GlpF at 2.2 Å resolution (residues 6–259), with its primary permeant substrate glycerol, elucidates the mechanism of selective permeability (Fu *et al.*, 2000). Three intermediate glycerol binding sites define an amphipathic transmembrane pathway that conducts glycerol and certain other linear carbohydrates. Glycerol bound close to the narrowest portion of the channel around conserved Arg-206 shows that glycerol molecules move through the channel in single file, with the alkyl backbone wedged against a two-sided hydrophobic corner. Two successive OH groups each form hydrogen bonds with a pair of acceptor and donor atoms on the opposite sides of a four-sided tripathic channel. The stereoselectivity of the channel is also revealed by functional assays of substrate conduction. This structure of the channel explains the preferential permeability for linear carbohydrates and absolute exclusion of ions and charged solutes.

Surrounding each monomeric, glycerol-conducting channel are six and two half-membrane-spanning α-helices arranged in a right-handed super-twisted fashion (Fig. 1). The anticipated location of the N terminus of GlpF was confirmed to be cytoplasmic (Fu *et al.*, 2000) by establishing that accessibility of a thrombin site inserted between the histidine tag and the mature N terminus is inaccessible to thrombin until cells are permeabilized. This has also been shown to be so in aquaporins (Preston *et al.*, 1994).

The protein is constructed as two segments representing the genetic duplication that are related by a quasi twofold axis that would pass through the center of the bilayer and almost intersect the fourfold axis of the tetramer. The first segment (residues 6–108) forms one side of the channel (Fig. 2). It begins with two membrane-spanning helical segments, M1 (residues 6–34) and M2 (40–60). These are followed by

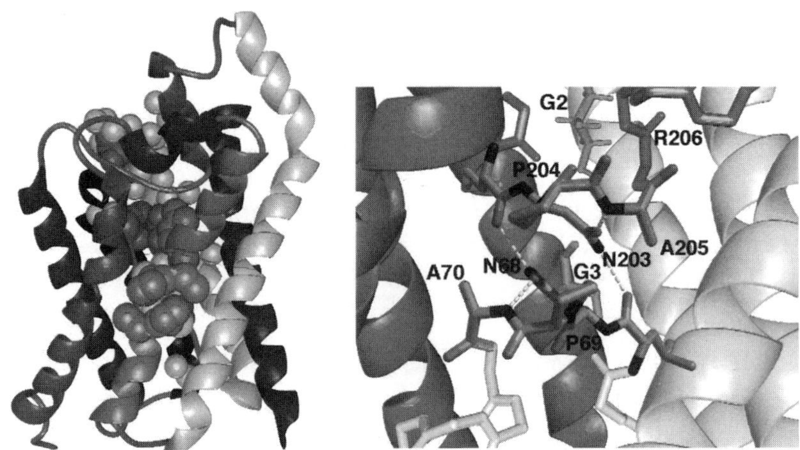

FIG. 1. (Left) The monomeric GlpF channel is vertical. GlpF and all AQP channels are composed of six and two half helices distinguished in different shades of gray. The view is from the lipid bilayer looking in to the quasi-twofold symmetry axis, which intersects at 90° with the fourfold axis of the tetramer that would be vertical, behind this image. These channels all present the NH groups of conserved side chains N68 and N203 into the center of the channel. The -NPA- regions are seen elaborated as space-filling representations in the center of the figure (medium gray). (Right) The detailed hydrogen bonding around the NPA conjugate pair is shown and orients the asparagine NH_2 groups to precisely position one NH from each of N68 and N203 to point directly into the lumen of the channel, at about the same height. There they determine the polarization of the central water molecule in the channel. Modified from Fu *et al.,* 2000, *Science* **290,** 481–486, with permission from the American Association for the Advancement of Science.

seven amino acids (61–67) that project into the center of the channel as an extended polypeptide chain that orients four carbonyls as hydrogen bond acceptors to the cytoplasmic vestibule (Fig. 3). This chain initiates M3 (68–78), the first of two short α-helices, that begins with the first -NPA- sequence near the center of the bilayer and returns to the cytoplasmic side. M3 packs laterally against M1 rather than its neighbor in the linear sequence, M2. This is followed by a third transmembrane helix M4 (85–108) that is packed between M3 and M2.

The second segment (144–254) reiterates a similar transmembrane topology, but beginning from the periplasmic side rather than from the cytoplasmic side. Two more spanning α-helices, M5 (144–168) and M6 (178–194), are followed by a second extended chain incursion into the transbilayer region from the outside. It presents four carbonyls as hydrogen bond acceptors to the periplasmic vestibule and initiates the second half-spanning helix M7 (203–216) (Fig. 3). Beginning with the second -NPA- sequence it returns to the outside. Finally M8 (232–253)

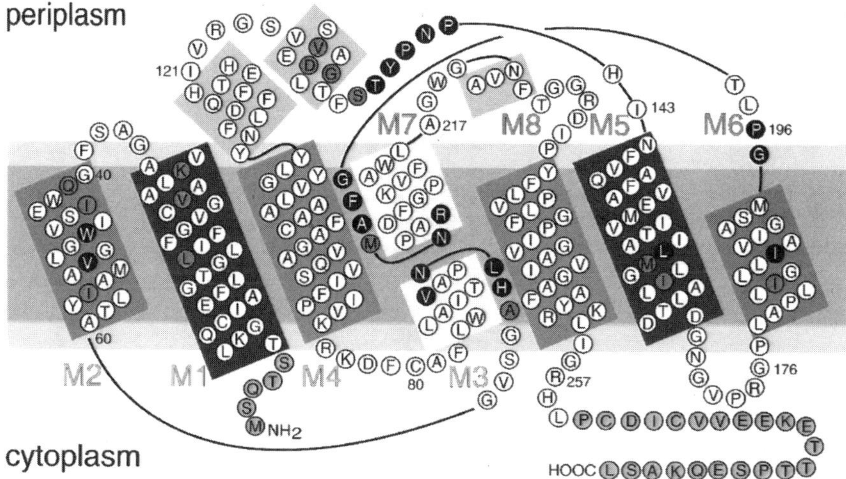

FIG. 2. The amino acid sequence of GlpF is arranged topologically as in the structure. The helices are viewed as if from inside the channel, viewed looking away from the fourfold axis out toward the lipids. The -NPA- motifs are in the center of the figure. Three and one half helices form each segment, labeled M1–M4 and M5–M8. Related helices M1, M5, M2, M6, etc., are boxed in similar shades of gray to represent their homologous roles in the structure. Residues in black circles interact with glycerol. Residues in gray circles contribute either carbonyl oxygen, or amides (NHs) to the channel, or contribute hydrocarbon to the channel. Light gray circles represent residues that are not seen in the structure. The deduced location of the hydrocarbon portion of the lipid bilayer is illustrated as a horizontal rectangle, and the head groups of the lipids are illustrated as light gray rectangles. Modified from Fu *et al.*, 2000, *Science* **290,** 481–486, with permission from the American Association for the Advancement of Science.

returns to the intracellular surface, placing both N and C termini on the inside of the plasma membrane.

Between the two genetic repeats the linking region forms two helices (109–120 and 126–134) as protrusions on the periplasmic side. These surround a concavity in the center of the tetramer where the shortest helices M2 and M6 meet near the fourfold axis of the tetramer, and on either side of the molecular quasi twofold axis. An extended chain (137–143) returns to the central channel and contributes three carbonyls to the entry vestibule. The two half-spanning helices M3 and M7, both in contact with the lipid-accessible exterior on the outside of the tetramer along one side, form an intimate contact between their N-terminal ends on the opposite side of the channel, related by the quasi twofold axis to form a coaxial bend. This unique, dimeric contact between N-terminal ends of two helices M3 and M7 plays a key role in maintaining a glycerol binding site in the center of the bilayer. The conserved "-NPA-" sequences (68–70 in M3, 203–205 in M7) create this interface, placing the proline

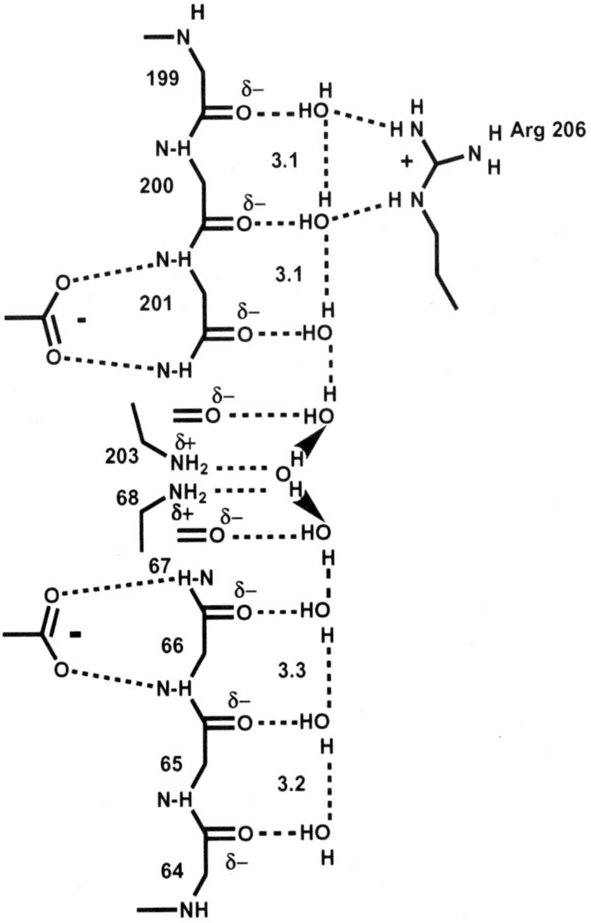

FIG. 3. The line of carbonyl acceptor groups that go through the GlpF channel is shown. The channel in GlpF is tripathic, with two of three sides being hydrophobic; one "side" is based on 8 carbonyl oxygens spaced at ~3.0 Å separation and supports a line of water, each with one OH acting as donor to the carbonyls through the channel. The central water position is acceptor from the conserved asparagine-68 and -203 NH_2 groups. This, along with the positively charged dipoles from half-spanning helices, polarizes the central water so that it donates hydrogen bonds to neighboring waters, thus polarizing them outward from the central position. This is the key to lack of proton conductance through the line of waters that are seen by crystallography to fill the channel with a line of polarized waters.

rings in van der Waals contact against each other, and cupping each prolyl side chain between the proline and alanine side chains of the opposite helix (Fig. 1). Each asparagine side chain is constrained by two hydrogen bonds from main chain NHs to the side chain Oδ1s, which thereby orient the side-chain NHs toward acceptors on the permeant substrate.

III. The Basis for Selectivity through the Channel

The channel pathway, defined by three glycerol molecules, has a wide vestibule ~15 Å across on the outer surface and reaches its constriction of ~3.8 Å × 3.4 Å in size, 8 Å above the central plane of the transmembrane region. This constriction lies at the start of a ~28 Å long selective tripathic channel ($r < 3.5$ Å) that subsequently widens out again toward the cytoplasmic surface. We discuss the glycerol sites in the order (G1–G3) seen by a glycerol molecule moving along its trajectory from the periplasmic side to the cytoplasmic side, and the glycerol OH moieties likewise in numbered order from most external to more internal.

Channel specificity is delineated by three glycerol molecules (G1, G2, G3) with single water molecules between them and several water molecules that are bound in the entry vestibule and down to the selectivity filter in the center of the bilayer. The O1–H of G1 remains hydrated by water molecules. G1 in the entry vestibule is oriented by G1 O2–H as hydrogen bond donor to the O of Tyr-138 (2.7 Å). G2 and G3 are both found in the selective region of the channel. G2 forms a hydrogen bond with one water molecule at its 1′-hydroxyl. Another bridging water molecule is hydrogen bonded to both G2 and G3, which suggests that glycerol and water are stoichiometrically cotransported.

The selectivity filter at G2 is strongly amphipathic, with two aromatic rings (Trp-48, Phe-200) forming a two-sided corner, two N–H donors from the guanidinium side chain of Arg-206 on another side, and two main-chain carbonyl oxygens on the fourth side. G2 binds in the Trp-Phe-Arg triad leaving no free space around it, such that Van der Waals, hydrogen-bond, and electrostatic forces each play a role. The O1H and O2H of G2 are each both hydrogen bond acceptors from NHs of the guanidinium group of Arg-206 (2.9 Å, 2.7 Å) and hydrogen bond donors to the carbonyl oxygens of Gly-199 (2.6 Å) and Phe-200 (2.8 Å), respectively. The alkyl backbone of G2 is tightly packed against the aromatic corner leaving no space for any substitutions at the C–H hydrogen positions.

The two main-chain amide nitrogens of residues 200–201 are pinned in place by hydrogen bonds to the buried carboxyl of Glu-152, which is invariant in all AQP family members and provides a powerful restraining force. The carboxyl group of Glu-152 orients the three adjacent carbonyls of Gly-199, Phe-200, Ala-201 in the entry vestibule and may increase negative charge on the carbonyl oxygens of Phe-200 and Ala-201. Thus the binding of glycerol in the selective filter creates the potential for the negative charge of Glu-152 to form an electrostatic interaction with the positive charge on Arg-206, via the amides of Phe-200, Ala-201, and the glycerol OHs. This could accentuate the polarization of the glycerol OHs and imply that a permeant molecule should also be polarizable in cross section.

The narrowest constriction (a putative "transition site" for transport) in the channel lies between G1 and G2, just above G2 O1H, and is barely large enough to allow passage of an HC–OH, the minimum cross section from the center of the glycerol molecule. At this narrowest position the hydrogen bond geometry from the donor NH2 of Arg-206 to an OH group becomes ideal, helping to specify the only possible orientation of the CHOH group in this putative transition site.

G3 straddles the highly constrained invariant Asn-203 and Asn-68, the two "N"s at the conjunction of the two NPA sequences between M3 and M7. The NH_2 of Asn203 is donor to O1 of G3 (3.0 Å) with ideal geometry. The NH_2 of Asn68 is donor to O2 of G3 (3.1 Å), also with ideal geometry. OH3 of G3 is hydrogen bond donor to the carbonyl of highly conserved His-66 (2.8 Å). Therefore glycerol molecules can only pass through the region of G2, G3 in single file, and with each CHOH within one molecule following its neighbor in single file.

A second highly conserved buried carboxyl of Glu-14 is hydrogen bonded to the amide nitrogens of Leu-67 and His-66. Thus the carbonyls of His-66, Ala-65, and Gly-64 are strongly oriented in the cytoplasmic vestibule of the channel by interactions with the carboxyl of Glu-14. The oriented carbonyls are the quasi twofold equivalent of those at residues 199–201 in the external vestibule. This creates a path of hydrogen bond acceptors that make a line down one wall of the channel. While glycerol is resolved at three sites through the translocation pathway, there may be other partially occupied intermediate sites that are interpreted here as water molecules.

The glycerols in the selectivity filter, G2, G3 have the same handedness, consistent with their lying on a pathway that conducts glycerol without rotation as it is transported. This emphasizes the evolution of a chiral pathway for conduction of organic molecules that have a handedness to them. During conductance, glycerol becomes progressively dehydrated in the external vestibule as carbonyl oxygens replace its water of hydration. Complete dehydration is not required since a water such as that seen between G2 and G3 can copermeate with each glycerol. Glycerol can move stepwise, in single file, through the selective filter by exchanging one set of H-bonds for another along the closely spaced, highly oriented set of eight carbonyl oxygen acceptors and four N–H donor atoms. Alternatively, glycerol diffusion may be visualized as a series of thermally activated transitions over the potential barriers that separate binding sites through the channel including those at G2, G3 (Lauger and Apell, 1982).

In two positions, G2 O1H, and G2 O2H, the donors and acceptors from the channel are oriented with ideal geometry, requiring the dual role of donor and acceptor ~120° apart from each of two successive groups in the permeant molecule. The hydroxyl moiety almost uniquely matches the requirement for hydrogen bond donor and acceptor from the same atomic center. Binding the OH of a permeant molecule at G2 may favor the local potential energy minimum by bridging a long distance electrostatic interaction (8.0–8.5 Å) between two charged side chains of Glu-152 and Arg-206 in the entrance to the selective filter.

IV. Roles of Conserved Residues: Functional and Structural

Conservation across the AQP family identifies conserved structural and functional roles. The 25 most conserved residues are classified according to their locations (Table I). Six residues are conserved only in the first segment (group I). These residues are all close packed together in a structural core of the first segment that is very close to Asn-203, suggesting that this core is conserved in structure and orients Asn-203 at the G3 site. Three conserved residues found only in the second genetic repeat (group II) are found in disparate regions, with one, Arg-206 as an essential functional determinant, at the beginning of the selectivity filter.

The gene-duplicated heritage represented by ~20% sequence identity between the two segments, residues 6–108 and 144–254, is found throughout the AQP family. This is represented in the structure as a quasi twofold relationship between these segments (Cheng *et al.*, 1997; Walz *et al.*, 1997; Fu *et al.*, 2000). Correspondingly several of the highly conserved residues between segments 1 and 2 (group III) have structural roles, and others have more functional roles. Of the structural class, the most conserved residues all lie near the center of the bilayer. Of eight pairs of residues conserved between both genetic repeats (group III), six pairs are located close to the plane through the center of the transmembrane region and related by the structural quasi twofold axis. They are found in pairs of side chains that contact each other, and these contacting pairs are twofold related to other similarly conserved pairs. This arrangement suggests a possible collective structural role that may be related to packing around the quasi twofold symmetry axis in GlpF.

TABLE I
Some Conserved Residues and Sequence–Structure Relationship in GlpF[a,b]

Residues	Location in structure
Group I	
S63	Core packing near N68
H66	Core packing near N68
V71	Core packing near N68
T72	Core packing near N68
Y89	Core packing near N68
Q93	Core packing near N68
Group II	
L159	In the channel, near G3
R206	Key selectivity filter residue G2
P240	Packing core near N203
Group III	
E14/E152	Orient amides in the vestibule
T18/T156	Packing in mid-membrane plane
G49/G184	Packing in mid-membrane plane
G96/G243	Packing in mid-membrane plane
G64/G199	Orients amides in the vestibules
NPA (68–70)/NPA (203–205)	Mid-membrane plane, and G3

[a] From Fu *et al.* 2000, *Science* **290,** 481–486, with permission from the American Association for the Advancement of Science.

[b] The roles of conserved residues are indicated. First the rank order of the sequence conservation was evaluated according to the informational entropy at every position in the sequence alignment of 46 representative AQP family members. The first group I includes residues only found in the first genetic repeat. The second group II includes residues conserved in the second repeat only, and group III includes residues that are present at equivalent positions and conserved in both repeats I and II. Groups I and II therefore may be related to structure and function specific to each respective repeat, while group III reflects the symmetric relationship between the two genetic repeats.

Four of the group III residues are conserved glycines located within helices. Far from any more usual role in disrupting helices of soluble proteins, these transmembrane helix glycines allow the closest approach in the helical bundle (Javadpour *et al.,* 1999) that occurs in the central plane. These glycines are invariant or highly conserved to fulfil their role in the close contacts at the helix crossing contacts (MacKenzie *et al.,* 1997).

The helices diverge outward from this central plane of closest contact to generate the vestibules, one on on each surface. The interhelix angles of $\sim+35^\circ$ to $+40^\circ$ are not in the categories most favored by knobs into holes packing (Walther *et al.*, 1996). However, helix packing within the bundle is under powerful constraints from the hydrophobic bilayer and from the interfaces between monomers in the tetramer where helices pack with more optimal knobs into hole packing angles of $\sim -20°$.

The two conserved, buried glutamates Glu-14, Glu-152 have clear functional roles, each orienting four carbonyls of successive amino acids in the vestibules. The conserved pair of group III Gly-64 and Gly-199 are symmetrically located at the extremities of the central selective filter and serve to structure the downturn of the extended chains from 65 to 68 and 200 to 203 that form the critical line of four carbonyls each that in turn provide the pitons for successive hydrogen bond donors on their way through the channel (Fig. 3).

The conservation of the -NPA- sequences at 68–70 and 203–205 can be rationalized in both structural and functional terms. Structurally it preserves a key signature contact between M3 and M7. There are plausible structural reasons why residues that are in contact around the quasi twofold axis such as the pair of NPA motifs and the Gly-49, Gly-184 pair would remain conserved, since each effectively resolves two packing situations at once. All these residues related by the quasi twofold within GlpF are among the most conserved residues throughout the entire AQP protein family. This therefore suggests that the GlpF structure is prophetic of all other AQP relatives.

V. Stereoselective Preferences of GlpF among Linear Alditols

Osmotic methods have been used to measure influx of carbohydrates into proteoliposomes, reconstituted using purified GlpF. The rate of influx of the tested molecules is indicated by recovery of the initial volume of the proteoliposome. Ribitol, with all OH groups having the same stereospecific relationship to the carbon backbone, generated a clear exponential reswelling with a time constant of $\sim 0.35\ s^{-1}$. In contrast D-arabitol, a chiral stereoisomer of ribitol with a mixed arrangement of hydroxyls, shows a $\sim$10-fold reduction in transport relative to ribitol, demonstrating stereoselectivity of the channel.

Membranes have a high intrinsic permeability for glycerol. When GlpF is present the reswelling rate increases such that the vesicles recover by a factor similar to that seen with ribitol. To ensure that this rate is due to GlpF, this rate was measured as a function of the amount of GlpF used in the reconstitution. The linear dependence demonstrates that the transport is directly proportional to the number of GlpF molecules per vesicle.

The rates demonstrate a stereoselective channel, consistent with the exact positioning of donors, acceptors, and hydrophobic contacts. The structure of the central selectivity filter shows that it can accommodate either prochiral carbon orientation, placing the hydrogen attached to the carbon in contact with the aromatic side chain either of Phe-200 or of Trp-48, depending on its chirality. Therefore the conductivity is in principle different depending on the enatiomer in the selectivity filter, or which way glycerol enters the channel.

Stereoselectivity is seen between different chain lengths and different chiralities. As an example there is a factor of 10 difference in rate between ribitol and its stereoisomer D-arabitol. The structure shows how this can be visualized since two successive –CHOH groups are oriented at the G2 site. The carbon backbone must be lined up along the channel axis. Depending on enantiomer, the hydrogens attached to the carbon are in contact with the aromatic ring either of Phe-200 or of Trp-48, that forms two sides of the four-sided channel in similar environments. However, any CHOH groups adjacent to these two will alternately place the carbon in one of two tetrahedrally disposed sites that have quite different environments. Of those tested, chiral molecules had lower rates than nonchiral molecules. There is no conductivity of cyclized pentoses and hexoses that can be rationalized as due to their size. The reswelling rates observed follow the same trend as was observed for GlpF upon induction in whole cells (Heller *et al.,* 1980), showing that *in vivo* characteristics are represented in our assay using recombinant protein reconstituted into vesicles.

VI. Simulations and Rates of Glycerol Passing through the Channel

Molecular mechanics provides a method of assessing various contributors to the rates of events. Though such methods are often criticized for leading structures to diverge from the experimentally observed, they remain the constant hope for a more rational and quantitative assessment of mechanisms. In all, 20 hydrogen bond acceptors and 23 hydrogen bond donors participate in the conductance of glycerol (Nollert *et al.,* 2001; Jensen *et al.,* 2002). The methods also can be held to account for the experimental values of measured properties, and if necessary rectified so that differences between one channel structure and another of different property can at least be suggested and used to redirect the next wave of experiments. In the context of GlpF, the first structure of an AQP at adequate resolution, such calculations seem to be quite robust such that after a 1 ns equilibration time the model remains within 1 Å rms of the crystal structure, leaving the channel structure essentially identical to the bounds of motion and structure obtained by X-ray crystallography.

The passage of glycerol evokes little if any change in the structure of the channel, as also seen by comparing the crystallographic structure in glycerol, versus that in water alone (Tajkhorshid *et al.,* 2002). Therefore there is theoretical and experimental support for the conclusion that there is no structural gating of the channel; it is a highly stable, open, selective channel.

For molecular mechanics, the GlpF tetramer is typically embedded in a model of 16:0/18:1c9-palmitoyl-oleyl-phosphatidylethanolamine (POPE) lipid bilayer fully hydrated by water, and subjected to molecular mechanics at 310K and atmospheric pressure.

To determine the rate of conductance of a permeant molecule, it is necessary to drive the molecule through the channel since it would take far too long for a single molecule to find its way from one side to the other without any directional assistance. In physiological terms the driving force is presented as a concentration gradient, such that the free energy seen by the permeant molecule has a net downhill slope to it. In molecular mechanics terms, a theoretical external force $f(t)$ is applied to move a glycerol molecule through the channel at constant velocity (Jensen *et al.,* 2002). The necessary force is related to the gradient of an energy profile through the channel at each particular position. The gradients are connected by a smooth mathematical function based on sums of sin wave functions to reveal an energy landscape through the channel. The potential of mean force provides a means of effectively mapping the energy landscape seen by a glycerol molecule in passage through the conducting pathway. Jarzynski's identity between free energy and irreversible work allows for reconstruction of the potential of mean force along the conduction pathway by sampling a series of time snapshots through the molecular dynamics trajectory. The experimentally observed binding sites were thus justified in molecular and thermodynamic terms. And barriers inside the channel could be associated with transition-state free energies, and hence rates of conduction. The energy profile also reveals a low-energy periplasmic vestibule suited for efficient uptake of glycerol from the environment.

VII. Simulation and Rates of Water Passage through the GlpF (an AQP) Channel

To understand the conduction of water through AQP channels such as the GlpF channel, the crystal structure of GlpF with only water in the channel at 2.7 Å that reveals the mean positions and probabilities of position can be compared with a molecular simulation in which 7–9 water molecules constitute a single file in the 20 Å constriction region of the channel. The calculated diffusion constant corresponds to a flux of 2.4×10^9 s^{-1}, which is 5 times larger than the deduced flux for GlpF (Borgnia *et al.,* 1999a,b; Borgnia and Agre, 2001) of 0.5×10^9 s^{-1}. The observed difference may be due to rate-limiting steps, such as the entrance of water into the channel, which cannot be included in the calculation of a one-dimensional diffusion constant. The agreement between simulation and the crystal structure is in general very good, indicating that the water molecules form a stable line of 7–9 water molecules through the entire channel, and corroborating that the force fields and assumptions necessary in molecular mechanics are a good representation of the situation in this type of channel.

Molecular mechanics is a rapidly evolving field that is now finding remarkable successes in the membrane channel field which may be well suited to the conditions applied. Thus it is most useful not only when molecular mechanics can rationalize the experimentally determined rates of conductance and selectivities, but when it also serves to allow predictive simulations that test hypotheses prior to experiment. In these cases the technology has reached a most useful prospective phase.

The structure and the simulation indicate that the water molecules proceed together through the channel driven by osmotic pressure. The water molecules are also polarized by the carbonyls such that they each are donors to the eight carbonyls that line the channel, but such that the central water molecule is oriented by the central asparagine side-chain NH donors to polarize the entire line of waters that face outward from that central position, as was postulated by Murata *et al.* (2000) from their 3.5–6 Å electron microscopic structure of AQP1. The helix dipoles of helices M3 and M7 also place their positively charged ends together near the NPA regions; thus they, too, have been suggested to play a role in orienting water molecules at the center, so "bipolarizing" the line of water. In addition it is suggested based on molecular simulations that the full charge present on Arg-206 may play a role in polarizing the central water molecule (de Groot and Grubmuller, 2001).

The relative contributions of these three contributors to polarizing the central water could be assessed for the degree of their contribution by "turning off" each one of the charges associated with the polarizing dipoles in turn (Tajkhorshid *et al.,* 2002). When the partial charges on the NH_2-groups of the side chains of Asn-68 and Asn-203 are turned off, the effect on water configuration is partial. As the partial charges on the amide backbone atoms of the half-helices M3 and M7 are turned off, which effectively eliminates the helix dipoles, a partial effect is also observed. And as this is combined with the NH_2 groups of the side chains of Asn-68 and Asn-203, the bipolar orientation of the single-file arrangement breaks down, potentially leading to a proton wire. Although the dipoles of the channel-lining carbonyl groups alone cannot maintain the bipolar orientation of water molecules, their absence significantly accelerates the relaxation toward orientational disorder. Therefore each of these factors contributes to the abrogation of any proton conductance through the channel.

VIII. Insulation against Proton Conduction in AQPs

The conduction of protons through water can be remarkably fast, proceeding by the Grotthus mechanism (Grotthuss, 1806) in which protons hop, perhaps cooperatively, throughout a line of hydrogen bonded water molecules, as is observed in the proton tunneling through ice. If the protons move cooperatively, no charge need ever be seen in the center of the channel, where the fully charged Arg-206 might become a major repulsive element, nor any bulk transfer of water take place, yet the proton could easily "leak" through the line of water. The molecular mechanics dissection of the contributors shows that the polarized central water, as implied by the crystallography (Fu *et al.*, 2000; Nollert *et al.*, 2001; Sui *et al.*, 2001; Fu *et al.*, 2002), is key to this insulation (Tajkhorshid *et al.*, 2002; de Groot and Grubmuller, 2001).

IX. Quaternary Structure of GlpF (and AQPs)

Each tetramer comprises four closely associated monomeric channels circled by a hydrophobic surface long enough to span the lipid bilayer (Fig. 4). Toward the cytoplasmic and periplasmic surfaces are layers that include side chains of tyrosine and tryptophan that can productively interact with the polar–nonpolar interface in the lipid head-group region as in other integral membrane proteins (Koeppe and Anderson, 1996). These layers are flanked by two outer layers of charged residues, 35 Å apart, that result in net positive charge on the cytoplasmic side.

Each tetramer has a central concave surface on the periplasmic side where the two central transmembrane helices M2 and M6 of each monomeric channel in the tetramer are not long enough to span the bilayer, consistent with the tetramer being the physiological quaternary structure. The interfaces between subunits are almost as hydrophobic as the exterior, suggesting that a monomer could plausibly lie in the membrane, though under stress due to mismatch of thickness that might lead to changes in the orientation of the short transmembrane helices, and therefore in conductance properties of such a postulated monomer.

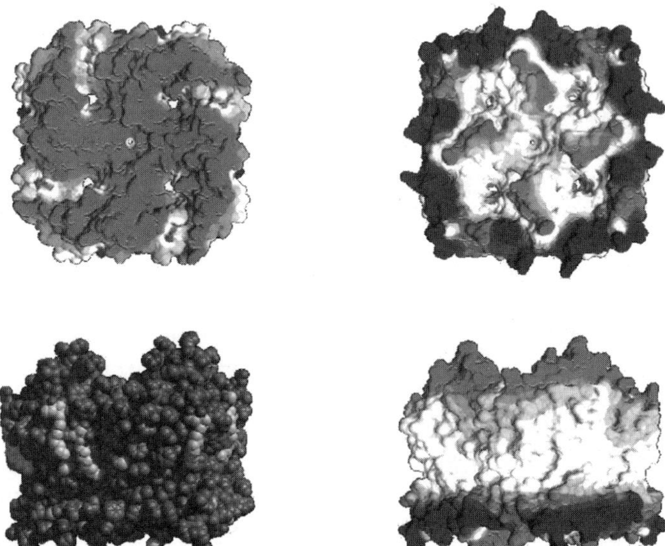

FIG. 4. A surface representation of the tetramer shaded according to electrostatic potential: Light gray represents negative, dark gray positive charge. Top left is a view from the periplasmic side that is predominantly negatively charged. The center is a fourfold symmetry axis that relates all four channels in the image. The centers of the channels can be seen to traverse the entire molecule, leaving a white zone seen clearly throughout the entire channel. At the top right the view is from the cytoplasmic side and is predominantly positively charged. Bottom left is a side view of a space-filling representation of the structure and shows ordered detergent molecules of octyl glucoside as they associate with the periplasmic leaflet interfacial zone. Bottom right is an electrostatic surface view of the same, showing the preponderance of negative charge on the periplasmic side of the membrane, and the positive charge on the cytoplasmic surface.

The tetramer seen in the GlpF crystal structure has close similarity to the tetramers that associate, alternate sides up, in reconstituted membranes to form two-dimensional crystals, as characterized by image reconstructions of AQP1 (Cheng *et al.,* 1997; Li *et al.,* 1997; Walz *et al.,* 1997; Hasler *et al.,* 1998a,b; Mitsuoka *et al.,* 1999; Ringler *et al.,* 1999). However, the GlpF tetramers are each surrounded by detergent, and the packing between neighboring tetramers is therefore different. For example, GlpF could reasonably be inserted into membranes as monomers during synthesis, and subsequently gain stability by tetramerization.

Twelve ordered octyl glucoside molecules are associated with the aromatic regions near the external side of each tetramer (Fig. 4). The pyranoside rings interact laterally with the external tryptophan/tyrosine layer, while the alkyl chains extend toward the hydrophobic core.

X. The Ion Channel in AQP6; a Possible Pore on the Fourfold Axis of AQPs?

Most ion channels are composed of multimers of proteins. Sometimes they are homomeric as with the MscL stretch-activated channel potassium channel (Chang *et al.*, 1998); sometimes they are composed of similar subunits as in the acetylcholine receptor (Stroud *et al.*, 1990). Therefore the tetramer in AQPs begs the question as to what properties pertain to the central fourfold axis. There are three density peaks on the fourfold axis near the external surface. One of these peaks, that refines to 12($\pm$3) electrons, is coordinated at 2.9 Å by four otherwise buried glutamates (Glu-43) in a square planar arrangement. Below this peak is a second, waterlike peak that is 2.2 Å from the first peak, implying that the first peak is an octahedrally coordinated cation, most likely Mg^{2+} (present at ~300 m*M*), that is bound in the fourfold axial pathway. Physiologically Mg^{2+} (~1 m*M*) or Na^{+} (~145 m*M*) could fulfill this role. The remainder of the fourfold transmembrane pathway is surrounded by exclusively nonpolar aliphatic side chains making it extremely hydrophobic, and seemingly closed to all hydrated ions and polar molecules larger than water. While this is the case in GlpF, there is a central cavity ~10.6 Å in diameter that resembles the structure in the KcsA potassium channel. And in AQP6, the constriction on the cytoplasmic side is significantly wider, suggesting that it then could be an ion-conducting channel.

XI. GlpF Channel Selectivity for Antimonite

Escherichia coli resistant to antimonite have a disruption in the glpf gene suggesting that GlpF conducts antimonite, the neutral pH form of which is $Sb(OH)_3$; therefore it is expected that the channel again provides adequately positioned hydrogen bond accepting carbonyls to support the polar side of the $Sb(OH)_3$ molecule.

XII. Selectivity against Passing Ions or an Electrochemical Gradient

A fascinating question pertains to the necessity for all plasma membrane channels to remain insulating to any electrical leakage of ions. How is this accomplished? Common to GlpF and AQPs is their absolute selectivity against ions and charged solutes. Therefore, comparison with known ion channels helps to define what is required to allow ion conductance in a channel. While the channel is large enough to hold an ion such as Na^+, K^+, Ca^{2+}, or Cl^-, it is too small to accommodate a hydrated ion, and by its amphipathic nature provides no replacement for water of hydration on one entire side. The energetic cost of removing even a single water of hydration from an ion can be prohibitive. Therefore cation conducting channels provide oriented carbonyl groups that carry a δ-charge. These can replace any water (s) of hydration displaced within any narrow selective region that intrudes on the first hydration shell (Doyle and Wallace, 1997; Chang *et al.*, 1998; Doyle *et al.*, 1998; Koeppe *et al.*, 1999; Roux and MacKinnon, 1999).

By requiring a group that is polarized with both $\delta+$ donor and $\delta-$ acceptor character, both anions and cations are excluded. The two charged side chains of Arg-206 and Glu-152 might impose an energy barrier to passage of a charged ion or solute molecule including OH- or H_3O^+. Since Arg-206 and Glu-152 are conserved across almost all AQPs and GlpF, electrostatic repulsion would play the same role in AQPs. That no fully charged counteranion is bound to Arg-206 where glycerol fills the site further emphasizes that removal of water of hydration from an anion without compensation is enough to prevent any passage of ions through this selectivity filter.

Removal of water and the fixing of a specific conformation of glycerol in the GlpF channel are also energetically costly. However, the structure specifically matches the donor plus acceptor and amphipathic nature of each of a series of C–OH groups, replacing the hydration shell in water by a precise match within the channel.

XIII. The Various Contributions to Rejection of Proton Conductance

A most interesting conundrum is how AQPs completely exclude conductance of protons that could be lethal to any cell. This is especially true in these channels that can support a line of water molecule positions through the channel region. The Grotthus mechanism involves the protonation of one water from a donor at one side of the mebrane, and the concerted passing of another proton from that water molecule to its neighbor through the channel. In principle, if concerted throughout the line of water molecules, this mechanism could leak protons from one side of the membrane to the other, without generating any charge within the 28 Å narrow region of the channel.

To analyze the basis for water conductance of GlpF we determined the crystal structure of GlpF in the absence of glycerol (Tajkhorshid *et al.*, 2002). This structure at 2.7 Å resolution showed that water molecules form a line through the channel in which each water molecule is a hydrogen bond donor to one of the eight carbonyls that line the channel, and also donor to a neighboring water. The only exception is for the central water molecule that is hydrogen bond acceptor from both conserved asparagines of the -NPA- box. Lying opposite the hydrophobic corner, its hydrogens point away from the center and become hydrogen bond donors to the neighboring waters on both sides, so bipolarizing the entire line of waters, making each water an acceptor from water on its inside relative to the center of the bilayer at W 522, opposite the conserved asparagines at the channel center.

There are two places at which there is no electron density for water molecules that we would expect to be good sites for water. One is triangulated opposite the terminal NH_2 group of conserved Arg-206 and the carbonyl of Gly-199 ($z \sim -3$ Å relative to $z = 0$ Å at the carbonyl of Ala-201, becoming positive toward the cytoplasmic side), leaving a distance of almost two hydrogen bonds, 5.5 Å, between the waters adjacent to this site. The second case where density is absent is opposite the carbonyl of the gene-duplicated Gly-64 at the other end of the channel, leaving a similar gap of 4.5 Å. The lack of density at these sites may be for several reasons: (1) There may be a break in the line of waters here. If so this would already provide some basis for the insulation toward passage of protons. However, this seems unlikely in light of the ideal sites, even though one of them lies at the narrowest region of the channel. (2) This reasoning put slightly differently could be restated that the probability of finding waters at these two sites is somewhat lower perhaps because they are somewhat higher in energy. Molecular mechanics tends to support this conclusion for the first of these waters that lies at the narrowest part of the channel. (3) There may be a somewhat disordered water molecule at each site, thus blurring and flattening the density. Even though the crystallography is carried out at -100°C where water is immobilized it still may be frozen differently in each molecule, thereby still being absent from the density.

XIV. Selectivity for Glycerol versus Water

GlpF also conducts water, but at a relative rate of 23 versus 154 per second for the water channel AQPZ (Borgnia and Agre, 2001). The conductivity of water preferring AQPs AQPZ and AQP-1 for glycerol, urea, or other uncharged small molecules is essentially undetectable. Conductivity for water in the absence of glycerol may be severely impaired by the amphipathic nature of the channel, since the hydration shell of water itself is quite stringent, with many more neighbors than found in carbohydrates. Several waters at one time, required to pass the filter in single file by the structure, could not retain more than two neighbors in the center of this channel. The dehydration of water would therefore make its transport energetically more costly than transport of the water molecules seen between OH groups on successive glycerols or alditols.

AQPs have a high permeability to water and not glycerol, whereas a subclass is permeant to both water and glycerol. The geometric requirements for hydration of water are stringent, with many four- and five-coordinate waters (Bernal and Fowler, 1933; Eisenberg and Kauzman, 1969). All residues throughout the selective filter are highly conserved between GlpF and AQPs except for Trp-48 (Phe or His in AQPs), Phe-200 (typically Ala, Thr, or Cys in AQPs), and Ala-201, all in the narrowest region of the filter. Therefore the larger size in the AQP filter will support an increased number of water molecules, so alleviating the barrier to water that would be imposed in GlpF by desolvation of several waters in single file, and allowing a greater degree of water coordination. At the same time removal of the indole of Trp-48 and the phenyl ring of Phe-200 will remove the hydrophobic corner that can no longer provide Van der Waals interaction with the alkyl chain of glycerol in AQPs.

In AQP1, C191 is responsible for mercury inhibition of transport (Prasad *et al.*, 1998). It is positioned similarly to its homologous residue Phe-200 in GlpF at the narrowest position of the channel. Thus mercury inhibition in aquaporins is brought about by a chemical modification of the thiol in the selectivity filter.

Mutagenesis of two residues, Tyr236Pro and Trp237Leu, in AQP1 allowed it to pass glycerol (Lagree *et al.*, 1999). In GlpF Pro-236 lies in the protein interior in contact with Ala-201 side chain behind G2. Leu-237 contacts Pro-204 of the second NPA sequence at the G3 site. Therefore mutational changes at these sites can be relayed to the filter to favor passage of glycerol.

Substitution of the conserved Arg-206 in a water channel, AQPZ (Arg189Val/Ser), inactivated water transport (Borgnia *et al.*, 1999b). A mutation of the conserved Arg-206 in the related aquaporin-2 (AQP2) (Arg-187) is known to cause nephrogenic diabetes insipidus in humans. It is believed that the disease is caused by an impaired intracellular membrane traffic as indicated by experiments demonstrating that human AQP2 Arg187Cys expressed in *Xenopus* oocytes is not transferred to the cell membrane (Deen *et al.*, 1995). However, the location of Arg-206 in the selectivity filter must alter the water transport of AQP2 in collecting duct principal cells in the kidney.

XV. Regulated Ion Channels Formed by Members of the AQP Family

Members of the AQP family are also thought to act as regulated ion channels in addition to their primary channel selectivity (Yool *et al.*,

1996; Agre *et al.*, 1997). However, anion conductance of AQP6 at low pH (Yasui *et al.*, 1999) and cGMP gated potassium conductance of AQP1 (Anthony *et al.*, 2000) have been demonstrated. In this regard the transmembrane pore formed between molecules around the fourfold axis deserves attention. This pore has a bound cation at the external surface, and the variations in pore diameter (Smart *et al.*, 1997) ($r_{\min} = 1.17$ Å) are closely similar to those of the (also tetrameric) KcsA potassium channel throughout its length ($r_{\min} = 0.53$ Å) (Andersen and Koeppe, 1992). In fact the diameters of the GlpF tetramer are greater except at one position near the cytoplasmic surface, and the narrow portion of the fourfold axis is shorter than in KcsA. Like the KcsA channel it also has a $r \sim 4$ Å cavity in the center of the bilayer, and it is ~10 Å shorter than the KcsA channel, making this a shorter track across the bilayer. The main impediment to regarding this as an ion channel is only that the walls are hydrophobic in GlpF (which is not an ion channel). The fourfold axis therefore deserves consideration as a candidate for becoming an ion channel in other AQPs where the structure responds to phosphorylation, ligand binding, pH, or interaction with other proteins.

References

Agre, P., Preston, G. M., Smith, B. L., Jung, J. S., Raina, S., Moon, C., Guggino, W. B., and Nielsen, S. (1993). Aquaporin CHIP: the archetypal molecular water channel. *Am. J. Physiol.* **265,** F463–476.

Agre, P., Lee, M. D., Devidas, S., and Guggino, W. B. (1997). Aquaporins and ion conductance [letter; comment]. *Science* **275,** 1490; discussion 1492.

Andersen, O. S., and Koeppe, R. E. d. (1992). Molecular determinants of channel function. *Physiol. Rev.* **72,** S89–158.

Anthony, T. L., Brooks, H. L., Boassa, D., Leonov, S., Yanochko, G. M., Regan, J. W., and Yool, A. J. (2000). Cloned human aquaporin-1 is a cyclic GMP-gated ion channel. *Mol. Pharmacol.* **57,** 576–588.

Baker, M. E., and Saier, M. H., Jr. (1990). A common ancestor for bovine lens fiber major intrinsic protein, soybean nodulin-26 protein, and *E. coli* glycerol facilitator. *Cell* **60,** 185–186.

Bernal, J. D., and Fowler, R. H. (1933). On the nature of water. *J. Chem Phys.* **1,** 515–520.

Borgnia, M. J., and Agre, P. (2001). Reconstitution and functional comparison of purified GlpF and AqpZ, the glycerol and water channels from Escherichia coli. *Proc. Natl. Acad. Sci. USA* **98,** 2888–2893.

Borgnia, M., Nielsen, S., Engel, A., and Agre, P. (1999a). Cellular and molecular biology of the aquaporin water channels. *Annu. Rev. Biochem.* **68,** 425–458.

Borgnia, M. J., Kozono, D., Calamita, G., Maloney, P. C., and Agre, P. (1999b). Functional reconstitution and characterization of AqpZ, the *E. coli* water channel protein. *J. Mol. Biol.* **291,** 1169–1179.

Braun, T., Philippsen, A., Wirtz, S., Borgnia, M. J., Agre, P., Kuhlbrandt, W., Engel, A., and Stahlberg, H. (2000). The 3.7 A projection map of the glycerol facilitator GlpF: a variant of the aquaporin tetramer. *EMBO Rep.* **1,** 183–189.

Chang, G., Spencer, R. H., Lee, A. T., Barclay, M. T., and Rees, D. C. (1998). Structure of the MscL homolog from *Mycobacterium tuberculosis:* a gated mechanosensitive ion channel. *Science* **282,** 2220–2226.

Cheng, A., van Hoek, A. N., Yeager, M., Verkman, A. S., and Mitra, A. K. (1997). Three-dimensional organization of a human water channel. *Nature* **387,** 627–630.

de Groot, B. L., and Grubmuller, H. (2001). Water permeation across biological membranes: mechanism and dynamics of aquaporin-1 and GlpF. *Science* **294,** 2353–2357.

de Groot, B. L., Engel, A., and Grubmuller, H. (2001). A refined structure of human aquaporin-1. *FEBS Lett.* **504,** 206–211.

Deen, P. M., Croes, H., van Aubel, R. A., Ginsel, L. A., and van Os, C. H. (1995). Water channels encoded by mutant aquaporin-2 genes in nephrogenic diabetes insipidus are impaired in their cellular routing. *J. Clin. Invest.* **95,** 2291–2296.

Doyle, D. A., and Wallace, B. A. (1997). Crystal structure of the gramicidin/potassium thiocyanate complex. *J. Mol. Biol.* **266,** 963–977.

Doyle, D. A., Morais Cabral, J., Pfuetzner, R. A., Kuo, A., Gulbis, J. M., Cohen, S. L., Chait, B. T., and MacKinnon, R. (1998). The structure of the potassium channel: molecular basis of K^+ conduction and selectivity [see comments]. *Science* **280,** 69–77.

Eisenberg, D., and Kauzman, W. (1969). In: *The Structure and Properties of Water.* Oxford University Press, Oxford, UK.

Engel, A., Fujiyoshi, Y., and Agre, P. (2000). The importance of aquaporin water channel protein structures. *Embo J.* **19,** 800–806.

Eze, M. O., and McElhaney, R. N. (1978). J. *Gen. Microbiol.* **105,** 233–242.

Finkelstein, A. (1987). In: *Water Movement through Lipid Bilayers, Pores and Plasma Membranes, Theory and Reality.* Wiley, New York.

Fischer, A. (1903). In: *Vorlesungen uber Bakterien,* Ch. III, pp. 24–25. Jena.

Fu, D., Libson, A., Miercke, L. J., Weitzman, C., Nollert, P., Krucinski, J., and Stroud, R. M. (2000). Structure of a glycerol-conducting channel and the basis for its selectivity. *Science* **290,** 481–486.

Fu, D., Libson, A., and Stroud, R. (2002). The structure of GlpF, a glycerol conducting channel. *Novartis Found. Symp.* **245,** 51–61.

Gorin, M. B., Yancey, S. B., Cline, J., Revel, J. P., and Horwitz, J. (1984). The major intrinsic protein (MIP) of the bovine lens fiber membrane: characterization and structure based on cDNA cloning. *Cell* **39,** 49–59.

Grotthuss, C. J. T. d. (1806). *Ann. Chim.* **LVIII.**

Hasler, L., Heymann, J. B., Engel, A., Kistler, J., and Walz, T. (1998a). 2D crystallization of membrane proteins: rationales and examples. *J. Struct. Biol.* **121,** 162–171.

Hasler, L., Walz, T., Tittmann, P., Gross, H., Kistler, J., and Engel, A. (1998b). Purified lens major intrinsic protein (MIP) forms highly ordered tetragonal two-dimensional arrays by reconstitution. *J. Mol. Biol.* **279,** 855–864.

Heller, K. B., Lin, E. C., and Wilson, T. H. (1980). Substrate specificity and transport properties of the glycerol facilitator of *Escherichia coli. J. Bacteriol.* **144,** 274–278.

Heymann, J. B., and Engel, A. (2000). Structural clues in the sequences of the aquaporins [In Process Citation]. *J. Mol. Biol.* **295,** 1039–1053.

Javadpour, M. M., Eilers, M., Groesbeek, M., and Smith, S. O. (1999). Helix packing in polytopic membrane proteins: role of glycine in transmembrane helix association. *Biophys. J.* **77,** 1609–1618.

Jensen, M. O., Park, S., Tajkhorshid, E., and Schulten, K. (2002). Energetics of glycerol conduction through aquaglyceroporin GlpF. *Proc. Natl. Acad. Sci. USA* **99,** 6731–6736.

Koeppe, R. E., 2nd, and Anderson, O. S. (1996). Engineering the gramicidin channel. *Annu. Rev. Biophys. Biomol. Struct.* **25,** 231–258.

Koeppe, R. E., 2nd, Greathouse, D. V., Providence, L. L., Shobana, S., and Andersen, O. S. (1999). Design and characterization of gramicidin channels with side chain or backbone mutations. *Novartis Found. Symp.* **225,** 44–55.

Lagree, V., Froger, A., Deschamps, S., Hubert, J. F., Delamarche, C., Bonnec, G., Thomas, D., Gouranton, J., and Pellerin, I. (1999). Switch from an aquaporin to a glycerol channel by two amino acids substitution. *J. Biol. Chem.* **274,** 6817–6819.

Lauger, P., and Apell, H. J. (1982). Jumping frequencies in membrane channels. Comparison between stochastic molecular dynamics simulation and rate theory. *Biophys. Chem.* **16,** 209–221.

Li, H., Lee, S., and Jap, B. K. (1997). Molecular design of aquaporin-1 water channel as revealed by electron crystallography [letter] [see comments]. *Nat. Struct. Biol.* **4,** 263–265.

Lin, E. C. C. (1976). Glycerol Dissimilation and its regulation in bacteria. *Annual Rev. Microbiol.* **30,** 535–578.

MacKenzie, K. R., Prestegard, J. H., and Engelman, D. M. (1997). A transmembrane helix dimer: structure and implications. *Science* **276,** 131–133.

Maurel, C., Reizer, J., Schroeder, J. I., and Chrispeels, M. J. (1993). The vacuolar membrane protein gamma-TIP creates water specific channels in *Xenopus* oocytes. *EMBO J.* **12,** 2241–2247.

Maurel, C., Reizer, J., Schroeder, J. I., Chrispeels, M. J., and Saier, M. H., Jr. (1994). Functional characterization of the Escherichia coli glycerol facilitator, GlpF, in Xenopus oocytes. *J. Biol. Chem.* **269,** 11869–11872.

Mitsuoka, K., Murata, K., Walz, T., Hirai, T., Agre, P., Heymann, J. B., Engel, A., and Fujiyoshi, Y. (1999). The structure of aquaporin-1 at 4.5-A resolution reveals short alpha-helices in the center of the monomer. *J. Struct. Biol.* **128,** 34–43.

Murata, K., Mitsuoka, K., Hirai, T., Walz, T., Agre, P., Heymann, J. B., Engel, A., and Fujiyoshi, Y. (2000). Structural determinants of water permeation through aquaporin-1. *Nature* **407,** 599–605.

Nollert, P., Harries, W. E., Fu, D., Miercke, L. J., and Stroud, R. M. (2001). Atomic structure of a glycerol channel and implications for substrate permeation in aqua(glycero)porins. *FEBS Lett.* **504,** 112–117.

Pao, G. M., Wu, L. F., Johnson, K. D., Hofte, H., Chrispeels, M. J., Sweet, G., Sandal, N. N., and Saier, M. H., Jr. (1991). Evolution of the MIP family of integral membrane transport proteins. *Mol. Microbiol.* **5,** 33–37.

Park, J. H., and Saier, M. H., Jr. (1996). Phylogenetic characterization of the MIP family of transmembrane channel proteins. *J. Membr. Biol.* **153,** 171–180.

Prasad, G. V., Coury, L. A., Finn, F., and Zeidel, M. L. (1998). Reconstituted aquaporin 1 water channels transport CO2 across membranes. *J. Biol. Chem.* **273,** 33123–33126.

Preston, G. M., and Agre, P. (1991). Isolation of the cDNA for erythrocyte integral membrane protein of 28 kilodaltons: member of an ancient channel family. *Proc. Natl. Acad. Sci. USA* **88,** 11110–11114.

Preston, G. M., Carroll, T. P., Guggino, W. B., and Agre, P. (1992). Appearance of water channels in Xenopus oocytes expressing red cell CHIP28 protein. *Science* **256,** 385–387.

Preston, G. M., Smith, B. L., Zeidel, M. L., Moulds, J. J., and Agre, P. (1994). Mutations in aquaporin-1 in phenotypically normal humans without functional CHIP water channels. *Science* **265,** 1585–1587.

Rees, D. C., Chang, G., and Spencer, R. H. (2000). Crystallographic analyses of ion channels: lessons and challenges. *J. Biol. Chem.* **275,** 713–716.

Reizer, J., Reizer, A., and Saier, M. H., Jr. (1993). The MIP family of integral membrane channel proteins: sequence comparisons, evolutionary relationships, reconstructed pathway of evolution, and proposed functional differentiation of the two repeated halves of the proteins. *Crit. Rev. Biochem. Mol. Biol.* **28,** 235–257.

Ringler, P., Borgnia, M. J., Stahlberg, H., Maloney, P. C., Agre, P., and Engel, A. (1999). Structure of the water channel AqpZ from *Escherichia coli* revealed by electron crystallography. *J. Mol. Biol.* **291,** 1181–1190.

Roux, B., and MacKinnon, R. (1999). The cavity and pore helices in the KcsA K^+ channel: electrostatic stabilization of monovalent cations [see comments]. *Science* **285,** 100–102.

Sanno, Y., Wilson, T. H., and Lin, E. C. C. (1968). Control of permeation to glycerol in cells of *E. coli. Biochem. Biophys. Res. Commun.* **32,** 344–349.

Smart, O. S., Breed, J., Smith, G. R., and Sansom, M. S. (1997). A novel method for structure-based prediction of ion channel conductance properties. *Biophys. J.* **72,** 1109–1126.

Smith, B. L., and Agre, P. (1991). Erythrocyte Mr 28,000 transmembrane protein exists as a multisubunit oligomer similar to channel proteins. *J. Biol. Chem.* **266,** 6407–6415.

Stahlberg, H., Braun, T., de Groot, B., Philippsen, A., Borgnia, M. J., Agre, P., Kuhlbrandt, W., and Engel, A. (2000). The 6.9-A structure of GlpF: a basis for homology modeling of the glycerol channel from *Escherichia coli. J. Struct. Biol.* **132,** 133–141.

Stroud, R. M., McCarthy, M. P., and Shuster, M. (1990). Nicotinic acetylcholine receptor superfamily of ligand-gated ion channels. *Biochemistry* **29,** 11009–11023.

Sui, H., Han, B. G., Lee, J. K., Walian, P., and Jap, B. K. (2001). Structural basis of water-specific transport through the AQP1 water channel. *Nature* **414,** 872–878.

Sweet, G., Gandor, C., Voegele, R., Wittekindt, N., Beuerle, J., Truniger, V., Lin, E. C., and Boos, W. (1990). Glycerol facilitator of *Escherichia coli:* cloning of glpF and identification of the glpF product. *J. Bacteriol.* **172,** 424–430.

Tajkhorshid, E., Nollert, P., Jensen, M. O., Miercke, L. J., O'Connell, J., Stroud, R. M., and Schulten, K. (2002). Control of the selectivity of the aquaporin water channel family by global orientational tuning. *Science* **296,** 525–530.

Verkman, A. S., Shi, L. B., Frigeri, A., Hasegawa, H., Farinas, J., Mitra, A., Skach, W., Brown, D., Van Hoek, A. N., and Ma, T. (1995). Structure and function of kidney water channels. *Kidney Int.* **48,** 1069–1081.

Walther, D., Eisenhaber, F., and Argos, P. (1996). Principles of helix–helix packing in proteins: the helical lattice superposition model. *J. Mol. Biol.* **255,** 536–553.

Walz, T., Hirai, T., Murata, K., Heymann, J. B., Mitsuoka, K., Fujiyoshi, Y., Smith, B. L., Agre, P., and Engel, A. (1997). The three-dimensional structure of aquaporin-1. *Nature* **387,** 624–627.

Weissenborn, D. L., Wittekindt, N., and Larson, T. J. (1992). Structure and regulation of the glpFK operon encoding glycerol diffusion facilitator and glycerol kinase of *Escherichia coli* K-12. *J. Biol. Chem.* **267,** 6122–6131.

Wistow, G. J., Pisano, M. M., and Chepelinsky, A. B. (1991). Tandem sequence repeats in transmembrane channel proteins. *Trends Biochem. Sci.* **16,** 170–171.

Yasui, M., Hazama, A., Kwon, T. H., Nielsen, S., Guggino, W. B., and Agre, P. (1999). Rapid gating and anion permeability of an intracellular aquaporin. *Nature* **402,** 184–187.

Yool, A. J., Stamer, W. D., and Regan, J. W. (1996). Forskolin stimulation of water and cation permeability in aquaporin 1 water channels [see comments]. *Science* **273,** 1216–1218.

AUTHOR INDEX

A

B

D

E

F

G

H

I

J

K

L

M

R

S

T

U

V

W

X

Y

Z

SUBJECT INDEX

F

G

H

I

K

L

M

N

O

P

Q

X

Y

ISBN 0-12-034263-4